Building Multiservice Transport Networks

Jim Durkin
John Goodman
Ron Harris
Frank Fernandez-Posse
Michael Rezek
Mike Wallace

Cisco Press

800 East 96th Street
Indianapolis, Indiana 46240 USA

Building Multiservice Transport Networks

Jim Durkin, John Goodman, Ron Harris, Frank Fernandez-Posse, Michael Rezek, Mike Wallace

Copyright © 2006 Cisco Systems, Inc.

Cisco Press logo is a trademark of Cisco Systems, Inc.

Published by:
Cisco Press
800 East 96th Street
Indianapolis, IN 46240 USA

Printed in the United States of America 1 2 3 4 5 6 7 8 9 0

First Printing March 2010

Library of Congress Cataloging-in-Publication Number: 2004114023

ISBN: 1-58714-245-7

Trademark Acknowledgments

All terms mentioned in this book that are known to be trademarks or service marks have been appropriately capitalized. Cisco Press or Cisco Systems, Inc. cannot attest to the accuracy of this information. Use of a term in this book should not be regarded as affecting the validity of any trademark or service mark.

Warning and Disclaimer

This book is designed to provide information about designing, configuring, and monitoring multiservice transport networks. Every effort has been made to make this book as complete and as accurate as possible, but no warranty or fitness is implied.

The information is provided on an "as is" basis. The authors, Cisco Press, and Cisco Systems, Inc. shall have neither liability nor responsibility to any person or entity with respect to any loss or damages arising from the information contained in this book or from the use of the discs or programs that may accompany it.

The opinions expressed in this book belong to the author and are not necessarily those of Cisco Systems, Inc.

Corporate and Government Sales

Cisco Press offers excellent discounts on this book when ordered in quantity for bulk purchases or special sales.

For more information please contact: **U.S. Corporate and Government Sales**   1-800-382-3419
corpsales@pearsontechgroup.com

For sales outside the U.S. please contact: **International Sales**   international@pearsoned.com

Feedback Information

At Cisco Press, our goal is to create in-depth technical books of the highest quality and value. Each book is crafted with care and precision, undergoing rigorous development that involves the unique expertise of members from the professional technical community.

Readers' feedback is a natural continuation of this process. If you have any comments regarding how we could improve the quality of this book, or otherwise alter it to better suit your needs, you can contact us through email at feedback@ciscopress.com. Please make sure to include the book title and ISBN in your message.

We greatly appreciate your assistance.

Publisher	Paul Boger
Cisco Representative	Anthony Wolfenden
Cisco Press Program Manager	Jeff Brady
Executive Editor	Elizabeth Peterson
Production Manager	Patrick Kanouse
Development Editor	Dan Young
Project Editor	Kelly Maish
Copy Editor	Krista Hansing
Technical Editor(s)	Gabriel Gutierrez, Rob Gonzalez
Editorial Assistant	Raina Han
Book Designer	Louisa Adair
Cover Designer	Louisa Adair
Composition	Interactive Composition Corporation
Indexer	Larry Sweazy

CISCO SYSTEMS

Corporate Headquarters
Cisco Systems, Inc.
170 West Tasman Drive
San Jose, CA 95134-1706
USA
www.cisco.com
Tel: 408 526-4000
      800 553-NETS (6387)
Fax: 408 526-4100

European Headquarters
Cisco Systems International BV
Haarlerbergpark
Haarlerbergweg 13-19
1101 CH Amsterdam
The Netherlands
www-europe.cisco.com
Tel: 31 0 20 357 1000
Fax: 31 0 20 357 1100

Americas Headquarters
Cisco Systems, Inc.
170 West Tasman Drive
San Jose, CA 95134-1706
USA
www.cisco.com
Tel: 408 526-7660
Fax: 408 527-0883

Asia Pacific Headquarters
Cisco Systems, Inc.
Capital Tower
168 Robinson Road
#22-01 to #29-01
Singapore 068912
www.cisco.com
Tel: +65 6317 7777
Fax: +65 6317 7799

Cisco Systems has more than 200 offices in the following countries and regions. Addresses, phone numbers, and fax numbers are listed on the
Cisco.com Web site at www.cisco.com/go/offices.

Argentina • Australia • Austria • Belgium • Brazil • Bulgaria • Canada • Chile • China PRC • Colombia • Costa Rica • Croatia • Czech Republic
Denmark • Dubai, UAE • Finland • France • Germany • Greece • Hong Kong SAR • Hungary • India • Indonesia • Ireland • Israel • Italy
Japan • Korea • Luxembourg • Malaysia • Mexico • The Netherlands • New Zealand • Norway • Peru • Philippines • Poland • Portugal
Puerto Rico • Romania • Russia • Saudi Arabia • Scotland • Singapore • Slovakia • Slovenia • South Africa • Spain • Sweden
Switzerland • Taiwan • Thailand • Turkey • Ukraine • United Kingdom • United States • Venezuela • Vietnam • Zimbabwe

About the Authors

Jim Durkin is a Senior Systems Engineer at Cisco Systems and is a specialist in optical transport technologies. Jim has more than 17 years of experience in the telecommunications industry, involving design and implementation of voice, data, and optical networks. He started his career at AT&T Bell Laboratories. Jim has a Bachelor's degree and a Master's degree in electrical engineering from the Georgia Institute of Technology. He holds the Optical Specialist, CCNA, and CCIP certifications from Cisco Systems.

John Goodman is a Senior Systems Engineer with Cisco Systems, supporting network solutions for service providers. He has spent 13 years in the planning, design, and implementation of optical transport networks. He has a Bachelor's degree in electrical engineering from Auburn University, and holds the Cisco Optical Specialist and CCNA certifications. John lives with his wife and two daughters in Tennessee.

Ron Harris is a Senior Systems Engineer at Cisco Systems and is a specialist in optical transport technologies. As a systems engineer for the last 6 years, Ron has worked with various service providers, Regional Bell Operating Companies, and energy/utilities company in the design and specifications of optical networks. He has amassed more than 18 years of experience in the telecommunications industry. Before joining Cisco in 2000, Ron worked as a technical sales consultant for Lucent Technologies where he led a team of sales engineers responsible for the sale of next-generation optical fiber and DWDM to transport providers. Before joining Lucent, he worked for several years in various engineering roles at a leading telecommunications provider in the Southeastern United States. Ron has earned an MBA from the University of Alabama at Huntsville, and a Bachelor's degree in computer and information sciences from the University of Alabama at Birmingham. He is presently Cisco certified as an Optical Specialist I, CCNP, and CCIP.

Frank Fernández-Posse has a diverse background in the telecommunications industry. Frank has been engaged in designing, validating, and implementing networks using various technologies. Given his broad background, he dedicated part of his career to validating technology/product integration, including data, ATM, optical, and voice technologies. Frank joined Cisco Systems in 2001 as a Systems Engineer and currently supports transport networking solutions for service providers; he is a certified Cisco Optical Specialist. Before joining Cisco Systems, he worked at Lucent Technologies.

Michael Rezek is an Account Manager at Cisco Systems and a specialist in optical transport technologies. Michael is a professionally licensed engineer in electrical engineering in North Carolina and South Carolina. He received his Master of Science degree in electrical engineering from the Georgia Institute of Technology. In 2001, Michael received his CCNP Voice Specialization in VoIP VoFR VoATM and has received his CCDA, CCNA, CCDP, and CCNP certifications in 2000. He graduated summa cum laude with a Bachelor of Engineering degree in electrical engineering from Youngstown State University. He has authored 32 patent disclosures, 10 of which Westinghouse pursued for patent. He sold his first ever Cisco Inter Office Ring (IOF) to a major ILEC. At Rockwell Automation Engineering, Michael designed, built, and tested hardware and software for a 15-axis robot for the fiber industry. As an engineer, he was commissioned to develop the intellectual property for a complex and proprietary fiber-winding technology which he then designed and tested.

Mike Wallace, a native of South Carolina, began his career in telecommunications with one of the largest independent telephone companies in South Carolina in January 1970. During his 21-year career there, he served in many technical positions; in 1984, he was promoted to central office equipment engineer, with a specialty in transmissions engineering. This special field required the tasks of planning, designing, and implementing optical transmission networks, while working closely with outside plant engineers to understand fiber-optic cable characteristics and specifications that would be the foundation for optical transmitters, receivers, and repeaters to come together to form optical transmission networks.

In 1991, Mike moved on from the telephone company to pursue other opportunities. He had a 14-month assignment with the University of North Carolina at Charlotte to engineer an optical network for the campus, and a 3-year assignment with ICG, Inc., a major CLEC with an optical network in the Charlotte, North Carolina market, where he provided technical support for the local sales teams. Mike had a 7-year assignment with Fujitsu Network Communications, Inc., a major manufacturer of optical transmission systems, where he served as a Sales Engineer for the Southeast territory. Mike has served as president of the local chapter of the Independent Telephone Pioneers Association, which is a civic organization that supports multiple charities in the Palmetto State.

About the Technical Reviewers

Rob Gonzalez, P.E., Cisco Optical Specialist, is a Member of Technical Staff for BellSouth's Technology Planning and Deployment, Transmission and Access lab. He is responsible for testing and evaluating Cisco optical products for use in the BellSouth network. Rob also is the Subject Matter Expert for Layer 1 and Layer 2 transport of data services using Packet-over-SONET. Rob has been with BellSouth for more than 11 years in different capacities, and has worked on the technical staff for almost 5 years.

Gabriel Gutierrez, CCNA, CCIP, COS-Optical, has worked in the telecommunications industry for over 10 years. He received his Bachelor's degree in Electrical Engineering from Southern Methodist University. Currently, Gabriel works at Cisco Systems as a System Engineer selling and supporting optical and data networking solutions.

Acknowledgments

Jim Durkin: I would like to thank Joe Garcia for his initial idea of writing this book and for his support and recognition during this time-consuming project. I also would like to thank John Kane and Dan Young for their outstanding support. Most of all, I want to thank my beautiful wife and children for their support and patience during the writing of this book. This book is dedicated to John Richards, my uncle, who has been a father figure and mentor to me in my life.

John Goodman: This book is dedicated to my wife, Teresa, and to Joe Garcia, who was instrumental in my participation in this project.

Ron Harris: This book could not have been possible without the tireless efforts of my editors and technical reviewers. I would like to personally thank Rob Gonzalez and Gabriel Gutierrez for their hard work and tremendous effort in technically reviewing the chapters covering MSTP. I also would like to thank Dan Young and his team of editors for their editorial spirit of excellence while preparing this book for publication. Most important, I owe a tremendous amount of gratitude to my wife and daughters for their support and patience during the compilation of this book.

Frank Fernández-Posse: I would like to thank my wife, Ana, for her patience and ongoing support, and my baby son, Alec, for putting a big smile on my face every day. I love you! I am also grateful for being part of a great team in which support is readily available from every member. Special thanks to Jim Durkin for kicking off and managing this effort.

Michael Rezek: I would like to acknowledge my wife for the sacrifices she has made to provide me with the time to write this book.

Mike Wallace: I'd like to acknowledge all of my co-authors for their patience and assistance in completing this project. I would especially like to acknowledge Jim Durkin for his vision to see the need for this project and for giving me the opportunity to participate. I'd like to thank all the technical reviewers for their diligence, comments, and dedication to make this book a value to those individuals interested in its subject matter. I'd like to dedicate this book to my wonderful wife, Rosanne, for her support and understanding, and also to all of the people (too many to mention) who have been a part of my telecommunications career and education. It has been a great ride!

This Book Is Safari Enabled

The Safari® Enabled icon on the cover of your favorite technology book means the book is available through Safari Bookshelf. When you buy this book, you get free access to the online edition for 45 days.

Safari Bookshelf is an electronic reference library that lets you easily search thousands of technical books, find code samples, download chapters, and access technical information whenever and wherever you need it.

To gain 45-day Safari Enabled access to this book:

- Go to http://www.ciscopress.com/safarienabled

- Complete the brief registration form

- Enter the coupon code DXDG-P64C-62EA-YYCL-PWJ7

If you have difficulty registering on Safari Bookshelf or accessing the online edition, please e-mail customer-service@safaribooksonline.com.

Contents at a Glance

Table of Contents

Icons Used in This Book

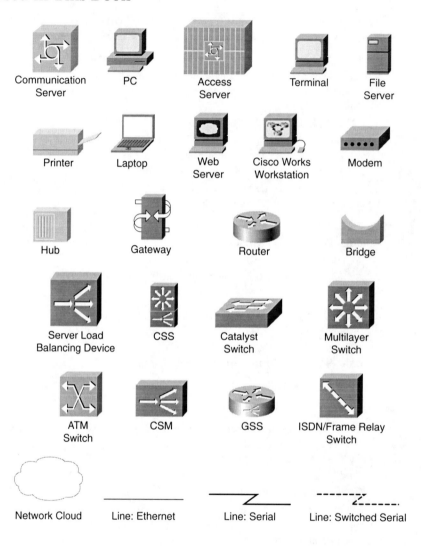

Command Syntax Conventions

The conventions used to present command syntax in this book are the same conventoins used in the IOS Command Reference. The Command Reference describes these conventions as follows:

- **Boldface** indicates commands and keywords that are entered literally as shown. In actual configuration examples and output (not general command syntax), boldface indicates commands that are manually input by the user (such as a **show** command).

- *Italics* indicate arguments for which you supply actual values.

- Vertical bars (l) separate alternative, mutually exclusive elements.

- Square brackets [ ] indicate optional elements.

- Braces { } indicate a required choice.

- Braces within brackets [{ }] indicate a required choice within an optional element.

Introduction

This book is a rare assemblage, in that it combines the best minds across a number of topics in one central repository. Books that are authored by one or two authors limit the depth and breadth of expertise to only that particular author(s). This book draws on the breadth and depth of each author as it pertains to each topic discussed, enhancing the book's overall value. The authors of this book are Cisco Systems Optical Engineers who have more than 75 years of combined optical networking expertise.

The authors of this book have seen a need to prepare those aspiring to grow their capabilities in multiservice transport networking. The result is this book, *Building Multiservice Transport Networks*. This book provides the reader with information to thoroughly understand and learn the many facets of MSPP and DWDM network architectures and applications with this comprehensive handbook. This includes topics such as designing, configuring, and monitoring multiservice transport networks. A multiservice transport network consists of MSPPs and MSTPs. Cisco's ONS 15454 is an example of a Multiservice Provisioning Platform (MSPP) and a Multiservice Transport Platform (MSTP).

It is important to understand that the Cisco ONS 15454 can be considered as two different products under one product family. The ONS 15454 MSPP is one product, and the other is the ONS 15454 MSTP. MSTP describes the characteristics of the ONS 15454 when used to implement either a fixed-channel OADM or a ROADM-based DWDM network. One of the unique capabilities of the ONS 15454 is that it remains one chassis, one software base, and one set of common control cards to support both MSPP applications and MSTP applications.

Service providers today understand the need for delivering data services—namely, Ethernet and SAN extension. However, most are uncertain of or disagree on the most economical network foundation from which these services should actually be delivered. When placed in newer environments, service providers instinctively leverage past knowledge of network deployments and tend to force-fit new technology into old design schemes. For example, some service providers have always used point-to-point circuits to deliver services, so when customers required Ethernet services, many immediately used private-line, point-to-point circuits to deliver them. Using the ONS 15454, this book shows you how to deliver basic private-line Ethernet service and how to deliver Ethernet multipoint and aggregation services using RPR to enable newer and more efficient service models.

This book also discusses how the MSPP and MSTP fit within the overall network architecture. This is important because many service providers are trying to converge and consolidate their networks. Service providers, such as ILECs, are looking to deliver more services, more efficiently over their network. This book can serve as a handbook that network designers and planners can reference to help develop their plans for network migration.

Goals and Methods

An important goal of this book is to help you thoroughly understand all the facets of a multiservice transport network. Cisco's ONS 15454 is addressed when discussing this because it is the leading multiservice transport product today. This book provides the necessary background material to ensure that you understand the key aspects of SONET, DWDM, Ethernet, and storage networking.

This book serves as a valuable resource for network professionals engaged in the design, deployment, operation, and troubleshooting of ONS 15454 applications and services, such as TDM, SONET/SDH,

DWDM, Ethernet, and SAN. By providing network diagrams, application examples, and design guidelines, this book is a valuable resource for readers who want a comprehensive book to assist in an MSPP and MSTP network deployment.

In summary, this book's goals are to

- Provide you with an in-depth understanding on multiservice transport networks
- Translate key topics in this book into examples of "why they matter"
- Offer you an end-to-end guide for design, implementation, and maintenance of multiservice transport networks
- Help you design, deploy, and troubleshoot ONS 15454 MSPP and MSTP services
- Provide real-life examples of how to use an MSPP and an MSTP to extend SAN networks
- Understand newer technologies such as RPR and ROADM, and how these can be deployed within an existing ONS 15454 transport architecture
- Review SONET and DWDM fundamentals

Who Should Read This Book?

This book's primary audience is equipment technicians, network engineers, transport engineers, circuit capacity managers, and network infrastructure planners in the telecommunications industry. Those who install, test, provision, troubleshoot, or manage MSPP networks, or who aspire to do so are also candidates for this book. Additionally, data and telecom managers seeking an understanding of TDM/data product convergence should read this book.

Business development and marketing personnel within the service-provider market can also gain valuable information from this book. This book should facilitate their understanding of how to market and price new services that can be delivered over their network.

How This Book Is Organized

The book provides a comprehensive view of MSPP and MSTP networks using the Cisco ONS 15454.

Chapters 1 through 15 cover the following topics:

Part I: "Building the Foundation for Understanding MSPP Networks"

- **Chapter 1, "Market Drivers for Multiservice Provisioning Platforms"**—This chapter builds the case for deploying a MSPP network. This chapter focuses on key reasons why MSPPs are needed and how MSPPs can reduce capital expenditures for service providers. It also discusses another important benefit for using an MSPP: the ease of operations, administration, maintenance, and provisioning (OAMP) of an MSPP.
- **Chapter 2, "Technology Foundation for MSPP Networks"**—This chapter provides an overview of key technologies that must be understood to successfully deploy an MSPP network. These include fiber optics, optical transmission, SONET principles, and synchronization and timing.
- **Chapter 3, "Advanced Technologies over Multiservice Provisioning Platforms"**—This chapter discusses three advanced technologies supported by MSPPs: 1) storage-area networking, 2) dense wavelength-division multiplexing, and 3) Ethernet. For each technology, this

chapter provides a brief history of the evolution of the service and then its integration into the MSPP platform.

Part II: "MSPP Architectures and Designing MSPP Networks"

- **Chapter 4, "Multiservice Provisioning Platform Architectures"**—This chapter describes various MSPP architectures. It reviews traditional network architectures and contrasts these with MSPP architectures. This comparison helps to point out the enormous benefits that MSPPs provide.

- **Chapter 5, "Multiservice Provisioning Platform Network Design"**—This chapter discusses how to design MSPP networks. It examines the key design components, including protection options, synchronization (timing) design, and network management. This chapter also discusses supported MSPP network topologies, such as linear, ring, and mesh configurations.

- **Chapter 6, "MSPP Network Design Example: Cisco ONS 15454"**—This chapter provides a realistic network design example of an MSPP network using the Cisco ONS 15454. It uses an example network demand specification to demonstrate an MSPP network design. The solution uses an ONS 15454 OC-192 ring. As part of the design, this chapter introduces the major components of the ONS 15454 system, including the common control cards, the electrical interface cards, the optical interface cards, the Ethernet interface cards, and the storage networking cards.

Part III: "Deploying Ethernet and Storage Services on ONS 15454 MSPP Networks"

- **Chapter 7, "ONS 15454 Ethernet Applications and Provisioning"**—This chapter discusses Ethernet architectures and applications supported on the ONS 15454, including Ethernet point-to-point and multipoint ring architectures. This chapter discusses the ONS 15454 Ethernet service cards: E Series, CE Series, G Series, and ML Series. Application examples are provided as well, including how to provision Ethernet services. As an example, this chapter discusses how to implement a resilient packet ring (RPR) using the ML-Series cards.

- **Chapter 8, "ONS 15454 Storage-Area Networking"**—This chapter discusses storage-area networking (SAN) extension using the Cisco ONS 15454. You can use 15454 networks to connect storage-area networks between different geographical locations. This is important today because of the need to consolidate data center resources and create architectures for disaster recovery and high availability.

Part IV: "Building DWDM Networks Using the ONS 15454"

- **Chapter 9, "Using the ONS 15454 Platform to Support DWDM Transport: MSTP"**—This chapter highlights the basic building blocks of the ONS 15454 MSTP platform. It describes the key features and functions associated with each ONS 15454 MSTP component, including fixed OADMs and ROADM cards, transponder/muxponder interface cards, and amplifier interface cards. This chapter provides network topology and shelf configuration examples. Each ONS 15454 MSTP shelf configuration example shows you the most common equipment configurations applicable to today's networks.

- **Chapter 10, "Designing ONS 15454 MSTP Networks"**—This chapter examines the general design considerations for DWDM networks and relays their importance for ONS 15454 Multiservice Transport Platform (MSTP) DWDM system deployment. Design considerations and design rules examples are included in this chapter. This chapter describes Cisco's MetroPlanner Design Tool, which you can use to quickly design and assist in turning up an ONS 15454 MSTP network.

- **Chapter 11, "Using the ONS 15454 MSTP to Provide Wavelength Services"**—This chapter discusses wavelength services using the ONS 15454 MSTP, and it explores the different categories and characteristics of wavelength services as they relate to ONS 15454 MSTP features and functions. You will understand how you can use the ONS 15454 MSTP to provide wavelength services, such as SAN, Ethernet, and SONET, while using different protection schemes. Both fixed-channel optical add/drop and ROADM based networks are discussed.

Part V: "Provisioning and Troubleshooting ONS 15454 Networks"

- **Chapter 12, "Provisioning and Operating an ONS 15454 SONET/SDH Network"**—This chapter describes how to install, configure, and power up the ONS 15454. It also discusses how to test, maintain, and upgrade software for the ONS 15454.

- **Chapter 13, "Troubleshooting ONS 15454 Networks"**—This chapter provides a high-level approach to troubleshooting ONS 15454 SONET networks. This chapter provides you with a general approach to troubleshooting the most common problems and issues found during turn-up of an ONS 15454 node, as well as ONS 15454 network-related issues.

Part VI: "MSPP Network Management"

- **Chapter 14, "Monitoring Multiple Services on an Multiservice Provisioning Platform Network"**—This chapter provides an overview of the fault- and performance-management capabilities of the ONS 15454. This chapter also includes a discussion of three key areas that are essential in managing MSPP networks: 1) SNMP MIBs, 2) TL1 support, and 3) performance management. The end of this chapter discusses the key differences in using the local Craft Interface application, called Cisco Transport Controller (CTC), versus an element-management system (EMS).

- **Chapter 15, "Large-Scale Network Management"**—This chapter provides a list of key functions supported by large-scale operational support systems (OSS). After discussing these functions, the following important question is asked and discussed: "Why use an element-management system (EMS)?" This chapter describes Cisco's EMS, called Cisco Transport Manager (CTM), and discusses how CTM provisions Layer 2 Ethernet Multipoint service step by step over an ONS 15454 ring equipped with ML-Series cards.

PART I

Building the Foundation for Understanding MSPP Networks

This chapter covers the following topics:

- Market Drivers
- Increased Demand for Bandwidth by LANs
- Rapid Delivery of Next-Generation Data and High-Bandwidth Services
- TCO
- OAM&P
- Capital Expense Reduction

CHAPTER 1

Market Drivers for Multiservice Provisioning Platforms

Multiservice Provisioning Platforms (MSPPs) are optical platforms into which you can insert or remove various ring and service cards. The interfaces on these cards deliver a variety of electrical and optical services, such as traditional time-division multiplexed (TDM) circuit-based and packet-based data services within the chassis. Because they are modular, MSPPs enable you to insert cards such as plug-ins or blades. This modularity accommodates the aggregation of traditional facilities such as DS1, DS3, STS1, OC-3, STM1, OC-12, STM4, OC-48, and STM64. MSPPs also support current data services such as 10-Mb Ethernet, 100-Mb Ethernet, and Gigabit Ethernet (GigE).

Emerging storage-area networking (SAN) services such as Fiber Connection (FICON), Fibre Channel (FC), and Enterprise Systems Connection (ESCON) can also be aggregated via STS-1, STS-3c, STS-6c, STS-9c, STS12-c, STS-24c, or STS-48c interfaces; these services can be transported over a single optical transport facility via an OC-3, OC-12, OC-48, OC-48 DWDM, OC-192, or OC-192 DWDM line interface. The service cards have multiple ports or connections per card. For example, a lower-density electrical DS1 card can drop out 14 DS1s, and higher-density cards can drop out up to 56 DS1s. This is true for all other service cards as well.

The platform flexibility translates into drastically improved efficiencies in the transport layer and dramatically increased savings in both the initial costs and the life-cycle costs of the deployment. Some of the latest technology in MSPP integrates dense wavelength-division multiplexing (DWDM) into the chassis and can deliver wavelengths of various types. The footprint is typically one quarter of a 19-inch or 23-inch standard rack or bay in size. Figure 1-1 shows an MSPP, and Figure 1-2 shows multiple MSPPs deployed in a single bay.

MSPPs are in use today by every major incumbent local exchange carrier (ILEC) and competitive local exchange carrier (CLEC), most independent carriers, and cable companies and large enterprise customers that utilize these MSPPs over leased fiber. Figure 1-3 shows a carrier (such as a service provider) deployment of MSPP to deliver services to multiple customers. Figure 1-4 shows a customer's private implementation over leased fiber. Figure 1-5 shows a CLEC implementation that uses the ILEC's network and delivers services to the customers.

Figure 1-1   *An MSPP*

Figure 1-2   *Standard 19-Inch Bay with Four MSPPs*

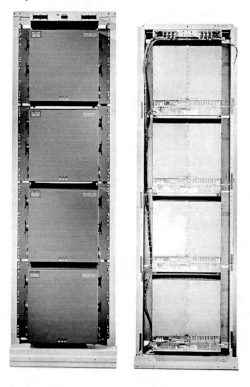

Figure 1-3   *Service Provider Network Implementation of MSPP*

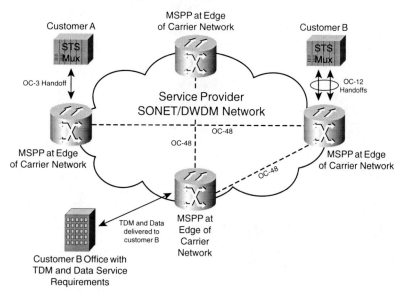

Figure 1-4   *Customer Private Network Implementation of MSPP*

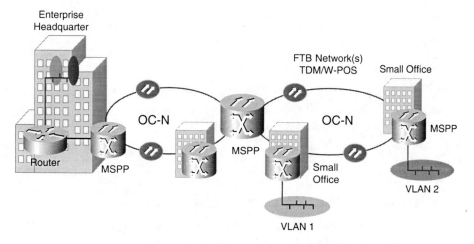

The emergence of MSPP in the late 1990s enabled CLECs to take advantage of regulatory changes that essentially forced incumbents to "unbundle" their networks. In other words, incumbents had to allow competitors to lease their network infrastructure. CLECs could very quickly and inexpensively establish a non-asset-based network because the ILEC owned the entire network infrastructure, except for the customer access MSPP shelf of the CLEC. CLECs then begin to sell services off the MSPP rapidly because they were not encumbered with legacy operating support systems (OSS). Figure 1-5 shows an example of this non-asset-based network.

Figure 1-5   *CLEC-to-ILEC Network Implementation of MSPP*

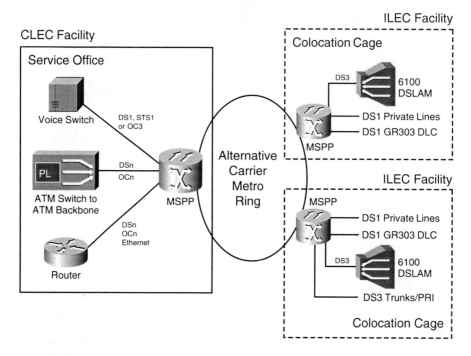

This also enabled CLECs to add network assets in proportion to revenue flow and thus grow their own asset-based network. They were able to wean themselves off the ILEC networks to become standalone asset-based CLECs.

Market Drivers

The mass proliferation of MSPPs is driven by the following three market drivers:

- An increase in demand for bandwidth by local-area networks (LANs) for connection to their service providers
- Rapid delivery of new next-generation data and high-bandwidth services
- Total cost of ownership (TCO)

Increased Demand for Bandwidth by LANs

An interesting phenomenon with regard to the proliferation of MSPPs is that, unlike PCs, which experience next-generation technology transformations thanks to increased processing speeds of their components (such as the microprocessor), MSPPs largely have emerged without increasing the optical data-transport ring speeds.

Before MSPPs emerged, optical platforms had already achieved speeds of OC-3, OC-12, OC-48, and OC-192. MSPPs did not increase these transport ring speeds, which would have been analogous to an increase in speed of the computer processor. So what led to the explosive growth in demand for MSPPs? The answer is the assimilation of numerous services within the same shelf, along with easy management of these services.

Legacy optical systems were limited to only a few different types of service drops, with a limited capacity for mixing these service types. MSPPs, on the other hand, integrate numerous optical and electrical interface cards that carry voice and data services within the same shelf. The port density of these cards is much greater than that of legacy platforms, so they consume much less rack space than legacy systems (see Figure 1-6).

Figure 1-6 *Example of Service Drops with Much Higher Densities*

Additionally, MSPPs deliver far more of each service type per MSPP shelf. This allows for a greater variety and quantity of each type of service (ToS).

Consequently, the forces behind the movement toward a new generation of Synchronous Optical Network (SONET) equipment in the metropolitan-area network (MAN) or wide-area network (WAN) include easy integration, substantial savings, and the capability to add new services. Along with these primary drivers is a need to improve the scalability and flexibility of SONET while maintaining its fault-tolerant features and resiliency.

The focus of first adopters was to integrate add/drop multiplexers (ADMs) and digital cross-connect systems (DCSs) with higher port density and more advanced grooming capabilities than those of the traditional standalone devices. Now the push is toward making SONET more efficient for data delivery. Next-generation data-networking services, such as Ethernet, are also integrated within the MSPPs. Next-generation data-networking services are covered later in this chapter.

What drove this demand for increased bandwidth in a smaller footprint? Well, just as more demanding computer applications such as voice and video drove the demand for faster processors, higher-speed LANs drove the need for speedier WANs.

The proliferation of the Internet created the demand for a higher-speed LAN to access content. Of course, the speed of accessing content over a LAN is throttled down to the speed at which the WAN can deliver it, as shown in Figure 1-7. With the backbone and core of the WAN as the optical network, the need for more high-speed T1s and T3s grew dramatically to keep up with the demands of the LAN.

Figure 1-7 *LAN Bottleneck*

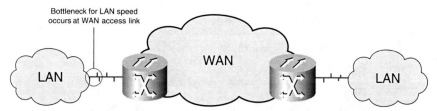

There is no question that the demand for data traffic is growing sharply throughout the public network. A wide variety of services continue to drive the increasing need for bandwidth. Data private lines, ATM, digital subscriber line (DSL), videoconferencing, streaming video, transparent LAN service (TLS), IP/Multiprotocol Label Switching virtual private networks (IP/MPLS VPNs), and other applications are all increasing in use. In addition, transport over delivery services is as different as DS1, DS3, STS-1, OC-3, OC-12, OC-48, 10-/100-Mbps Ethernet, and GigE. Ubiquitous Internet and intranet applications, coupled with the significant rescaling of the interexchange network that has taken place in recent years, further spurs the demand for data services both in MANs and across WANs.

Caught between the increased long-haul capacity and the growing access demand from customers, MANs have now become the bottleneck in the overall network. Despite the growing deployment of optical fiber in cities, MAN expansion has not kept up with the increased demand for data transport and value-added IP services that continue to drive bandwidth consumption higher. MAN networks not only need to add capacity, but they also need to be flexible enough to offer cost-effective aggregation of a variety of services at multiple layers, such as TDM packet and wavelength services, to supply the services needed by customers.

Along with the emergence of demand for Internet content and the transmission of large documents, such as video clips and photographs, came a cascading effect of demand for larger WAN connections. The surge for increased WAN speeds meant that offices and businesses were looking at speeds of DS3 and even OC-3, OC-12, and OC-48 to connect to their service providers. Additionally, business customers who ordered dedicated rings from their service providers to connect multiple buildings within a metropolitan area desired various service drops to meet the demands of various types of LAN equipment. Thus, a mixture of DS3, DS1, and Ethernet services are all desired in the various locations, as shown in Figure 1-8.

Figure 1-8   *Customer Application Using Diverse Service Drops*

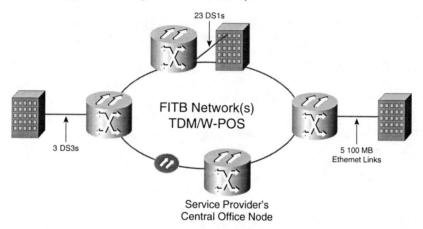

One of the interesting aspects of increasing the available speed on the WAN to support higher LAN speeds is that this increase in speed drove the development of new applications, such as voice and video over data. This, in turn, created an even greater demand for WAN speeds to support these LAN applications. This ripple, or domino, effect of technology is very often seen in computing: An enhancement in the computer video card might enable software writers to write more sophisticated new applications. As the limits of computing technology are pushed, PC developers must again enhance the computer.

Rapid Delivery of Next-Generation Data and High-Bandwidth Services

Next-generation MSPP platforms accelerate service-provider return on investment (ROI) of services such as Ethernet in several ways. First, the startup cost is low because the service provider can install the platform using only the currently required capacity. Then, as demand increases, such as for an OC-48, the node can be scaled to OC-192 and finally to OC-192 DWDM channels. In addition, aggregation of all types of services into each wavelength maximizes bandwidth efficiency and reduces incremental service costs. Efficient DWDM utilization saves 70 to 80 percent of bandwidth, which increases as the network scales. Fewer wavelengths have to be activated, and, as the network expands, carriers do not have to move from the C-band into the more costly L-band to add metro color wavelengths. Overall, service providers can realize a significant ROI in a brief amount of time (such as a single quarter) instead of in 1 to 2 years, as they did previously.

With a single platform to deploy instead of many platforms, service providers also limit their operating costs. A one-card interface designed into a multiservice platform saves space so that a single bay can more often handle all the broadband multiplexing and demultiplexing that a network node requires. In fact, when thousands of network interfaces are involved, as in a central office, the savings in vertical rack space from using multiservice platforms can be measured in miles. Additional savings come from an eightfold reduction in power consumption with next-generation multiservice systems.

With only one type of platform in use, fewer spare interface cards must be held in inventory. In addition, less training is involved for installation and maintenance. A single network OSS can be used to configure and manage all services, thus minimizing the difficulties involved in administering a network with multiple platforms with multiple bit rates, topologies, and operating systems. A more advanced end-to-end service-provisioning design has been developed for provisioning and restoration across Layers 1 to 3 of the network through the introduction of the Unified Control Plane. All these features work to simplify the network infrastructure, helping to save operating costs and increase profitability.

Ethernet Services

Ethernet and other data services actually cause service providers to rethink how they design and architect their local transport networks. These networks were originally designed to carry circuit-switched voice traffic. However, with the emergence of data services, such as Ethernet over the MAN, these networks can be inefficient when it comes to data.

Because customers demand a variety of different Ethernet services carriers are finding that a "one size fits all" architecture doesn't work. As these carriers adopt Ethernet technologies that are still new to the OSS, such as to personnel and carriers, they face a delicate balancing act. Should they wait until standards and network-management interfaces are mature and tested before they adopt key enhancements? Or do the benefits outweigh the cost penalties of using a nonstandard solution?

The answers tend to be different across carriers and enhancements. For example, Regional Bell Operating Companies (RBOC) are deploying different Ethernet offerings that drive the specifications of the infrastructure that they're using to support them. The great news for these carriers is that MSPP can support all the three major offerings.

Traditional Ethernet Service Offerings

Traditional Ethernet services fall into three basic types (see Figure 1-9):

- Ethernet Private Line, or point-to-point (sometimes also called E-line)
- Ethernet relay/wire services, or point-to-multipoint
- Ethernet multipoint, or multipoint-to-multipoint (sometimes also called E-LAN services)

Figure 1-9   *Various Ethernet Deployment Models*

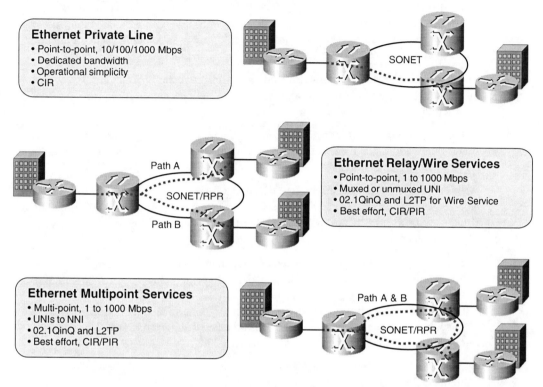

Ethernet Private Line
- Point-to-point, 10/100/1000 Mbps
- Dedicated bandwidth
- Operational simplicity
- CIR

SONET

Path A

SONET/RPR

Path B

Ethernet Relay/Wire Services
- Point-to-point, 1 to 1000 Mbps
- Muxed or unmuxed UNI
- 02.1QinQ and L2TP for Wire Service
- Best effort, CIR/PIR

Ethernet Multipoint Services
- Multi-point, 1 to 1000 Mbps
- UNIs to NNI
- 02.1QinQ and L2TP
- Best effort, CIR/PIR

Path A & B

SONET/RPR

Ethernet Private Line (EPL) services are equivalent to traditional dedicated bandwidth leased-line services offered by service providers. EPL service is a dedicated point-to-point connection from one customer-specified location to another, with guaranteed bandwidth and payload transparency end to end.

Ethernet Wire Service (EWS) is a point-to-point connection between a pair of sites. It usually is provided over a shared switched infrastructure within the service provider network, can be shared with one or more other customers, and is offered with various choices of committed bandwidth levels up to the wire speed. To help ensure privacy, the service provider separates each customer's traffic by applying virtual LAN (VLAN) tags. This service is also a great alternative for customers who do not want to pay the expensive cost of a dedicated private line.

Ethernet multipoint services are sometimes TLSs because their support typically requires the carrier to create a metro-wide Ethernet network, similar to the corporate LANs that have become a staple of the modern working world.

TLSs provide Ethernet connectivity among geographically separated customer locations and use virtual local-area networks (VLANs) to span and connect those locations. Typically, enterprises deploy TLS within a metro area to interconnect multiple enterprise locations. However, TLS also can be extended to locations worldwide. Used this way, TLS tunnels wide-area traffic through VLANs so that the enterprise customer does not need to own and maintain customer premises equipment (CPE) with wide-area interfaces. Customers are unchained from the burden of managing or even being aware of anything with regard to the WAN connection that links their separate LANs.

TLS is significantly less expensive and easier to deploy over an Ethernet infrastructure than on a Frame Relay or ATM infrastructure. These lower costs are derived primarily from lower equipment costs.

Deploying TLS on Ethernet also provides the increased flexibility carriers obtain in provisioning additional bandwidth, with varying quality of service (QoS) capabilities and service-level agreements (SLAs).

With TLS being a low-cost service, the carriers can use it to encourage customers to use a bundled service arrangement. This increases margins and strengthens the customer bonds. Typical value-added services include an Ethernet interface to the Internet, SAN over SONET, and data-center connectivity.

To support a shared multipoint offering, carriers traditionally had to install Ethernet switches around a metro fiber ring. These carriers have implemented shared multipoint services directly over fiber, which means that those services do not include SONET restoration capability; this effectively limits them to noncritical traffic. However, carriers that use the metro Ethernet network over MSPP take advantage of SONET's restoration capabilities: redundant fiber paths and 50-ms switching.

One leading vendor's multilayer ethernet cards(that is, the card integrates Open Systems Interconnection [OSI] Model Layer 1, Layer 2, and Layer 3 functionality into cards) enables customers to experience actual LAN speed across the metro ring by simply provisioning this card in the MSPP chassis. It offers the added benefits of SONET diversity and rapid restoration technology.

This application targets customers who want to interconnect multiple locations, offering a less complicated alternative to the mesh network that they would otherwise need. Unlike with traditional private lines, data can travel across the metro network as fast as it does on a company's internal LAN.

Customers connect their own Ethernet equipment to the metro network. All they need is a router interface, which is analogous to Frame Relay architecture.

Resilient Packet Ring

Some carriers refer to this transparent LAN service as resilient packet ring (RPR) architecture for restoration. Unlike SONET architecture, in which half of all available bandwidth typically sits idle, RPR uses the backup, or protection, facilities to carry traffic even under normal conditions. If a failure occurs, traffic is rerouted on a priority basis, as shown in Figure 1-10.

This is attractive to customers who have SONET with extra bandwidth because they can now dedicate part of that bandwidth to the LAN.

Figure 1-10 *RPR Using Both Rings Under Normal Conditions*

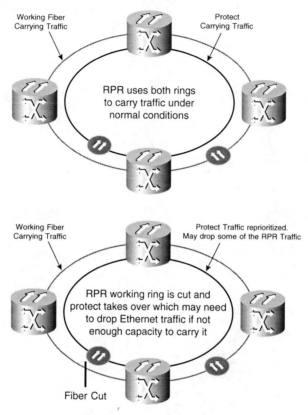

DWDM Wavelength Services

For customers who want Ethernet connectivity at speeds of 1 Gbps or higher, carriers most likely will use a delivery platform based on DWDM in the future. Currently, however, that market is limited to the very largest corporate customers.

An enterprise customer's investment in a DWDM network can't be justified merely by Ethernet alone. Today 90 percent of all DWDM sales are for SAN service types of protocol. The price of Ethernet over SONET is three times cheaper than a DWDM-based solution.

Over time, however, DWDM-based solutions are likely to become more popular as enterprise customer bandwidth needs increase and as vendors drive equipment costs down.

Many vendors are making Ethernet cheaper with course wavelength-division multiplexing (CWDM).

DWDM-based offerings can be point-to-point or multipoint. To date, these have been dedicated services, with an entire wavelength used by just a single customer. But some carriers reportedly are experimenting with a shared offering.

Figure 1-11 shows one type of DWDM implementation: a variation of a SONET platform called Multi Service Transport Platform (MSTP). In an MSTP application, the transponders and muxponders are integrated into the MSPP shelf itself.

Figure 1-11 *An MSPP DWDM Application*

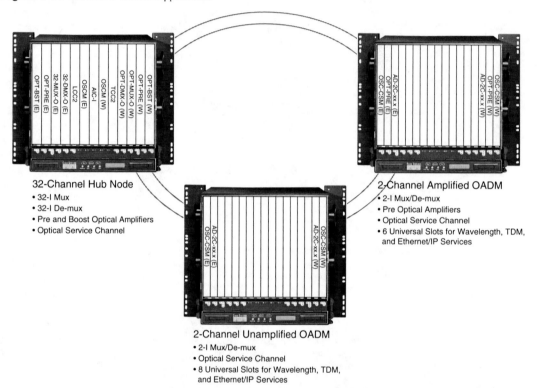

32-Channel Hub Node
- 32-I Mux
- 32-I De-mux
- Pre and Boost Optical Amplifiers
- Optical Service Channel

2-Channel Amplified OADM
- 2-I Mux/De-mux
- Pre Optical Amplifiers
- Optical Service Channel
- 6 Universal Slots for Wavelength, TDM, and Ethernet/IP Services

2-Channel Unamplified OADM
- 2-I Mux/De-mux
- Optical Service Channel
- 8 Universal Slots for Wavelength, TDM, and Ethernet/IP Services

In the other implementation of DWDM, the muxponders are located in a separate device and merely have a specific ITU-T optics wavelength card placed in the MSPP shelf to convert the wavelength of light. These optics wavelengths are then multiplexed into a separate device, sometimes referred to as a filter, and then carried over the fiber to a demultiplexor. There the light signals are redistributed into the corresponding optical cards that read the specific various wavelengths.

DWDM over MSPP brings enhanced photonic capabilities directly into the service platform, extends geographical coverage, ensures flexible topology support, and delivers on the requirements of today's multiservice environment. Integrating DWDM into MSPPs, such as the Cisco ONS 15454, provides operational savings and allows the platform to be tailored for MSPP architectures, pure DWDM architectures, or a mixed application set. With the DWDM over MSPP strategy, customers can use a single solution to meet metro transport requirements, ranging from a campus environment to long-haul applications, with significantly more flexibility in the DWDM layer than traditional long-haul solutions.

The MSTP strategy consists of three phases: integrated DWDM, flexible photonics, and dynamic services. This MSTP technology allows for any or all wavelengths to be provisioned at any or all the MSTP nodes at any time.

DWDM provides unprecedented density in a shelf that has a footprint of one quarter of a 19-inch bay. These wavelength services can be handed to customers and terminated on their customer premises equipment. Figure 1-12 shows a DWDM fiber with several wavelengths carrying various traffic types.

Figure 1-12 *DWDM Fiber with Several Wavelengths Carrying Various Traffic Types*

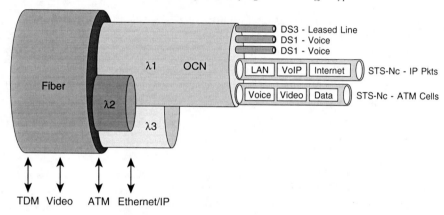

SAN Services

A number of applications are driving the need for MSPPs to deliver storage-area networking. These include applications for disaster recovery, data backup, SAN extension, and LAN extension. These applications are achieved with services that are delivered over SONET.

Storage represents the largest component of enterprise IT spending today, and it is expected to remain so for the foreseeable future. Resident on enterprise storage media is a business's most valuable asset in today's environment of electronic business and commerce information. The complexity of storing and managing this information, which is the lifeblood of corporations in almost every vertical segment of the world economy, has brought about significant growth in the area of SANs.

Alongside the growth of SAN implementations is the desire to consolidate and protect the information within a SAN for the purposes of business continuance and disaster recovery (BC/DR) by transporting storage protocols between primary and backup data centers. Enterprises and service providers alike have found that SONET is one of the technologies that best facilitates the connectivity of multiple sites within the MAN and the WAN.

MSPP vendors have recognized the need to transport storage protocols over SONET/ Synchronous Digital Hierarchy (SDH) networks and have developed the technology and plug-in cards for MSPP. This enables customers to transparently transport FC, FICON, and ESCON.

FC technology has become the protocol of choice for the SAN environment. It has also become common as a service interface in metro DWDM networks, and it is considered one of the primary drivers in the DWDM market segment. However, the lack of dark fiber available for lease in the access portion of the network has left SAN managers searching for an affordable and realizable solution to their storage transport needs.

Thus, service providers have an opportunity to create revenue streams to efficiently connect to and transport the user's data traffic through FC handoffs. Service providers must deploy metro transport equipment that will enable them to deliver these services cost-effectively and with the reliability their SLAs require—hence the need for MSPPs. Industry experts expect this growth to mirror the growth in Ethernet-based services and recommend following a similar path to adoption.

Voice and Video Applications

Although voice and video are not optical platform data services (instead, these applications ride upon the underlying protocol, such as Ethernet), it is important to cover them here because, in many ways, they are responsible for driving the demand of the underlying metro service, such as Ethernet over SONET.

A communications network forms the backbone of any successful organization. These networks serve as a transport for a multitude of applications, including delay-sensitive voice and bandwidth-intensive video. These business applications stretch network capabilities and resources, but also complement, add value to, and enhance every business process. Networks must therefore provide scaleable, secure, predictable, measurable, and sometimes guaranteed services to these applications. Achieving the required QoS by managing the delay, delay variation (jitter), bandwidth, and packet-loss parameters on a network, while maintaining manageability, simplicity, and scalability, is the recipe for maintaining an infrastructure that truly serves the business environment from end to end.

One of the benefits that MSPP provides is a very reliable MAN infrastructure upon which to deploy IP telephony and IP video, as shown in Figure 1-13. The 50 ms switch time of SONET provides reliability, and the high data speeds allow thousands of calls to be carried across the MSPP network. Customers and service providers can rapidly deploy these services. Thus, customers who are deploying their own private networks within the metropolitan area using MSPP can quickly turn up service and begin realizing the cost savings of IP-based telephony (IPT) and IP-based video. Likewise, service providers can deploy these services and generate new revenue streams.

Figure 1-13 *MSPP Infrastructure Used to Deploy Voice over IP (VoIP)*

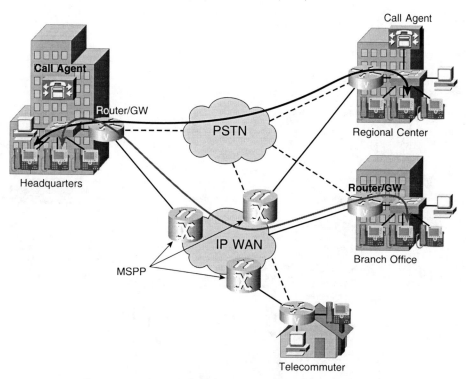

Some cards that can be used within MSPPs use Layer 2 (switching) and Layer 3 (routing) with built-in technology. Cisco calls this card the ML card (for "multiLayer"). This simply suggests that the card contains Layer 1, Layer 2, and Layer 3 features and functionality. Thus, MSPP is an enabling underlying infrastructure for these emerging, high-demand services. Layer 2 functionality handles the switching, thereby creating a truly switched network between the nodes of the optical ring. The Layer 3 service that this ML card provides prioritizes voice and video traffic over data traffic, known as QoS, as shown in Figure 1-14. (This MSPP infrastructure was referred to earlier as RPR.) This virtual LAN feature enables voice and video to flow seamlessly across the MAN. It is called virtual because traditionally the "locality" of LANs has been the premises or campus. Now local is the optical ring, which could be 50 or more miles long, not just what is physically "local" in the traditional use of the word.

It is important to point out that point-to-point Ethernet over SONET, or Ethernet Private Line, is another means of enabling voice and video, even without RPR. Today's LANs almost always consist of Ethernet as the LAN technology. Thus, when a customer receives an Ethernet handoff from the MSPP, the MSPP is easily integrated right into the LAN infrastructure. The customer or service provider can connect multiple sites with Ethernet Private Lines and then use this underlying infrastructure to transport the higher-layer voice or video.

Figure 1-14 *MSPP Multilayer Card Handles Layer 2 Switching and Layer 3 QoS*

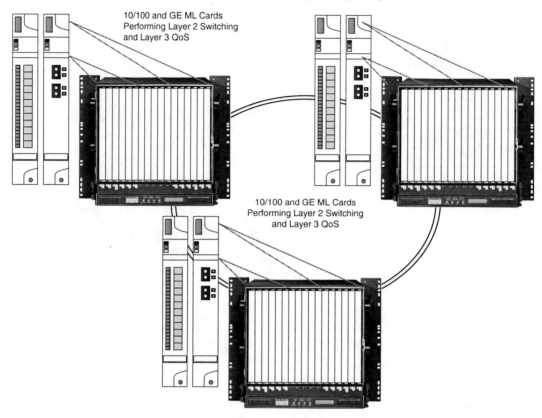

TCO

Reducing capital and operational expenditures is critically important to the long-term survivability of service providers and enterprise customers. Delivering new services over existing infrastructure enables the service provider to improve top-line revenue numbers and enterprise customers to support additional business capabilities without forklift upgrades. Thus, next-generation platforms must not only deliver advanced multiservice capabilities, but also help reduce the overall costs of operating a network.

Legacy Optical Platforms

To better understand the benefits of next-generation optical MSPP, you should review legacy optical networking architectures. In the next section, you will learn about three different legacy platforms: a legacy OC-3 platform, a legacy OC-12 platform, and a legacy OC-48 platform. Then you will compare them to today's MSPP technology, which delivers the same services. Although numerous vendors provided these legacy platforms, the focus remains on architecture that reflected the leading vendors' legacy optical platforms, to compare MSPP with the best legacy equipment.

OC-12 and Below-High-Speed Optics

The legacy optical platform shown in Figure 1-15 uses OC-3 or OC-12 optical cards for the high-speed cards. The footprint of this platform allows the customer to place six of these in a bay. Not much flexibility exists regarding what services can be delivered or dropped from this OC-3 or OC-12 configured platform.

Figure 1-15 *OC-12 or Below Platform*

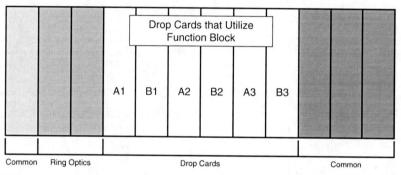

The cross-connecting is done within the high-speed optical cards, OC-3 or OC-12. The system controller, timing, communication, and alarming (user panel) cards reside in the section of the shelf labeled "Common0."

The DS1s, DS3s, and other services specific to the vendor are delivered in the section of the shelf labeled "Drop Cards." Different groupings in the shelf are shown as A, B, and C. Each group must consist of either 28 DS1s, 1 DS3, or 1 STS1. Unfortunately, you cannot mix different service cards within the same group: Even if a slot was available in a given group, such as group A, group B, or group C, you could use that slot for only the same service type, such as DS1 or DS3. This means that slots might have to go unused if the service type is not required in any given group.

If you are using OC-12 cards on the high-speed slots of this shelf, it could drop out an OC-3. However, this shelf is not upgradeable in service from an OC-3 to an OC-12, and it could never be upgraded to an OC-48 or OC-192. On the other hand, MSPP allows such upgrades simply by taking a 50 ms SONET switch hit.

High-bit digital subscriber line (HDSL) is also an option for certain vendors' products in the "Drop Card" section. This shelf can drop out up to 84 DS1s, but in so doing it leaves no room for DS3 or other service drops. This shelf also can drop up to three DS3s per shelf, but, again, nothing else can be dropped.

Notice that next-generation service drops, such as Ethernet and SAN protocols (FC, FICON, and ESCON) are available from this platform. There also is no transmux functionality, which allows DS1s to be multiplexed and carried in a DS3 payload through the ring, demultiplexed on the other end, and handed back off as DS1s. Finally, no DWDM functionality exists on this platform.

Given the shelf drops, Table 1-1 shows how many services can be dropped from a shelf using this platform.

Table 1-1   *Drops Available for a Typical OC-3/OC-12 Legacy Optical Platform*

Service Type	Number of Drops	Conditions
DS1	28	If no DS3s
DS3/ EC-1	3	If no DS1s
HDSL	2	If no data
OC-3	1	If OC-12 ring speed
Data	1	If no HDSL

OC-12 High-Speed Optics Only

The legacy platform shown in Figure 1-16 uses OC-12 optical cards only for the high-speed optics. The footprint of this platform allows the customer to place four of these in a bay. Not much flexibility exists for which services can be delivered or dropped from the OC-12 ring.

Figure 1-16 *Legacy OC-12-Only Platform*

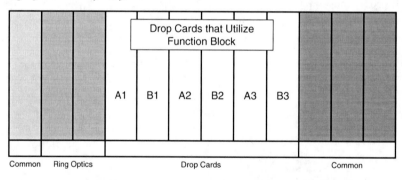

For example, in the OC-3 or OC-12 platforms, you can drop DS1s. This OC-12 can drop only DS3s or EC1s, but not services more granular than a DS3, such as a DS1. You can drop up to four DS3s or EC1s per shelf, and you can mix them as well. OC-1s and OC-3s can also be dropped.

Cross-connecting is done in the high-speed optical cards, OC-3 or OC-12. The system controller, timing, communication, and alarming (user panel) cards reside in the section of the shelf labeled "Common."

The DS3s or EC1s, and other services specific to the vendor are delivered in the section of the shelf labeled "Drop Cards."

Again, as with the OC-3/OC-12 shelf, this shelf is not upgradeable to any other optical ring speed other than OC-12. MSPP allows such upgrades simply by taking a 50 ms SONET switch hit.

Just as in the previous case, notice that advanced services, such as Ethernet and SAN protocols (FC, FICON, or ESCON) are available from this platform. No transmux functionality or DWDM functionality on this platform is integrated into the SONET platform.

Given the shelf drops, Table 1-2 shows how many services can be dropped from a shelf using a typical legacy OC-12 optical platform.

Table 1-2   *Drops Available for a Typical OC-12 Legacy Optical Platform*

Service Type	Number of Drops	Conditions
DS1	0	—
DS3/ EC-1	4 protected	If no LAN cards
HDSL	2	If no data
OC-3	4	If no DS3s or EC-1s
Data	4	If no HDSL

OC-48 High-Speed Optics Only

The legacy platform as shown in Figure 1-17 uses OC-48 optical cards only for the high-speed optics. The footprint of this platform allows the customer to place two of these in a bay. Not much flexibility exists regarding which services can be delivered or dropped from the OC-48 ring.

Figure 1-17   *Legacy OC-48-Only Platform*

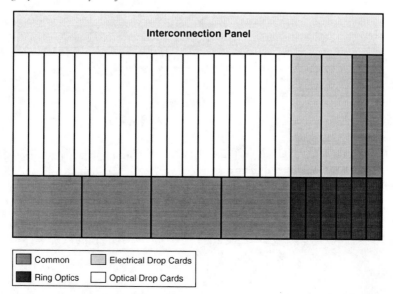

For example, in the OC-12 platform, you cannot drop DS1s. This OC-48 can drop only DS3s or EC1s, but not services which are more granular than a DS3, such as a DS1. You can drop up to 48 DS3s or EC1s per shelf, and you also can mix them. Again, just as with the OC-12-only shelf, OC-1s and OC-3s can be dropped.

OC-12s can be dropped from the OC-48-only platform, but keep in mind that these cards weigh more than 25 pounds each.

The cross-connections are performed on the high-speed optical cards, OC-3 or OC-12. The system controller, timing, communication, and alarming (user panel) cards reside in the section of the shelf labeled "Common."

The DS3s or EC1s or OC-12, or other services specific to the vendor are delivered in the section of the shelf labeled "Drop Cards."

Again, as with the OC-3/OC-12 shelf, this shelf is not upgradeable to any other optical ring speed than OC-48.

Just as in the previous cases, notice that advanced services, such as Ethernet and SAN protocols (FC, FICON, or ESCON) are available from this platform. There also is no transmux functionality or DWDM functionality on this platform that is integrated into the SONET platform.

Given the shelf drops, Table 1-3 shows how many services can be dropped from a shelf using a typical legacy OC-48 optical platform.

Table 1-3   *Drops Available for a Typical OC-48 Legacy Optical Platform*

Service Type	Number of Drops	Conditions
DS1	0	—
DS3/ EC-1	48 protected	—
HDSL	—	—
OC-3	4 protected	If no OC-12
OC-12	1 protected	—

MSPP

To compare the OC-3/OC-12, OC-12-only, and OC-48-only legacy platforms shown in the last section with the same implementation of OC-3, OC-12, and OC-48 on MSPPs, take a look at Figures 1-18, 1-19, and 1-20.

Figure 1-18 shows an MSPP configured for OC-3 high-speed optics.

Figure 1-18 *MSPP Configured for OC-3*

Figure 1-19 shows an MSPP configured for OC-12 high-speed optics.

Figure 1-19 *MSPP Configured for OC-12 High-Speed Optics*

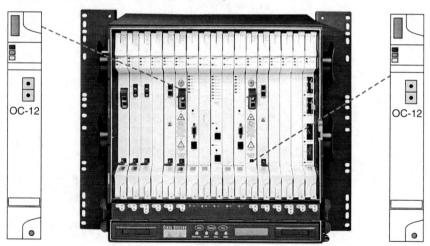

Figure 1-20 shows an MSPP configured for OC-48 high-speed optics.

MSPP offers so many advantages that it is difficult to know where to begin. Not only do MSPPs provide flexibility within the shelf, but they also allow for very flexible topologies, as shown in Figure 1-21.

Figure 1-20 *MSPP Configured for OC-48 High-Speed Optics*

Figure 1-21 *Various Topologies Available for MSPPs Integrated within a Network*

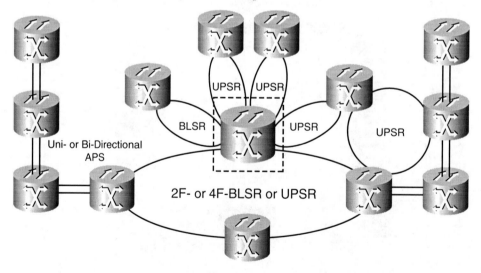

With MSPPs, it is easy to configure the topology of the MSPP network by simply pointing and clicking on the desired topology choice within the graphical user interface (GUI). Choices include 2F or 4F bidirectional line switched ring (BLSR), unidirectional path switched ring (UPSR), and unidirectional or bidirectional APS.

An MSPP must support a wide variety of network topologies. In SONET, the capability to support UPSR as stated by the Telcordia GR-1400, 2-Fiber BLSR and 4-Fiber BLSR as stated by the Telcordia GR-1230, and 1+1 APS is essential.

Figure 1-22 shows the flexibility of Path Protected Mesh Networks (PPMN) and a path-protection scheme. A path-protection scheme based on meshing is similar to a routing environment: Not only does it offer ease of management and provisioning, but it also can provide significant cost savings. Because of its strict adherence to SONET standards, a PPMN logical ring is similar to the Telecordia-specified standard for UPSR. These cost savings are realized when network span needs to be scaled to a higher bandwidth. Without PPMN functionality, a complete UPSR ring must be upgraded with new higher-rate optics, even if only a fraction of the ring needs the additional bandwidth.

Figure 1-22 *Path-Protected Meshed Network Topology and Protection Scheme*

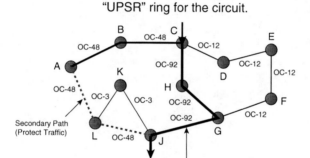

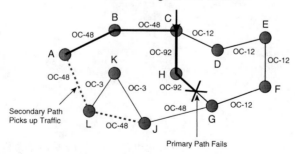

With PPMN, only spans that require the additional bandwidth are upgraded, thus reducing the overall cost to support additional higher-bandwidth services. Unfortunately, RBOCs that are encumbered with legacy OSS cannot take advantage of the benefits of PPMN. Therefore, this is seen predominately in the Enterprise and CLEC networks.

High-Speed Optical Cards

When using the high-speed optics of MSPPs, notice that the shelf is the same, regardless of which set of high-speed slots cards you are using (such as OC-3, OC-12, OC-48, or even OC-192 optics).

NOTE   The MSPP used throughout this book is a leading vendor's platform; sections are labeled as "Common," "Main"- High Speed Optics, and "Drop Cards."

Instead of building cross-connect functionality into the high-speed slot optical cards, as in numerous legacy platforms, this functionality resides within dedicated cross-connect cards. The division of tasks thus is reflected in the architecture, making it easy to upgrade optics.

High-speed optics use a dedicated transmit/receive port. They also weigh nothing close to what legacy cards did because of the exponential improvement in electronics and optical technology. Next-generation MSPP optics cards use Small Form Factor Pluggables (SFP), similar to the way Gig E uses Gigabit Interface Connectors (GBICs). The user thus requires only a multirate card for sparing because that card can use the SFP technology. Again, this is similar to the GBICs, which, though they all provide Ethernet protocol service drops, have varying span budgets and loss specifications (such as short reach, long reach, and extra long reach), as shown in Figure 1-23.

Figure 1-23   *Gigabit Ethernet Using Varying GBICs for Different SPAN Budget Requirements*

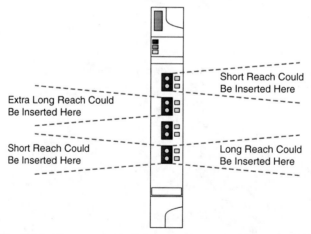

MSPP offers added features, including in-service optical line rate upgrades (or downgrades, if necessary). These are quite simple to perform. The following example illustrates performing an upgrade from OC-12 to OC-48:

Step 1   Manually switch any traffic that is riding the span that you are going to upgrade. Remove the OC-12 card, and change the provisioning of the slot to an OC-48. Install the OC-48 card in the slot. Repeat this step on the opposite side of the span. Manually switch the working card to the newly installed OC-48 card using the software on both sides of the span. Now the ring traffic is riding on the new OC-48 card, as shown in Figure 1-24. Figure 1-25 shows the traffic being rolled to the new OC-48 card.

Figure 1-24 *Replacing an OC-12 Protection Card with an OC-48 Card*

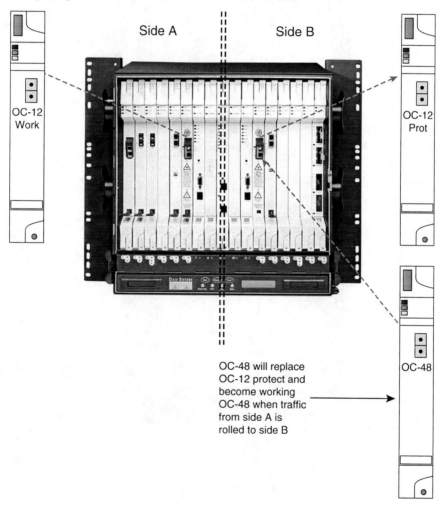

OC-48 will replace OC-12 protect and become working OC-48 when traffic from side A is rolled to side B

Step 2 Change the provisioned slot to an OC-48 and install the card (on both ends of the span). Remove all switches that might be up (see Figure 1-26).

Figure 1-25 *Working Traffic Placed on New OC-48 Card*

Figure 1-26 *Replacing an OC-12 Card with an OC-48 in Slot Provisioned as "Protect"*

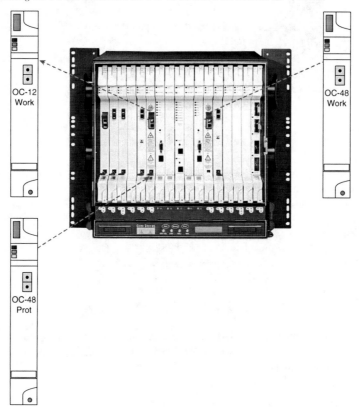

The upgrade is complete. The traffic can stay on the OC-48 card as is or can be rolled back onto the second OC-48 card that was installed. This is up to the user and procedures used.

As mentioned, DWDM capabilities are also available on the high-speed optical cards. Various wavelengths, as defined by the ITU, can be multiplexed over the fiber, thus improving the overall bandwidth of the fiber by orders of magnitude. For example, you could multiplex up to 32 wavelengths of OC-48 over a single fiber, providing an incredible OC-1536 bandwidth, as shown in Figure 1-27. Figure 1-28 shows DWDM delivered off MSPP shelves.

Figure 1-27 *Eighteen DWDM Wavelengths Multiplexed over a Single Fiber*

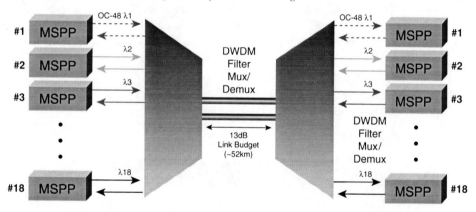

Figure 1-28 *DWDM off MSPP*

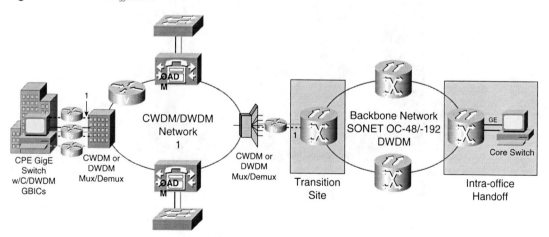

Table 1-4 lists the DWDM wavelengths as defined by the ITU for OC-48 and OC-192.

Table 1-4 *ITU Wavelengths in Nanometers for OC-48*

1528.77
1530.33
1531.12
1531.90

The value of MSPPs is found in the utilization, density, and flexibility of drops or service cards. Both legacy and next-generation MSPPs require high-speed optical cards and use them in a similar manner; the exceptions (as previously shown) include upgradeability, flexibility, and ring optic shelf space requirements. However, as you will see throughout the rest of this chapter, the density of these services has increased in MSPPs in many cases.

Looking at Figure 1-29, you can see that a number of slots available for drop cards can deliver various services. This varies depending on the vendor.

Figure 1-29 *Drop Cards off MSPP*

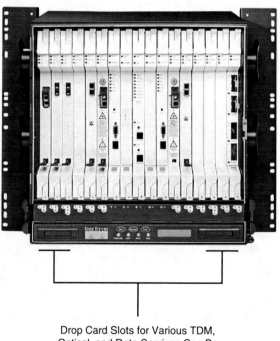

Drop Card Slots for Various TDM,
Optical, and Data Services Can Be
Inserted in These Slots

Again, the common cards and high-speed slot optical cards are in the central seven slots of the shelf. This leaves the rest of the shelf available for drop cards; these can be electrical, optical, or data services cards, or any combination of these.

Electrical Service Cards

Electrical cards are sometimes referred to as TDM-based cards. Depending on the vendor, each card is commonly referred to as low density or high density. An OC-48, for example, can carry 48 DS3s, and an OC-192 can carry 192 DS3s on the ring. If low-density cards are used, as shown in Figure 1-30, you could fill the entire left side of the shelf and drop out 48 DS3s if each working card has 12 ports. Figure 1-31 shows a high-density application in which there are 48 ports on each working and protection card.

Figure 1-30 *MSPP Shelf Filled with Electrical Cards*

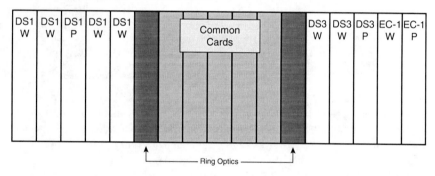

Figure 1-31 *MSPP Filled with 48 DS3s on Left Side of Shelf*

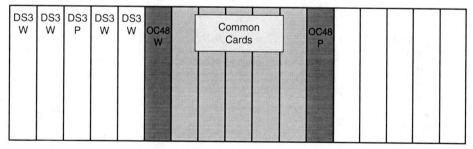

This brings up another advantage of MSPP over many legacy platforms: protection schemes. MSPPs allow what is referred to as either 1:1 (1 by 1) or 1:*n* (1 by *n*) protection for electrical DS1, DS3, and EC-1 cards, as Figure 1-32 shows. In this example, DS3 cards have 12 ports each, and DS1 cards have 14 ports each.

1:1 protection means that every card that will carry live traffic must have another card that will protect it if it fails.

1:*n* protection means that a protection card can protect two or more cards ("n" cards) where *n* is 1 to 5.

Figure 1-32 *1:1 and 1:n Protection Schemes for Low-Density Cards*

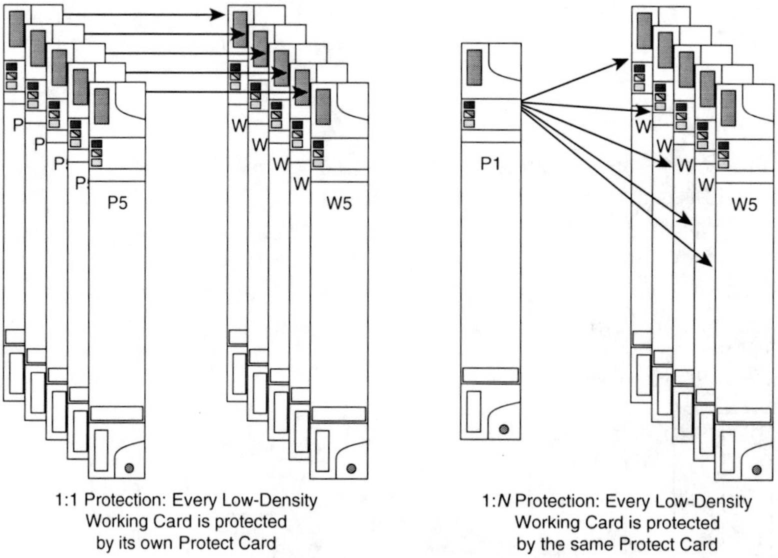

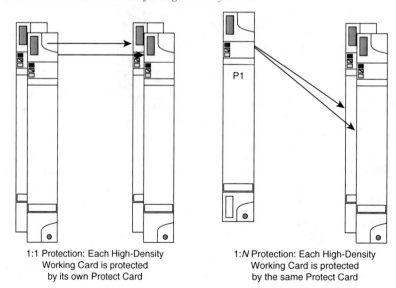

1:1 Protection: Every Low-Density
Working Card is protected
by its own Protect Card

1:*N* Protection: Every Low-Density
Working Card is protected
by the same Protect Card

If high-density cards are used in this leading vendor's MSPP, as shown in Figure 1-33, the user can drop up to 48 DS3s out of each card. This requires only two working cards and one protection card, totaling three slots, to drop out 96 DS3s per side. With five service or drop slots per side, this leaves two slots per side available for other service cards, such as optical or data cards. (Keep in mind that DS3 cards have 48 ports each and DS1 cards have 56 ports each.) This also allows all 192 DS3s to be dropped from this shelf, if necessary.

Figure 1-33 *1 and 1:n Protection Schemes for High-Density Cards*

1:1 Protection: Each High-Density
Working Card is protected
by its own Protect Card

1:*N* Protection: Each High-Density
Working Card is protected
by the same Protect Card

A similar scenario exists for DS1 cards. DS1 cards come in the same two flavors, low density and high density. Low-density cards have about 14 ports per card. High-density cards will have up to 56 DS1 ports per card. Again, this is a quantum leap over legacy equipment. Just one DS1 working card and one DS1 protection card, using just two slots in one MSPP shelf, can replace two entire OC-3/OC-12 legacy shelves, as shown in Figure 1-34.

Figure 1-34 *Two MSPP High-Density DS1 Cards Replace Two Legacy Platform Shelves*

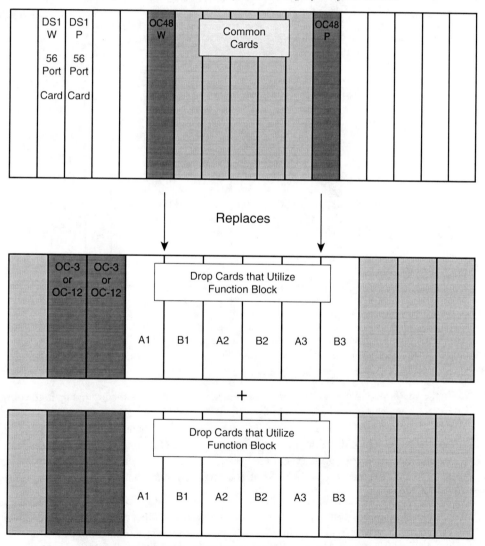

Now consider that an MSPP shelf filled with four working DS1 high-density cards (two on each side) and two protection DS1 high-density cards (one on each side) can drop 224 DS1s and leave four slots available (two on each side) for other optical or data service cards. One MSPP shelf can drop out more DS1s than an entire bay of legacy shelves.

Optical Service Cards

Figure 1-35 shows some typical optical cards and port densities, (densities vary by vendor). OC-3 cards come in various port densities, such as one-port, four-port, and eight-port cards. OC-12 cards come in single-port and four-port cards. OC-48 and OC-192 come in single-port cards. Today there are even multirate cards that offer 12 optical ports that can accept SFPs for OC-3, OC-12, and OC-48 optics, for incredible density and variety in a single slot. Various wavelength SFPs also exist for these optical cards, for various distance requirements based on signal loss over the fiber. This is often referred to as span budget, and it is noted in decibels (dB).

Figure 1-35 *Typical Optical Cards Used in MSPP*

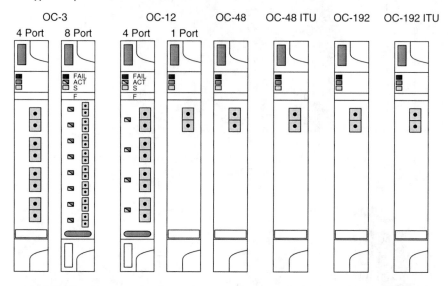

1+1 (1 plus 1) protection is used to protect the optical ports when protection is required. Thus, any given port on a four-port OC-3 card (protection port) will protect another port on a four-port OC-3 card (working port).

In the shelf under consideration, there are ten service slots. This means that up to ten optical cards of any sort can be used, as long as the required STSs (1 STS-1 is equivalent to an OC-1) required to be carried on the ring do not exceed the bandwidth of the high-speed optical cards. Thus, if the high-speed optical cards are OC-48, you can mix any type of optical card based on service requirements as desired. However, you cannot carry more than a combined number of 48 STSs on the ring (assuming that a UPSR is used).

Also, if every optical card required protection, that would allow six pair of optical cards, with each pair consisting of one working card and one protect card. Again, these can be mixed based on drop requirements. With the recent emergence of the multirate optical card, the distinction between high-speed optics and low-speed optics on a slot-by-slot basis is no longer a factor: Both high-speed and low-speed cards can be used in the same slot. Thus, you can see the enormous leap in density and flexibility of these optical cards, as well as a greatly reduced space requirement. Optics cards can also be mixed with electrical service cards and data service cards, as discussed in the next section.

Data Service Cards

The primary data cards in use today are Ethernet cards. They typically come in two basic speed categories, 10/100 Mb and 1 Gb speeds. The gigabit cards can have fixed optics ports and can be open-slotted (for the insertion of GBIC) to allow different optical wavelengths for different distance requirements based on signal loss over the fiber. The 10-/100-Mb cards typically come in 8-port to 12-port densities; the gigabit cards come in single-port to four-port cards. The gigabit cards can also be provisioned at a subrate speed, such as less than 1 Gb. This allows service providers to sell subrate gigabit and metered services, and gives flexibility to their product offerings: If a customer needs more than 100 Mb of bandwidth but less than a gigabit, bandwidth can be provisioned so as to not require full utilization of 21 to 24 STSs, depending on the vendor's equipment (1 STS or OC-1 is approximately 51.84 Mb), to be dedicated per gigabit port.

Along with Layer 1 Ethernet over SONET implementations, MSPP has the capability to use Layer 2 and Layer 3 routing features to provide multilayer Ethernet. Multilayer Ethernet cards enable the user to use Layer 2 switching and Layer 3 routing features. One of the most significant is that of QoS, which is used to classify, prioritize, and queue traffic based on traffic type. As mentioned earlier in this chapter, certain traffic, such as voice and video, cannot tolerate latency, so it must be given priority when competing with data traffic for the available bandwidth.

These multilayer cards also allow carriers to provide, or customers to privately deploy, a transparent LAN.

In addition, this service must allow any number of customer VLANs to be tunneled together so that the metro network is run by the service provider and so that the customer is not limited by number or forced to use specific assigned VLANs.

Figure 1-36 illustrates how the metro Ethernet services overlay on the metro optical transport network and the proposed architecture to deliver TLSs. The TLS flow is end to end between the switches.

SAN Service Cards

As mentioned earlier, certain cards deliver SAN protocol services.

These are some of the key features and functions of the SAN cards:

- Line-rate support for 1 Gbps and 2 Gbps
- FC or FICON
- SAN extension over both SONET and synchronous digital hierarchy infrastructures
- Use of generic framing procedure-transparent (GFP-T) to deliver low-latency SAN transport

Figure 1-36 *Ethernet LAN Services from End to End*

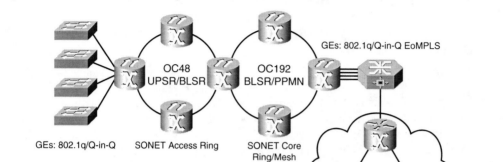

Some of the advantages of SAN over SONET include these:

- Capability to take advantage of SONET resiliency schemes
- Support for extended distances with innovative business-to-business credit management
- Transport efficiency through "subrating" the service and virtual concatenation (VCAT)
- ESCON support

Additionally, MSPP SAN features include the following:

- **Operational ease**—Storage IT experts do not have to become optical transport experts; they can simply purchase FC services from a managed service provider.
- **Disaster recovery/business continuance (DR/BC) planning**—IT personnel are not restricted by the unavailability of dark-fiber availability between data centers. If dark fiber is unavailable, SONET interconnect likely can be quickly turned up.
- **Multiservice/platform convergence**—MSPPs provide enterprise customers with the capability to deliver multiple voice, data, video, and storage services from a single, high-availability platform.
- **Single point of contact**—When MSPP-based SAN is deployed as customer premises equipment (CPE), the customer has a single point of contact, which is the MSPP, for transport of Ethernet, storage, and TDM services.

OAM&P

One of the most important aspects of any piece of equipment is the ongoing costs associated with provisioning or commissioning the equipment, operating it, administering the installed base of that equipment, and maintaining it. Industry pundits refer to this as OAM&P, for operations, administration, maintenance, and provisioning.

Some of the OAM&P advantages that MSPPs have over legacy platforms to which we have already alluded are listed here:

- Significant footprint and power savings, compared to legacy SONET platforms
- Integration of data and TDM traffic on a single scalable platform
- Faster installation and service turn-up of metro optical Ethernet and TDM services, which facilitates accelerated time to revenue
- Integrated optical network-management solution that simplifies provisioning, monitoring, and troubleshooting
- One multiservice network element instead of many single-service products

GUI

GUIs are used with MSPP and allow point-and-click interaction, rapidly improving the speed of provisioning.

GUI interfaces have gained acceptance among service providers, allowing technicians to perform OAM&P functions intuitively and with less training. In addition to the GUI, the MSPP should offer traditional command-line interfaces (CLIs) such as the widely accepted Telcordia TL1 interface.

A leading MSPP provides network, element, and card views of SONET. This enables the user to point to a node on a map and click on it to bring up the node view. You can then click on a card in the element view and provision card-level settings. This is a time saver because it enables the user to literally turn up a multinode ring in just minutes or hours instead of days. With IP as the communication protocol, you can access the elements from literally anywhere in the world if you have access to the Internet and know the appropriate passwords.

End-to-End Provisioning

With cross-network circuit provisioning, circuits can be provisioned across network elements without needing to provision the circuit on a node-by-node basis. The MSPP can reduce the required time to provision a circuit to less than 1 minute, as shown in Figure 1-37. Different views of the map, shelf, and card enable users to quickly navigate through screens during provisioning, testing, turn-ups, and troubleshooting.

Figure 1-37 *Map View and Shelf View of MSPPs*

Card View   Network View of Ring

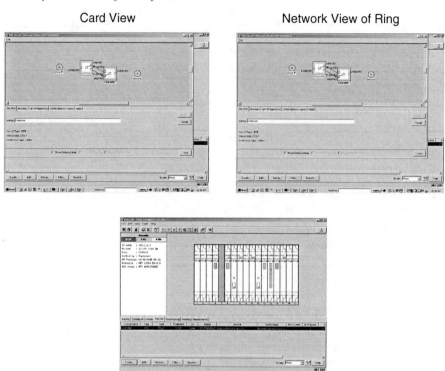

Shelf View

Wizards

Procedure wizards guide users step by step through complicated functions (see Figure 1-38). By employing wizards, features such as span upgrade, software installations, and circuit provisioning are available. These MSPP wizards dramatically reduce the complexity of many OAM&P tasks.

Figure 1-38 *An End-to-End Provisioning Sequence of Screens*

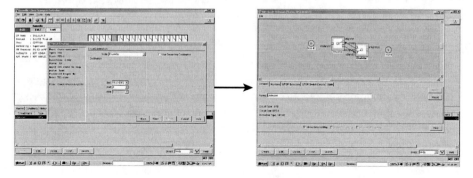

Alarms

Alarming screens are also advanced as compared to legacy platforms. Alarm logs can be archived more easily because LAN equipment using the IP protocol is connected to the MSPP. In addition, alarms are displayed automatically as they arise on the craft interface and show when they have been acknowledged and cleared, with different colors for easy recognition. In legacy platform craft interfaces, alarms had to be manually retrieved to be seen: If an alarm occurred, it sometimes took minutes or hours to be identified because the alarms were not automatically displayed. Additionally, there was no color indication to help facilitate the process of identifying which state alarms were in. Alarms could be logged but not viewed as easily when looking for patterns and diagnosing recurring problems.

Software Downloads

MSPP has simplified software downloads. In the past, what called for a craft person to be on site is now a simple function of a point-and-select function of the software from any location with TCP/IP access. Thanks to IP management enabled on MSPP, these updates can be scheduled from servers to be done automatically during off-peak usage hours. This takes advantage of server and data-networking technology, reducing the personnel required to perform them and also the time it takes to execute the update.

Event Logging

Event logging has existed in legacy optical platforms; however, the logs had to be manually retrieved. The data in the log could not be color-coded for easy recognition. Imagine looking at event log history that dated back perhaps months without a color code classifying certain event types—it could take hours to find the data you need. With MSPP GUI color codes, the search for data can be accelerated dramatically, enabling you to complete a search in minutes versus what could potentially take hours.

Capital Expense Reduction

It goes almost without saying that the capital expenditure (capex) reduction is significant for those who use MSPPs. A ring can be deployed today from 25 to more than 50 percent of the cost of a legacy ring because of the increase of port density per card and the overall shelf density. If there were no operational advantages, the savings in capex alone would be enough to justify the migration to MSPP.

Not only have the technology improvements allowed for greater port densities and numerous service drops on each shelf, but MSPP component costs have also greatly been reduced: More MSPP manufacturers are in the market today than when only legacy platforms existed. The rapid deployment of services on MSPP and the reduced cost barrier to entry into the market have increased the number of service providers who offer services off MSPPs. In the past, only a few service providers (perhaps even only one service provider) were in any one market because of either regulatory constraints on competition or cost barriers to entry. The expense of building a large network is enormous. No revenue can be realized until the network is in place, and even when the network is complete, there is no guarantee that the revenue will be adequate to pay for the investment. With deregulation of the telecommunications industry, competitive carriers could lease the incumbent networks. They would thus gain a more cost-effective pseudo-network infrastructure and bring in revenue to offset the costs of gradually building their own networks.

Another emerging phenomenon in the electronics industry that blossomed about the time MSPPs began to proliferate is that of outsourcing electronic component manufacturing by MSPP vendors. Electronic component manufacturing suppliers (EMS) have had to become extremely price-conscious to keep from losing business from their key MSPP customers. A brief review of the evolution of the EMS industry will shed a little light on the overall component price-reduction phenomenon.

Outsourcing in the electronics industry has evolved dramatically during the past decade. In its earliest and most basic form, the MSPP manufacturers made "manufacture vs. outsource" decisions based largely on opportunities to reduce costs or meet specialized manufacturing needs. The EMS companies took on manufacturing for specific products on a contract-by-contract basis. Through economies of scale across many such contracts, EMSs leveraged operational expertise, lower-cost labor, and buying power (or some combination of these) to lower the costs to the MSPP manufacturer. For instance, an MSPP manufacturer that made a broad range of optics products could achieve greater economics by working with an EMS that specialized in building these components instead of maintaining that capability in-house. Because the EMS built far more optics products than the MSPP manufacturer ever would, the EMS received significant volume discounts from its own suppliers—savings that the EMS passed along in part to its customers. Also, because the EMS specialized in manufacturing, it was more efficient in reducing setup and changeover times.

Such conventional outsourcing served electronics legacy platform manufacturers well into the mid-1990s, when demand was relatively predictable, competition was less fierce, and products were simpler. Then things began to change. Products grew more complex. As MSPP manufacturers emerged, they found that they had to dramatically boost investments in capital equipment to keep up with new manufacturing requirements, which ate into profits. The pace of innovation increased dramatically, leading to shorter product life cycles and increased pressure to decrease time to market.

In response, a number of MSPP manufacturers have used outsourcing to quickly and cost-effectively enter new markets. By teaming with an experienced partner, an MSPP manufacturer can significantly cut the time and cost involved in developing new products. Some MSPP manufacturers found that a move toward more collaborative outsourcing arrangements could improve their planning accuracy and capability to respond more quickly to changing market conditions. By outsourcing manufacturing and some of their "upstream" supply chain activities, MSPP manufacturers could free themselves to focus on their core competencies, tighten planning processes, and be more responsive to customer demand, whether this was a service provider or enterprise customer. Instead of having to ramp up or down the workforce, or start and shut down operations, the MSPP manufacturer could simply adjust the fee structure of the outsourcing agreement.

This increased demand for EMS products created an increased competition among electronic component manufacturing suppliers, thus driving down overall component prices (as competition always does) and greatly reduced the production cost of the MSPPs themselves.

This increased demand for MSPPs meant that component suppliers to the MSPP manufacturers could ramp up production and thus decrease the per-unit production costs of the components—another case of "demand creating demand."

Thus, MSPP suppliers, now more numerous, were forced to be more competitive and pass on this cost savings to the service providers themselves. This then reduced the overall capital costs for the service provider.

Thus, service providers receive a less expensive MSPP that can deliver more services, which are far more diverse from a reduced footprint. This, of course, means that the end customer of the TDM, optical, and data services has a choice of service providers and can negotiate more fiercely and get more bandwidth for the dollar.

Now you can see the supply chain effect from beginning to end: MSPP suppliers receive better component pricing because of the volumes these component manufacturers produce. These savings are passed along to MSPP customers in the form of more competitive pricing. This, in turn, is passed along to the customers who buy the services from the service providers.

Summary

MSPPs are changing the economics of service delivery for a number of reasons. The increased demand for bandwidth in the LAN caused by an explosion of data technology and services has prompted service providers to transform themselves. Service providers must rapidly provision a variety of services from traditional TDM circuits to next-generation data services such as metro Ethernet and storage. The high service density, flexibility, scalability, and ease of provisioning of MSPPs are allowing service providers and enterprise customers (who deploy their own private optical networks) to experience lower operational, administrative, and maintenance costs. Additionally, the service providers and enterprise customers are capable of decreasing the time required to turn up service, known as service velocity. This allows service providers to experience a lower TCO.

This enables service providers to offer customers next-generation data, storage, and DWDM technology integrated into the MSPPs, without the added cost and management of separate equipment. Thus, service providers benefit from higher margins on the services they deliver and an increase in the variety and quantity they can sell. Customers also receive more and varied bandwidth for their dollar. Thus, all parties can rejoice in the emergence of MSPPs.

This chapter covers the following topics:

- What Is an MSPP Network?
- Fiber Optic Basics
- SONET/SDH Principles
- SONET Overhead
- Synchronization and Timing

CHAPTER **2**

Technology Foundation for MSPP Networks

Multiservice Provisioning Platform (MSPP) networks are currently being deployed by service providers and in private networks worldwide to enable flexible, resilient, high-capacity transport of voice, video, and data traffic. MSPP systems have overtaken traditional time-division multiplexing (TDM)–only transport systems because the need for networks to provide far more than voice services has dramatically increased. This chapter introduces you to several foundational topics that help you grasp the role of the MSPP in modern networks. This includes a fundamental description of the manner in which light travels through optical fiber, and an introduction to the Synchronous Optical Network (SONET) and Synchronous Digital Hierarchy (SDH) standards upon which MSPPs are based. Finally, this chapter discusses the importance of precise synchronization in an MSPP network.

What Is an MSPP Network?

MSPP networks are relatively new to the telecommunications industry. These networks are the result of the need to combine traditional TDM networks and data networks on the same physical platform or equipment. These platforms are sometimes referred to as next-generation SONET or SDH systems. The typical MSPP can operate at any of the SONET OC-*n* rates, from OC-192 to OC-3, and also can provide various drop ports of electrical or optical nature. These drop ports can be DS-1, DS-3, EC-1 (STS-1), OC-3, OC-12, OC-48, OC-192, Ethernet, Fibre Channel (FC), Fiber Connectivity (FICON), or Enterprise Systems Connection (ESCON), and can also include dense wavelength-division multiplexing (DWDM) capabilities.

MSPP networks can be deployed in point-to-point, linear add/drop, unidirectional path-switched ring (UPSR), and bidirectional line switch ring (BLSR) configurations.

Fiber Optic Basics

MSPP systems use optical fibers to carry traffic through the network. In this section, you explore the basic construction of optical fiber and learn how the reflection and refraction of light affect its propagation through a fiber cable. This section also discusses the differences between the two major classes of optical fibers, multimode and single mode.

Optical Fiber

Two fundamental components of optical fiber allow it to transport light: the core and the cladding. Most of the light travels from the launch point to the end of the fiber segment in the core. The cladding is around the core to confine the light. Figure 2-1 illustrates the typical construction of an optical fiber.

Figure 2-1   *Typical Construction of an Optical Fiber*

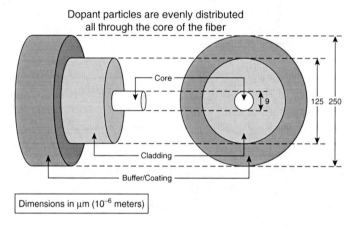

The diameters of the core and cladding are shown, but the core diameter might vary for different fiber types. In this case, the core diameter of 9 µm is very small, considering that the diameter of a human hair is about 50 µm.

The cladding's outer diameter is a standard size of 125 µm. The uniformity in cladding diameter allows for ease of manufacturing and ubiquity among optical component manufacturers.

The core and the cladding are made of solid glass. The only difference between the two is the method by which the glass was constructed. Each has different impurities, by design, which change the speed of light in the glass. These speed differences confine the light to the core.

The final element in this picture is the buffer/coating, which protects the fiber from scratches and moisture. Just as a glass pane can be scratched and easily broken, fiber-optic cable exhibits similar physical properties. If the fiber were scratched, in the worst case, the scratch could propagate, resulting in a severed optical fiber.

Light Propagation in Fiber

Reflection and refraction are the two phenomena responsible for making optical fiber work.

Reflection and Refraction

The phenomenon that must occur for light to be confined within the core is called reflection. Light in the core remains in the core because it is reflected by the cladding as it traverses the optical fiber. Thus, reflection is a light ray bouncing off the interface of two materials. It is most familiarly illustrated as light from an object being returned (reflected) as your image in a mirror.

Refraction, on the other hand, occurs when light strikes the cladding at a different angle (compared to the angle of reflection). Light undergoes refraction when it exits the core and proceeds into the cladding. This degrades optical transmission effects because key parts of the optical signal pulse are lost in the fiber cladding. Thus, refraction is the bending of the light ray while going from one material to another.

Refraction is probably less familiar, but you can see its effect in day-to-day living. For example, a drinking straw placed in a glass of clear liquid looks as if it is bent. You know that it is not bent, but the refraction properties of the liquid cause the straw to appear as bent.

Reflection and refraction are described mathematically by relating the angle at which they intersect the material surface and the angle of the resultant ray. In the case of reflection, the angles are equal.

Figure 2-2 illustrates a typical refraction/reflection scenario. Where light is predominantly reflected and not refracted, the index of refraction for the core is greater than the index of refraction for the cladding. The index of refraction is a material property of the glass and is discussed in the next section.

Figure 2-2   *Refraction and Reflection*

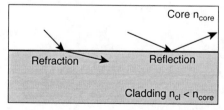

Two possible outcomes when light is in a fiber:
- Refraction – light leaks out of the fiber core (this is bad)
- Reflection – the light remains in the fiber core (this is good)

Core n_{core}

Refraction   Reflection

Cladding $n_{cl} < n_{core}$

In the case of refraction, Snell's Law relates the angles as detailed in the next section.

Index of Refraction (Snell's Law)

The speed of light varies depending upon the type of material. For example, in glass, the speed of light is about two-thirds the speed of light in a vacuum. The relationship shown in Figure 2-3 defines a quantity known as the index of refraction.

The index of refraction is used to relate the speed of light in a material substance to the speed of light in a vacuum. Glass has an index of refraction of around 1.5, although the actual number varies slightly from one type of glass to another. In comparison, air has an index of refraction of about 1, and water has an index of refraction of about 1.33.

Figure 2-3 *Index of Refraction*

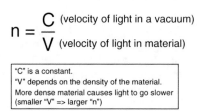

$$n = \frac{C}{V} \begin{array}{l} \text{(velocity of light in a vacuum)} \\ \text{(velocity of light in material)} \end{array}$$

"C" is a constant.
"V" depends on the density of the material.
More dense material causes light to go slower
(smaller "V" => larger "n")

The index of refraction (*n*) is a constant of the material at a specific temperature and pressure. The index of refraction is a fixed value. In another material at those same conditions, *n* would be different.

In glass, *n* is controlled for the glass by adding various dopant elements during the fiber-manufacturing process. Adding controlled amounts of dopants enables fiber manufacturers to design glass for different applications, such as single-mode or multimode fibers. For optical fiber, *n* is engineered slightly differently for the core and the cladding.

Types of Optical Fiber: Multimode and Single Mode

Two basic types of optical fibers exist—multimode fiber (MMF) and single-mode fiber (SMF). The most significant difference lies in their capabilities to transmit light long distances at high bit rates. In general, MMF is used for shorter distances and lower bit rates than SMF. For long-distance communications, SMF is preferred.

Figure 2-4 shows the basic difference between MMF and SMF.

Figure 2-4 *Optical Fiber—Multimode and Single Mode*

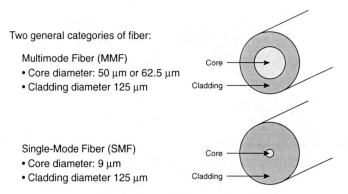

Two general categories of fiber:

Multimode Fiber (MMF)
• Core diameter: 50 µm or 62.5 µm
• Cladding diameter 125 µm

Core
Cladding

Single-Mode Fiber (SMF)
• Core diameter: 9 µm
• Cladding diameter 125 µm

Core
Cladding

Notice the physical difference in the sizes of the cores. This is the key factor responsible for the distance/bit rate disparity between the two fiber types. Figure 2-5 illustrates the large core effect of MMF.

In a nutshell, the larger core diameter allows for multiple entry paths for an optical pulse into the fiber. Each path is referred to as a mode—hence, the designation multimode fiber. Because MMF has the possibility of many paths through the fiber, it exhibits the problem of modal dispersion. Some of the paths in the MMF are physically longer than other paths. The light moves at the same speed for all the paths, but because some of the paths are longer, the light arrives at different times. Consequently, optical pulses arrive at the receiver with a spread-out shape. Not only is the shape affected, but the overall duration of the pulse is increased. When the pulses are too close together in time, they can overlap. This leads to confusion by the receiver and is interpreted as unintelligible data.

For SMF, only a single entrance mode is available for the optical signal to traverse the fiber. Thus, the fiber is appropriately referred to as single-mode fiber.

Figure 2-5 *Light Propagation*

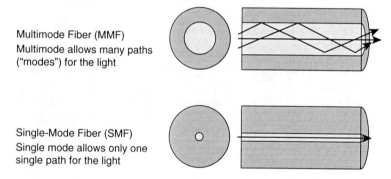

Multimode Fiber (MMF)
Multimode allows many paths
("modes") for the light

Single-Mode Fiber (SMF)
Single mode allows only one
single path for the light

SONET/SDH Principles

SONET/SDH defines a family of transmission standards for compatible operation among equipment manufactured by different vendors or different carrier networks. The standards include a family of interface rates; a frame format; a definition of overhead for operations, administration, and protection; and many other attributes.

Although SONET/SDH is typically associated with operations over single-mode fiber, its format can be transmitted over any serial transmission link that operates at the appropriate rate. For example, implementations of SONET/SDH over wireless (both radio and infrared) links have been deployed.

SONET/SDH has roots in the digital TDM world of the telecommunications industry, and its initial applications involved carrying large numbers of 64-kbps voice circuits. SONET/SDH was carefully designed to be backward compatible with the DS1/DS3 hierarchy used in North America, the E1/E3/E4 hierarchy used internationally, and the J1/J2/J3 hierarchy used in Japan. The initial SONET/SDH standards were completed in the late 1980s, and SONET/SDH technology has been deployed extensively since that time.

Digital Multiplexing and Framing

Framing is the key to understanding a variety of important SONET functions. Framing defines how the bytes of the signal are organized for transmission. Transport of overhead for management purposes, support of subrate channels including DS1 and E1, and the creation of higher-rate signals are all tied to the framing structure.

It's impossible to understand how signals are transported within SONET/SDH without understanding the basic frame structure.

Digital time-division networks operate at a fixed frequency of 8 KHz. The 8 KHz clock rate stems from the requirement to transmit voice signals with a 4 KHz fidelity. Nyquist's theory states that analog signals must be sampled at a rate of twice the highest frequency component, to ensure accurate reproduction. Hence, sampling of a 4-KHz analog voice signal requires 8000 samples per second. All digital transmission systems operating in today's public carrier networks have been developed to be backward compatible with existing systems and, thus, operate at this fundamental clock rate.

In time-division transmission, information is sent in fixed-size blocks. The fixed-size block of information that is sent in one 125-microsecond ([1/8000] of a second) sampling interval is called a frame. In time-division networks, channels are delimited. First locate the frame boundaries and then count the required number of bytes to identify individual channel boundaries.

NOTE   Although the underlying goal is similar, time-division frames have several differences when compared to the link layer frame of layered data communications.

The time-division frame is fixed in size, does not have built-in delimiters (such as flags), and does not contain a frame check sequence. It is merely a fixed-size container that is typically subdivided into individual fixed-rate channels.

All digital network elements operate at the same clock rate. Combining lower-rate signals through either bit or byte interleaving forms higher-rate signals. As line rates increase, the frame rate of 8000 times per second remains the same, to maintain compatibility with all the subrate signals. As result, the number of bits or bytes in the frame must increase to accommodate the greater bandwidth requirements.

The North American digital hierarchy is often called the DS1 hierarchy; the international digital hierarchy is often called the E1 hierarchy.

Even though both hierarchies were created to transport 64-kbps circuit-switched connections, the rates that were chosen were different because of a variety of factors. In North America, the dominant rates that are actually deployed are DS1 and DS3. Very little DS2 was ever deployed. What DS4 has existed has been replaced by SONET. Internationally, E1, E3, and E4 are most common.

NOTE	A DS2 is roughly four times the bandwidth of a DS1, a DS3 is roughly seven times the bandwidth of a DS2, and a DS4 is roughly six times the bandwidth of a DS3. But the rates are not integer multiples of one another. The next step up in the hierarchy is always an integer multiple plus some additional bits. Thus, TDM has to be done asynchronously.

Prior to a fully synchronized digital network, digital signals had to be multiplexed asynchronously. Figure 2-6 shows the signal rates in the North American and international digital hierarchies. An example of asynchronous multiplexing can be seen in the process of combining 28 DS1s to form 1 DS3 (North American hierarchy). Because each of the 28 constituent DS1s can have a different source clock, the signals cannot be bit- or byte-interleaved to form a higher-rate signal. First, their rates must be adjusted so that each signal is at exactly the same rate. This is accomplished by bit-stuffing each DS1 to pad it up to a higher common rate. In addition, control bits are inserted to identify where bit stuffing has taken place.

Figure 2-6 *Digital Transmission Before SONET/SDH Digital Signal Hierarchy*

North America		International	
Signal	Rate	Signal	Rate
DS1	1.544 Mb/s	E1	2.048 Mb/s
DS2	6.312 Mb/s	E2	8.448 Mb/s
DS3	44.736 Mb/s	E3	34.368 Mb/s
DS4	274.176 Mb/s	E4	139.264 Mb/s

When all 28 DS1s are operating at the same nominal rate, the DS3 signal is formed by bit-interleaving the 28 signals. The insertion of stuffing and control bits is why 28×1.544 Mbps = 43.232 Mbps, although a DS3 operates at 44.736 Mbps.

To demultiplex the signals, the control bits and stuffing bits must be removed. Because the control and stuff bits aren't fully visible at the DS3 level, the only way to demultiplex or remove one DS1 from the DS3 stream is to demultiplex the entire DS3 into its 28 constituent DS1s. For example, consider an intermediate network node in which you need to drop one DS1. First, the entire DS3 is demuxed into 28 DS1s, then the target DS1 is dropped, and finally the remaining 27 DS1s (and perhaps a 28th DS1 that is added) are remultiplexed to re-create the DS3 for transport to the next node. Figure 2-7 shows an example of multiplexing before SONET/SDH.

Figure 2-7 *Multiplexing Before SONET/SDH*

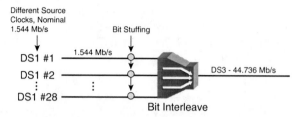

- Individual DS1s within the DS3 are not visible, access to any DS1 requires demuxing all DS1s
- A similar process is required in the E1 hierarchy
- SONET/SDH uses synchronous byte interleaving, individual signals can be demuxed without demuxing the entire signal

DS1 Frame

The most common time-division frame in the North American hierarchy is the DS1 frame. One DS1 carries twenty-four 64-Kbps channels. Figure 2-8 shows an individual frame of a DS1 and indicates that it contains a single 8-bit sample from each of the 24 channels. These twenty-four 8-bit samples ($8 \times 24 = 192$) dominate the DS1 frame.

Each frame also contains a single bit called the framing bit. The framing bit is used to identify frame boundaries. A fixed repetitive pattern is sent in the framing bit position. The receiver looks for this fixed framing pattern and locks on it. When the framing bit position is located, all other channels can be located by simply counting from the framing bit. Thus, in a TDM network, there is no requirement for special headers or other types of delimiters. Channels are identified simply by their position in the bit stream.

Each DS1 frame has 193 bits. The frame is repeated every 125 microseconds (8000 times per second), leading to an overall bit rate of 1.544 Mbps. All other digital time-division signals operate in a similar fashion. Some bits or bytes will always uniquely identify frame boundaries. When frame boundaries are established, individual channels can be located by a combination of counting and pre-established position in the multiplex structure.

Figure 2-8 *DS-1 Frame*

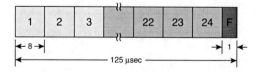

- Eight bits for each of the 24 channels sent every 125 μsec
- Framing bit used for channel alignment, error detection, and embedded operations channel
- $(8 \times 24) + 1 = 193$ bits/frame
- 193 bits/frame × 8000 frames/second = 1.544 Mb/s

STS-1 Frame

The basic building block of a SONET signal is the STS-1 frame. To facilitate ease of display and understanding, SONET and SDH frames are usually described using a matrix structure, in which each element of a row or column in the matrix is 1 byte. The matrix always has nine rows, but the number of columns depends on the overall line rate. In the case of an STS-1, the frame is presented as a 9-row-by-90-column matrix, which results in a total of 810 bytes per frame. The bytes are transmitted from left to right, top to bottom. In Figure 2-9, the first byte transmitted is the one in the upper-left corner. Following that byte are the remaining 89 bytes of the first row, which are followed by the first byte in the second row, and so on until the right-most byte (column 90) of the bottom row is sent. What follows the last byte of the frame? Just as in any other TDM system, the first byte of the next frame.

The transmission of 810 bytes at a rate of 8000 times per second results in an overall line rate of 51.84 Mbps.

Thirty-six bytes of the 810 bytes per frame, or roughly 2.3 Mbps, are dedicated to overhead. This results in a net payload rate of 49.536 Mbps.

This might seem like an odd transmission rate. It's not an ideal match to either the DS1 or E1 hierarchy in terms of backward compatibility. However, as you'll see, it's a relatively good match as a single-rate compromise that's compatible with both the E1 and DS1 hierarchies.

If SONET rates were chosen simply for operation in North American networks, a different rate would have been chosen. Similarly, if SDH had been developed strictly for Europe, a rate that allows more efficient multiplexing from the E1 hierarchy would have been chosen. But for SONET and SDH to adopt common rates, compromises were made.

Figure 2-9   *STS-1 Frame*

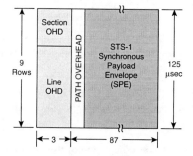

- $9 \times 90 = 810$ bytes per frame
- 810 bytes/frame $\times$ 8 bits/byte $\times$ 8000 frames/sec = 51.84 Mb/s
- 36 bytes per frame for section, line, and path overhead

STS-1 Frame and the Synchronous Payload Envelope

Figure 2-10 illustrates the relationship between the STS-1 frame and the Synchronous Payload Envelope (SPE). The SPE is the portion of the STS-1 frame that is used to carry customer traffic. As the figure shows, the position of the SPE is not fixed within the frame. Instead, the SPE is allowed to "float" relative to the frame boundaries.

This doesn't mean that the SPE varies in size. The STS-1 SPE is always 9×87 bytes in length, and the first column of the SPE is always the path overhead. "Floating" means that the location of the SPE, as indicated by the first byte of the path overhead, can be located anywhere within the 783 payload bytes of the frame.

Because the location of the beginning of the SPE is not fixed, a mechanism must be available to identify where it starts. The specifics of the overhead bytes have yet to be presented, but they are handled with a "pointer" in the line overhead. The pointer contains a count in octets from the location of the pointer bytes to the location of the first byte of the path overhead.

Several benefits are gained from this floating relationship. First, because the payload and the frame do not need to be aligned, the payload does not need to be buffered at the end nodes or at intermediate multiplexing locations to accomplish the alignment. The SPE can be immediately transmitted without frame buffering.

A second benefit, related to the first, occurs when creating higher-rate signals by combining multiple STS-1s to form an STS-N, such as an STS-12, at a network node.

The 12 incoming STS-1s might all derive timing from the same reference so that they are synchronized, but they might originate from different locations so that each signal has a different transit delay. As a result of the different transit delays, the signals arrive at the multiplex location out of phase. If the SPE had to be fixed in the STS-1 frame, each of the 12 STS-1s would need to be buffered by a varying amount so that all 12 signals could be phase-aligned with the STS-12 frame. This would introduce extra complexity and additional transit delay to each signal. In addition, this phase alignment would be required at every network node at which signals were processed at the Synchronous Transport Signal (STS) level. By allowing each of the STS-1s to float independently within their STS-1 frame, the phase differences can be accommodated and, thus, the associated complexity can be reduced. Reducing the requirement for buffering also reduces the transit delay across the network.

A final advantage of the floating SPE is that small variations in frequency between the clock that generated the SPE and the SONET clock can be handled by making pointer adjustments.

The details of pointer adjustments are beyond the scope of this chapter, but this basically involves occasionally shifting the location of the SPE by 1 byte to accommodate clock frequency differences. This enables the payload to be timed from a slightly different clock than the SONET network elements without incurring slips. The pointer adjustments also allow the payload clock to be tunneled through the SONET network.

Figure 2-10 *STS-1 SPE Relative to Frame Boundary*

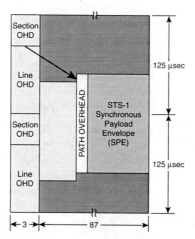

SONET/SDH Rates and Tributary Mapping

The SONET and SDH standards define transmission rates that are supported in MSPP systems. These rates are specific to each of these two standards, yet (as you will see) they are closely related. In this section, you will learn about these rates of transmission and how they are used to carry a variety of customer traffic.

SONET Rates

Figure 2-11 shows the family of SONET rates and introduces the terminology that is used to refer to the rates. The base SONET rate is 51.84 Mbps.

Figure 2-11 *SONET Rates*

Level		Line Rate	Payload Capacity	Capacity in T1s	Capacity in DSOs
Electrical	Optical				
STS-1	OC-1	51.84 Mb/s	50.112 Mb/s	28	672
STS-3	OC-3	155.52 Mb/s	150.336 Mb/s	84	2016
STS-12	OC-12	622.08 Mb/s	601.334 Mb/s	336	8064
STS-48	OC-48	2.488 Gb/s	2.405 Gb/s	1344	32256
STS-192	OC-192	9.952 Gb/s	9.621 Gb/s	5376	129024
STS-768	OC-768	39.808 Gb/s	38.486 Gb/s	21504	516096

This rate is the result of the standards body compromises between the SDH and SONET camps, which led to equally efficient (or inefficient, depending on your point of view) mapping of subrate signals (DS1 and E1) into the signal rate. All higher-rate signals are integer multiples of 51.84 Mbps. The highest rate currently defined is 39.808 Gbps. If traditional rate steps continue to be followed, the next step will be four times this rate, or approximately 160 Gbps.

The SONET signal is described in both the electrical and optical domains. In electrical format, it is the STS. In the optical domain, it is called an Optical Carrier (OC). In both cases, the integer number that follows the STS or OC designation refers to the multiple of 51.84 Mbps at which the signal is operating.

Sometimes confusion arises regarding the difference between the STS and OC designations. When are you talking about an OC instead of an STS? The simplest distinction is to think of STS as electrical and OC as optical. When discussing the SONET frame format, the assignment of overhead bytes, or the processing of the signal at any subrate level, the proper signal designation is STS. When describing the composite signal rate and its associated optical interface, the proper designation is OC. For example, the signal transported over the optical fiber is an OC-N, but because current switching fabric technology is typically implemented using electronics (as opposed to optics), any signal manipulation in an add/drop or cross-connect location is done at the STS level.

SDH Rates

Figure 2-12 shows the rates that SDH currently supports. The numbers in the Line Rate and Payload Capacity columns should look familiar: They are exactly the same as the higher rates defined for SONET. SDH does not support the 51.84-Mbps signal because no international hierarchy rate maps efficiently to this signal rate; that is, E3 is roughly 34 Mbps and E4 is roughly 140Mbps. So the SDH hierarchy starts at three times the SONET base rate, or 155.52 Mbps, which is a fairly good match for E4.

Figure 2-12 *SDH Rates*

Level	Line Rate	Payload Capacity	Capacity in E1s	Capacity in DSOs*
STM 1	155.52 Mb/s	150.336 Mb/s	63	2016
STM 4	622.08 Mb/s	601.334 Mb/s	252	8064
STM 16	2.488 Gb/s	2.405 Gb/s	1008	32256
STM 64	9.952 Gb/s	9.621 Gb/s	4032	129024
STM 256	39.808 Gb/s	38.486 Gb/s	16128	516096

*On a typical E1 time slots 0 and 16 are reserved for network information. For the number of working DSOs multiply the number in this column by .9375

SDH calls the signal a Synchronous Transport Module (STM) and makes no distinction between electrical and optical at this level. The integer designation associated with the STM indicates the multiple of 155.52 Mbps at which the signal is operating.

You can see how well the standards body compromise on SONET/SDH rates worked by comparing the capacity in DS0s column of this chart with the similar column of the SONET chart. The DS0 capacity is equivalent for each line rate. This implies that the efficiency of mapping E1s into SDH is equivalent to the efficiency of mapping DS1s into SONET.

When you look at the two charts, notice that even though all the terminology is different, the rate hierarchies are identical. Also note that capacity in DS0s is the same, so the two schemes are equally efficient at supporting subrate traffic, whether it originates in the DS1 or E1 hierarchies. This compatibility in terms of subrate efficiency is part of the reason for the SONET base rate of 51.84 Mbps; it was a core rate that could lead to equal efficiency.

Even though the rates are identical, the SDH and SONET standards are not identical. Differences must be taken into account when developing or deploying one technology versus the other. The commonality in rates is a huge step in the right direction, but you still need to know whether you're operating in a SONET or SDH environment.

Transporting Subrate Channels Using SONET

When SONET and SDH standards were developed during the 1980s, the dominant network traffic was voice calls, operating at 64 kbps. Any new transmission system, such as SONET/SDH, must be backward compatible with these existing signal hierarchies. To accommodate these signals, SONET has defined a technique for mapping them into the SONET synchronous payload envelope. Mappings for DS1, E1, DS1C, DS2, and DS3 signals have been defined. The mappings involve the use of a byte-interleaved structure within the SPE. The individual signals are mapped into a container called a virtual tributary (VT). VTs are then mapped into the SPE using a structure called a virtual tributary group (VTG); Figure 2-13 shows an example of a VTG. VTs define the mechanisms for transporting existing digital hierarchy signals, such as DS1s and E1s, within the SONET payload. Understanding the VT structure and its mapping into the SONET payload enables you to understand how DS1 and E1 can be accommodated for transport within SONET. This also clarifies the flexibility for transporting these signals and how channel capacity must be sized to meet the customers transport needs.

The basic container for transporting subrate traffic within the SONET SPE is the VTG. The VTG is a subset of the payload within the SPE. The VTG is a fixed time-division multiplexed signal that can be represented by a 9-row-by-12-column matrix, in which where the members of each row and column are bytes, just as in the previous example of the SONET frame. If you do the arithmetic, you'll find that each VTG has a bandwidth of 6.912 Mbps, and a total of seven VTGs can be transported within the SPE. An individual

VTG can carry only one type of subrate traffic (for example, only DS1s). However, different VTGs within the same SPE can carry different subrates. No additional management overhead is assigned at the VTG level, but as you'll see, additional overhead is assigned to each virtual tributary that is mapped into a VTG.

The value of the VTG is that it allows different subrates to be mapped into the same SPE. When an SPE of an STS-1 is defined to carry a single VTG, the entire SPE must be dedicated to transporting VTGs (that is, you cannot mix circuit and packet data in the same SPE except by using the VTG structure). However, different VTGs within the same SPE can carry different subrates.

Figure 2-13   *VTGs*

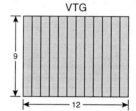

- Each VTG is 9 rows by 12 columns
  - Bandwidth is (9 × 12) bytes/frame × 8 bits/byte × 8000 frames/sec = 6.912 Mb/s
  - 7 VTGs fit in an STS-1 SPE
- There is no VTG level overhead
- VTGs are byte interleaved into SPE
- Virtual tributaries are mapped into VTGs

Now that you know about the structure of an individual VTG, let's see how the VTGs are multiplexed into the STS-1 SPE. Figure 2-14 illustrates this. As with all the other multiplexing stages within SONET/SDH, the seven VTGs are multiplexed into the SPE through byte interleaving. As discussed previously, the first column of the SPE is the Path Overhead column. This byte is followed by the first byte of VTG number 1, then the first byte of VTG number 2, and so on through the first byte of VTG number 7. This byte is followed by the second byte of VTG 1, as shown in Figure 2-14. The net result is that the path overhead and all the bytes of the seven VTGs are byte-interleaved into the SPE.

Note that columns 30 and 59 are labeled "Fixed Stuff." These byte positions are skipped when the payloads are mapped into the SPE, and a fixed character is placed in those locations. The Fixed Stuff columns are required because the payload capacity of the SPE is slightly greater than the capacity of seven VTGs. The SPE has 86 columns after allocating space for the path overhead. But the seven VTGs occupy only 84 columns (7 × 12). The two Fixed Stuff columns are just a standard way of padding the rate so that all implementations map VTGs into the SPE in the same way.

Individual signals from the digital hierarchy are mapped into the SONET payload through the use of VTs. VTs, in turn, are mapped into VTGs. A VT mapping has been defined for each of the multiplexed rates in the existing digital hierarchy.

For example, a DS1 is transported by mapping it into a type of VT referred to as VT1.5s. Similarly, VT mappings have been defined for E1 (VT2), DS1C (VT3), and DS2 (VT6) signals. In the current environment, most implementations are based on VT1.5 and VT2.

Because each of the subrates is different, the number of bytes associated with each of the VT types is also different.

Figure 2-14 *VTG Structure*

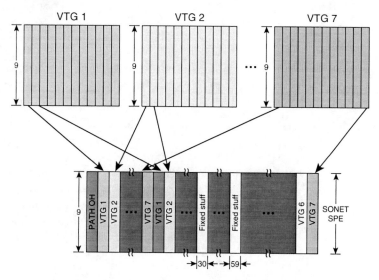

As we said, VTGs are fixed in size at 9 × 12 = 108 bytes per frame. Because the size of the individual VTs is different, the number of VTs per VTG varies.

A VTG can support four VT1.5s, three VT2s, two VT3s, or one VT6, as shown in Figure 2-15. Only one VT type can be mapped into a single VTG, such as four VT1.5s or three VT2s; however, you cannot mix VT2s and VT1.5s within the same VTG. Different VTGs within the same SPE can carry different VT types. For example, of the seven VTGs in the SPE, five might carry VT1.5s and the remaining two could carry VT2s, if an application required this traffic mix.

As a reminder, VT path overhead is associated with each VT, which can be used for managing the individual VT path. In addition, a variety of mappings of the DS1 or E1 signal into the VT have been defined to accommodate different clocking situations, as well as to provide different levels of DS0 visibility within the VT.

The most common VT in North America is the VT1.5. The VT1.5 uses a structure of 9 rows × 3 columns = 27 bytes per frame (1.728 Mbps) to transport a DS1 signal. The extra bandwidth above the nominal DS1 signal rate is used to carry VT overhead information. Four VT1.5s can be transported within a VTG. The four signals are multiplexed using byte interleaving, similar to the multiplexing that occurs at all other levels of the SONET/SDH hierarchy. The net result of this technique in the context of the SONET frame is that the individual VT1.5s occupy alternating columns within the VTG. Figure 2-15 shows an example of four DS-1 signals mapped into a VTG.

Figure 2-15 *Four DS1s Mapped into a VTG*

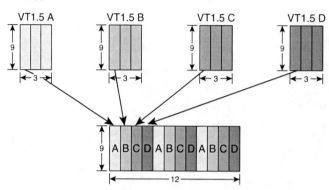

Outside North America, the 2.048-Mbps E1 signal dominates digital transport at the lower levels. The VT2 was defined to accommodate the transport of E1s within SONET. The VT2 assigns 9 rows × 4 columns = 36 bytes per frame for each E1 signal, which is 4 more bytes per frame than the standard E1. As is the case for VT1.5s, the extra bandwidth is used for VT path overhead. Because the VT2 has four columns, only three VT2s can fit in a VTG. The VTG is again formed by byte-interleaving the individual VT2s.

Signals of Higher Rates

In this section, you'll learn about the creation of higher-rate signals, such as STS-48s or STS-192s, and concatenation.

Remember that rates of STS-N, where N = 1, 3, 12, 48, 192, or 768 are currently defined. An STS-N is formed by byte-interleaving N individual STS-1s. Except for the case of concatenation, which is discussed shortly, each of the STS-1s within the STS-N is treated independently and has its own associated section, line, and path overhead.

At any network cross-connect or add/drop node, the individual STS-1s that forms the STS-N can be changed. The SPE of each STS-1 independently floats with respect to the frame boundary.

In an OC-48, the 48 SPEs can each start at a different byte within the payload. The H1 and H2 pointers (in the overhead associated with each STS-1) identify the SPE location. Similarly, pointer adjustments can be used to accommodate small frequency differences between the different SPEs.

When mapping higher-layer information into the SPE (such as VTGs or Packet over SONET), the SPE frame boundaries must be observed. For example, an STS-48 can accommodate 48 separate time-division channels of roughly 50 Mbps each. The payload that's mapped into the 48 channels is independent, and any valid mapping can be transported in the channel. However, channels at rates higher than 50 Mbps cannot be accommodated within an OC 48. For higher rates, concatenation is required.

Byte Interleaving to Create Higher-Rate Signals

Figure 2-16 shows an example of three STS-1s being byte-interleaved to form an STS-3. The resultant signal frame is now a 9-row-by-270-column matrix. The first nine columns are the byte-interleaved transport overhead of each of the three STS-1s. The remaining 261 columns are the byte-interleaved synchronous payload envelopes. Higher-rate signals, such as STS-48s or STS-192s, are formed in a similar fashion. Each STS-1 in the STS-N adds 3 columns of transport overhead and 87 columns of payload. All the individual STS-1s are byte-interleaved in a single-stage process to form the composite signal. Each of the STS-1s is an independent time-division signal that shares a common transport structure. The maximum payload associated with any signal is roughly 50 Mbps. A technique called concatenation must be used to transport individual signals at rates higher than 50 Mbps.

Figure 2-16 *Creating Higher-Rate Signals*

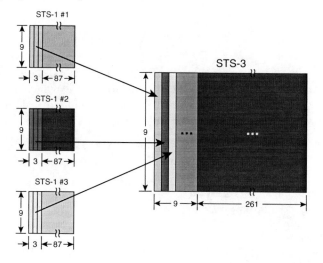

Concatenation

Increasingly, data applications in the core of the network require individual channels to operate at rates much greater than the 50 Mbps that can be accommodated in a single STS-1. To handle these higher rate requirements, SONET and SDH define a concatenation capability.

Concatenation joins the bandwidth of N STS-1s (or N STM-1s) to form a composite signal whose SPE bandwidth is N multiplied by the STS-1 SPE bandwidth of roughly 50 Mbps. Signal concatenation is indicated by a subscript c following the rate designation. For example, an OC-3c means that the payload of three STS-1s has been concatenated to form a single signal whose payload is roughly 150 Mbps.

The concatenated signal must be treated in the network as a single composite signal. The payload mappings are not required to follow the frame boundaries of the individual STS-1s.

Intermediate network nodes must treat the signal as a contiguous payload. Only a single path overhead is established because the entire payload is a single signal. Many of the transport overhead bytes for the higher-order STSs are not used because their functions are redundant when the payload is treated as a single signal.

Concatenation is indicated in the H1 and H2 bytes of the line overhead. The bytes essentially indicate whether the next SPE is concatenated with the current SPE. It is also possible to have concatenated and nonconcatenated signals within the same STS-N. As an example, an STS-48 (OC-48) might contain 10 STS-3cs and 18 STS-1s.

No advantage is gained from concatenation if the payload consists of VTs containing DS1s or E1s. However, data switches (for example, IP routers) typically operate more cost-effectively if they can support a smaller number of high-speed interfaces instead of a large number of lower-rate interfaces (if you assume that all the traffic is going to the same next destination).

So the function of concatenation is predominantly to allow more cost-effective transport of packet data.

Figure 2-17 shows an example of a concatenated frame. It's still nine rows by $N \times 90$ columns. The first $3N$ columns are still reserved for overhead, and the remaining $N \times 87$ columns are part of the SPE. The difference is that, except for the concatenation indicators, only the transport overhead associated with the first STS-1 is used in the network.

In addition, the payload is not N byte-interleaved SPEs: The payload is a single SPE that fills the full $N \times 87$ columns of the frame. A single path overhead is associated with the concatenated signal. The 9 bytes per frame of path overhead that are normally associated with the remaining STS-1s are available for payload in the SPE. A variety of mappings have been identified to transport packet protocols within the SPE.

Figure 2-17 *Concatenated Frames*

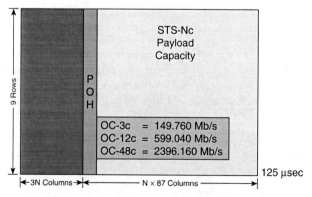

Even though concatenation does allow greater flexibility and efficiency for higher-rate data communications, there are several limitations to its use.

First is the signal granularity. Each rate in the hierarchy is four times the preceding rate. That makes for very large jumps between successive rates that are available. The signal granularity issue is compounded when looking at rates common to data communications applications. For example, Ethernet and its higher-speed variants are increasingly popular data-transport rates for the metropolitan-area networks (MANs) and wide-area networks (WANs). But in the Ethernet rate family of 10 Mbps, 100 Mbps, 1 Gbps, and 10 Gbps, only the 10 Gbps signal is a close match to an available concatenated rate.

An additional problem is that network providers are not always equipped to handle concatenated signals, especially at the higher rates, such as OC-192c. Implementations that operate at these high rates today are often on a point-to-point basis, and the signal does not transit through intermediate network nodes. This problem of availability of concatenated signal transport is especially an issue if the signal transits multiple carrier networks.

In an attempt to address some of these limitations, you can use the virtual concatenation technique.

As the name implies, virtual concatenation means that the end equipment sees a concatenated signal, but the transport across the network is not concatenated. The service is provided by equipment at the edge of the network that is owned by either the service provider or the end user. The edge equipment provides a concatenated signal to its client (for example an Internet Protocol (IP) router with an OC-48c interface), but the signals that transit the network are independent STS-1s or STS3cs. The edge equipment essentially uses an inverse multiplexing protocol to associate all the individual STS-1s with one another to provide the virtually concatenated signal. This requires the transmission of control bytes to provide the association between the various independent STS channels on the transmit side.

The individual channels can experience different transit delays across the network, so at the destination, the individual signals are buffered and realigned to provide the concatenated signal to the destination client.

Virtual concatenation defines an inverse-multiplexing technique that can be applied to SONET signals. It has been defined at the VT1.5, STS-1, and STS-3c levels. At the VT1.5 level, it's possible to define channels with payloads in steps of 1.5 Mbps by virtually concatenating VT1.5s. Up to 64 VT1.5s can be grouped. For example, standard Ethernet requires a 10-Mbps channel; this can be accomplished by virtually concatenating seven VT1.5s.

Similarly, a 100-Mbps channel for Fast Ethernet can be created by virtually concatenating two STS-1s. Virtual concatenation of STS-3cs provides the potential for several more levels of granularity than provided by standard concatenation techniques.

SONET/SDH Equipment

SONET/SDH networks are typically constructed using four different types of transmission equipment, as shown in Figure 2-18. These are path-terminating equipment, regenerators, add/drop multiplexers, and digital cross-connects. Each equipment type plays a slightly different role in supporting the delivery of services over the SONET/SDH infrastructure. All are necessary to provide the full range of network capabilities that service providers require.

Figure 2-18 *SONET/SDH Equipment*

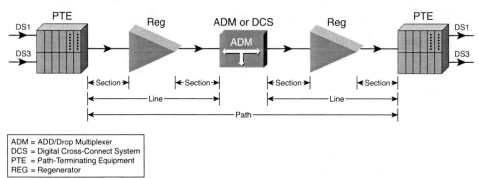

ADM = ADD/Drop Multiplexer
DCS = Digital Cross-Connect System
PTE = Path-Terminating Equipment
REG = Regenerator

Path-terminating equipment (PTE), also sometimes called a terminal multiplexer, is the SONET/SDH network element that originates and terminates the SONET/SDH signal. For example, at the originating node, the PTE can accept multiple lower-rate signals, such as DS1s and DS3s, map them into the SONET payload, and associate the appropriate overhead bytes with the signal to form an STS-*N*. Similarly, at the destination node, the PTE processes the appropriate overhead bytes and demultiplexes the payload for distribution in individual lower-rate signals.

When digital signals are transmitted, the pulses that represent the 1s and 0s are very well defined. However, as the pulses propagate down the fiber, they are distorted by impairments such as loss, dispersion, and nonlinear effects. To ensure that the pulses can still be properly detected at the destination, they must occasionally be reformed to match their original shape and format. The regenerator performs this reforming function.

The beauty of digital transmission is that, as long as the regenerators are placed close enough together that they don't make mistakes (that is, a 1 is reformed to look like a clean 1, not a zero, and vice versa), digital transmission can be essentially error free.

The regenerator function is said to be 3R, although there is sometimes disagreement over exactly what the three R's stand for.

These three functions are performed:

- Refresh or amplify the signal to make up for any transmission loss

- Reshape the signal to its original format to offset the effects of dispersion or other impairments that have altered the signal's pulse shape

- Retime the signal so that its leading edge is consistent with the timing on the transmission line

In today's world, the full 3R functionality requires O-E-O conversion. Because it is tied to the electrical signal format, the regenerator is unique to the line rate and signal format being used. Upgrades in line rate—say, from OC-48 to OC-192—require a change-out of regenerators. As such, it is a very expensive process, and network operators try to minimize the number of regenerators that are required on a transmission span. One of the benefits of optical amplifiers has been that they have significantly reduced the number of 3R regenerators required on a long-haul transmission span.

The add/drop multiplexer (ADM) is used, as the name implies, to "add" and "drop" traffic at a SONET/SDH node in a linear or ring topology. The bandwidth of the circuits being added and dropped varies depends on the area of application. This can range from DS1/E1 for voice traffic up to STS-1/STM-1 levels and, in some cases, even concatenated signals for higher-rate data traffic.

The functions of an ADM are very similar to the functions of a terminal multiplexer, except that the ADM also handles pass-through traffic.

Although this book is primarily related to MSPP, the next network element of a SONET network bears some mention and explanation.

The Digital Cross-Connect System (DCS) exchanges traffic between different fiber routes. The key difference between the cross-connect and the add/drop is that the cross-connect provides a switching function, whereas the ADM performs a multiplexing function. The cross-connect moves traffic from one facility route to another.

A cross-connect is also used as the central connection point when linear topologies are connected to form a mesh. There might be no local termination of traffic at the cross-connect location. In fact, if there is, the traffic might first be terminated on an add/drop multiplexer or terminal multiplexer, depending on the signal level at which the cross-connect operates.

DCSs are generally referred to as either wideband digital cross-connect, narrowband cross-connect, or broadband digital cross-connect.

A narrowband digital cross-connect is designed for interconnecting a large number of channels at the DS0 level.

A wideband digital cross-connect is designed for interconnecting a large number of channels at the DS1 basic level.

A broadband digital cross-connect is designed for interconnecting a large number of channels at the DS3 and higher levels.

SONET Overhead

In the SPE, 36 bytes (roughly 4 percent) of the STS-1 frame are used to transport overhead information. This overhead is multiplexed with the payload by byte interleaving.

The SONET overhead channels provide transport of the information required to manage and maintain the SONET network. Without the use of the overhead channels, SONET networks would be unreliable and very difficult to maintain. You can't understand how the network operates, how to isolate troubles, or what information is available for network tests without understanding SONET overhead.

Embedded in the SONET frame is the SPE. The SPE contains the client signal that is being transported. The first column of the SPE contains the path overhead, is one of three fields containing overhead for operations and management purposes. The section and line overhead maintain a fixed position in the first three columns of the matrix.

The location of the path overhead within an individual frame does not have a fixed position. Section, line, and path overhead are multiplexed with the payload using byte interleaving. The section and line overhead are inserted in the frame/signal 3 bytes at a time, and the path overhead is inserted 1 byte at a time.

SONET/SDH Transmission Segments

When referencing SONET overhead, it is imperative to have an understanding of the transmission segments.

For management purposes, SONET/SDH transmission systems are divided into different segments. In SONET, these segments are called sections, lines, and paths.

Sections

The lowest-level transmission segment is the section. The section exists between any adjacent O-E-O processing points on the transmission facility. All the SONET network elements you've learned about so far perform O-E-O processing; they also all terminate sections. It's not shown in Figure 2-19, but on long spans in which there are several consecutive regenerators, even the segment between adjacent regenerators is considered a section. The section is the shortest transmission segment that is visible from a management perspective.

On long-haul systems that employ optical amplifiers, section overhead is not terminated at the optical amplifier. In SDH, the equivalent terminology is a regenerator section.

In Figure 2-20, you can see the 9 bytes of section overhead. Recall that a section is the transmission segment between adjacent O-E-O processing points. The overhead represented in this figure is thus terminated and processed at essentially every SONET network element, whether it is a regenerator, cross-connect, add/drop mux, or terminal multiplexer. The section overhead provides the lowest level of granularity for management visibility of the SONET network.

Figure 2-19 *Examples of Sections*

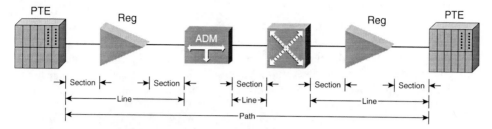

- Transmission segment between adjacent O-E-O processing points
- All SONET network elements, including regenerators,
  originate/terminate sections and process section overhead
- Section management information carried in overhead of SONET frame

Several functions are supported by these bytes of overhead information. The A1 and A2 bytes are used for framing; they indicate the beginning of the STS-1 frame. All other byte positions are determined by counting from these bytes.

The J0/Z0 byte has different meanings, depending on whether the STS-1 is the first STS-1 of an STS-*N* or one of the second through *N* STS-1s. In the first STS-1, the J0 byte, called the trace byte, is transmitted. A unique character is transmitted in the J0 byte; at any downstream location, the identity of the STS-1/STS-*N* can be verified by comparing the J0 byte received to the one that was transmitted.

The B1 byte carries the result of a parity check. The parity check occurs over all bits of the entire STS-*N* signal of the previous frame. Because of this, the use of the B1 byte is defined for only the first STS-1 of an STS-*N*.

The E1 byte is a section orderwire. It can be used for voice communications between regenerator section locations. Recall that each byte of overhead provides a 64-Kbps channel so that standard pulse code modulation (PCM) voice communications are possible over this channel.

F1 is a "user" byte. Its use is not standardized. Vendors can use this byte to provide special features within a single vendor network.

The D1, D2, and D3 bytes form a 192-Kbps data-communications channel. It is used for communication between network-management system elements.

Figure 2-20 *Section Overhead Bytes*

- Used for management on
  regenerator section by
  regenerator section basis
- Functions include framing,
  trace, parity check, order
  wire, and data communications
  channel

A1	A2	J0/Z0
B1	E1	F1
D1	D2	D3

Lines

The next SONET transmission segment is the line. The line exists between consecutive network elements that process the signal at the STS level. Any SONET node that does multiplexing or cross-connecting terminates the line. As discussed, these nodes also terminate a section. Management overhead at the line level is used for functions such as protection switching, error detection, synchronization status, and functions related to the position of the SONET payload within the SONET frame. Figure 2-21 defines the endpoints of the line segment.

In SDH, this segment is called a multiplexer section.

Figure 2-21 *Line Example*

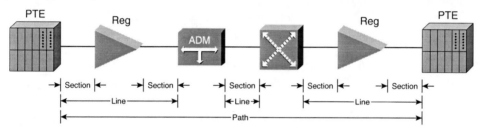

- Transmission segment between adjacent SONET level mux/demux points
- Any processing of the STS-*N* level signal terminates the line (e.g., ADM or DCS) and processes line level overhead
- Any line termination point also terminates section overhead
- Line management information carried in overhead of SONET frame

In Figure 2-22, you can see 18 bytes of line overhead. Table 2-1 documents the functions of the bytes in the line and section overhead.

Figure 2-22 *Line Overhead Bytes*

- Used for management of multiplexed signals
- Functions include payload pointers, pointer action, parity check, protection switching indicators, order wire, data communications channel, and synchronization status

H1	H2	H3
B2	K1	K2
D4	D5	D6
D7	D8	D9
D10	D11	D12
S1/Z1	M0/M1	E2

Table 2-1 *SONET Overhead Bytes for Section and Line*

Overhead Byte	Function	Description
A1, A2	Frame synchronization	These bytes indicate the beginning of an STS-1 frame.
B1, B2	Section and line parity bytes	The parity of each particular frame section is formed within a group of 2, 8, or 24 bits. These bit groups are arranged in columns, and the parity of each individual bit in the vertical direction is calculated.
D1 to D3	Section DCC	The data-communication channels (DCC) allow the transmission management and status information.
D4 to D12	Line DCC	
E1, E2	Section and line orderwire bytes	These bytes are allocated as orderwire channels for voice communication.
F1	Section user's data channel	These bytes are allocated for user purposes.
H1, H2	Payload pointer bytes	These bytes indicate offset in bytes from the pointer to the first byte of the STS-1 SPE in that frame. They are used in all STS-1s of an STS-M. They are also used as concatenation indicators.
H3	Pointer action byte	This byte is used in all STS-1s of an STS-N to carry extra SPE bytes when negative pointer adjustments are required.
J0 (C1)	Section trace	The J0 byte contains a plain-text sequence.
K1, K2	Automatic protection switching (APS) control	These bytes are used to control APS in the event of extreme communications faults.
S1/Z1	Synchronization status byte	The S1 byte indicates the signal clock quality and clock source.
M0, M1	Remote error indication	These bytes contain the number of detected anomalies (M1 only for STS-1/OC-1).

Paths

The final transmission segment is the path, the end-to-end trail of the signal. The path exists from wherever the payload is multiplexed into the SONET/SDH format to wherever demultiplexing of the same payload takes place. As the name implies, this function is generically performed at the PTE. A PTE can function as a terminal multiplexer or an add/drop multiplexer. Unlike sections and lines, which deal with the composite signal, paths are associated with the client signal that is mapped into the SONET/SDH payload. In addition, when supporting subrate multiplexing, such as transporting multiple DS1s, each individual DS1 has an associated path overhead. Figure 2-23 shows an example of a path.

The use of the term *path* is common in SONET and SDH.

Figure 2-23 *Path Example*

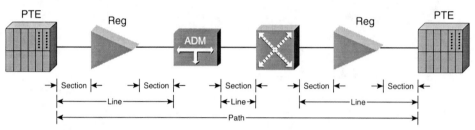

- End-to-end SONET segment between origination/mux point and termination/demux point
- May transit multiple network elements
- May be muxed/demuxed with other paths at network nodes (e.g., an OC-48 may contain 48 independent STS-1 paths)
- Any path termination also terminates section and line overhead
- Path management information carried within information payload

In Figure 2-24, you can see 9 bytes of path overhead. Each is described as follows:

- **J1**—The J1 byte is the path trace byte. Its operation is similar to the J0 byte, except that it's at the path level. The unique pattern in the J1 byte allows both source and destination to verify that they have a continuous path between them and that no misconnections have been made. All paths within an STS-N have a unique J1 byte.

- **B3**—The B3 byte is a parity check over all the bytes of the previous frame of the SPE.

- **C2**—The C2 byte is a label byte. It is used to indicate the type of payload that has been mapped into the SPE.

- **G1**—The G1 byte is the path status byte. It is used to relay information about the status of the path termination back to the originating location. Indicators, such as remote error indications and remote defect indications, are transmitted using this byte.

- **F2**—The F2 and F3 bytes provide a "user" channel between PTEs. The specific implementation is not standardized.

- **H4**—The H4 byte is a multiframe indicator byte. It is used only when a particular structured payload is transported in the synchronous payload envelope.

- **K3**—The K3 byte (formerly Z4) is used to transport automatic protection switching information at the path level.

- **N1**—The N1 byte (formerly Z5) is used for tandem connection maintenance. Network operators can use the N1 byte to perform path-level maintenance functions over intermediate segments of the path without overwriting the end-to-end path information.

Figure 2-24 *Path Overhead Bytes*

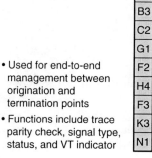

Now that you've seen all the bytes for the SONET overhead, Figure 2-25 serves as a visual reminder of the structure of the SONET overhead and SPE.

Figure 2-25 *SONET Overhead*

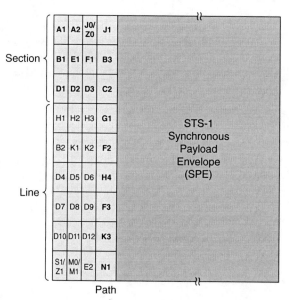

Synchronization and Timing

Synchronization is an important part of any digital time-division network. If network elements are not synchronized, entire frames of the SONET/SDH signal will occasionally be lost.

Losing a frame means that all the data bits or voice samples carried within the frame are lost. Clearly, slips must be minimized to provide high-quality transmission.

Digital time-division networks operate at a fundamental frequency of 8 KHz. This frequency was derived from the desire to support voice communication with a 4-KHz bandwidth. All network elements that perform switching or multiplexing have an internal clock that operates at 8 KHz.

As a general rule, clocks don't keep perfect time. They have a nominal operating frequency, but they drift over time. The better the quality of the oscillator in the watch, the less the clock drifts. The same is true of the clock in network elements. Two adjacent switches operating with independent clocks (called free running) will drift relative to one another. If they drift too much, a "slip" occurs. A slip results in dropping or duplicating a time-division frame that contains voice or data. To avoid slips, network clocks must be synchronized, such as with two time-division network elements—for example, these could be switches, multiplexers, or cross-connects. All network elements operate at a nominal frequency of 8000 samples per second, or 1 sample every 125 microseconds. If two network elements are operating independently with their own internal clocks, inevitably the two clock rates will drift relative to one another, and one clock will be slightly faster than the other.

This faster operation means that data is being sent at a higher rate than the other clock is processing it (because it has a slower clock). The receiving node buffers the excess bits that arrive until it has an entire time-division frame (that is, a DS1 frame or an STS-1 frame) of information that it hasn't yet processed. At that point, to realign the clocks and avoid falling even further behind, the receiving node discards the extra frame. This frame discard is called a slip.

In the opposite direction, the "faster" switch is receiving the incoming signal at a slower rate than its clock rate. Eventually, the switch gets to the point at which its incoming frame buffer is empty. At that point, to realign the two nodes, the switch repeats the previous frame of information. No data is lost, but the same data is sent twice. This overwriting of information is also called a slip.

Slips lead to loss or duplication of a time-division frame. This is an obvious problem for digital data. Occasional slips are tolerated, but each slip can lead to one or more retransmissions of data. Excessive slips not only affect the performance of the application, but they also can lead to network-congestion problems if the volume of retransmissions is too great.

Voice is surprisingly tolerant of slips. Unofficial subjective tests have been completed to show that users will tolerate slip rates of as high as one slip per 100 samples before they complain about call quality. This is an extremely high slip rate. When slips do become noticeable, they tend to produce audible pops and clicks that can become annoying.

Timing slips need to be eliminated. Since digital switching was introduced to the public switched digital network in 1976, a synchronization plan has been in place to ensure that network elements can trace their timing reference to a common clock. The plan has evolved over the years, but it's still the primary defense against slips.

To address this situation, SONET and SDH have defined a clever pointer-adjustment scheme using the H3 byte of the line overhead. The pointer adjustments allow the network to tolerate small frequency differences without incurring slips.

Timing

Every SONET/SDH network element has its own internal clock. When the element operates using its own clock, it is said to be free running. The highest-quality clock used in communications networks is called a Stratum 1 clock. It is also referred to as the primary reference source. Two free-running switches with Stratum 1 clocks would experience about five slips per year. The simple solution to the slip problem is to design every network node with a Stratum 1 clock.

The problem with this idea has been cost. Stratum 1 clocks are expensive, so the Bell System of the 1970s had one Stratum 1 clock located in the geographic center of the country (Missouri). The alternative plan was to design each network element with its own lower-quality internal clock and tie the clock back to a Stratum 1 clock. As long as the internal clock is tied to the Stratum 1, it will operate at the Stratum 1 frequency. If it loses its link to Stratum 1, it will eventually free-run.

Four quality or accuracy levels for timing sources have been defined for digital networks. As you go down the chain from Stratum 1 to Stratum 4, the clock quality and cost go down. Any clock will operate at the Stratum 1 frequency, as long as it has a timing chain or connection to a Stratum 1 clock. The difference in clock quality shows up when the timing chain is broken and the clocks go into free-running mode.

Typically there are two measures of clock quality: accuracy and holdover capability.

Accuracy is the free-running accuracy of the clock. Holdover is essentially a measure of how well the clock "remembers" the Stratum 1 clock frequency after the timing chain is broken. It is usually measured in terms of clock accuracy during the first 24 hours after a timing failure.

The idea is that if timing can be restored within the first 24 hours, the network is operating at the holdover accuracy instead of the free-running accuracy.

Table 2-2 describes the different quality levels or stratum values of clocks.

Table 2-2 *Synchronization Clock Stratum Levels*

Stratum Level	Characteristics
Stratum 1	Accuracy of 1×10^{-11} Stratum 1 is a network's primary reference frequency. Other network clocks are "slaves" to Stratum 1. Two free-running Stratum 1 clocks will experience five or fewer slips per year. Today the most common source is Global Positioning System receivers (GPSs).
Stratum 2	"Holdover" accuracy of 1×10^{-10} is required during the first 24 hours after the loss of primary reference frequency. Stratum 2 involves a free-running accuracy of 1.6×10^{-8}. Two free-running Stratum 2 clocks will experience 10 or fewer slips per day. This stratum is typically used in core network nodes, such as a large tandem switch.
Stratum 3	Stratum 3 has a holdover requirement of less than 255 slips during the first 24 hours after the loss of the primary reference frequency. This stratum has a free-running accuracy of 4.6×10^{-6}. Two free-running Stratum 3 clocks would experience 130 or fewer slips per hour. Typically, this stratum is used in local switches, cross-connects, and large private branch exchanges (PBXs). Stratum 3E has an accuracy of 1×10^{-6}.
Stratum 4	No holdover requirements exist. Stratum 4 has a free-running accuracy of 3.2×10^{-5}. Two free running Stratum 4 clocks will experience 15 or fewer slips per minute. Typically, this stratum is used in channel banks, PBXs, and terminal multiplexers.

So where do you find a Stratum 1 clock that is affordable enough to deploy over a large network? Try using a Global Positioning System receiver.

Global Positioning System

Thanks to the current availability of the Global Positioning System (GPS), networks can take advantage of a cheap source for Stratum 1 timing. Inexpensive GPS receivers make it possible to have a Stratum 1 clock source in nearly every location.

Timing distribution is accomplished through the use of Building Integrated Timing Source (BITS) and the BITS synchronization method. Figure 2-26 displays how this method is deployed.

BITS normally provides two Stratum 1 clocks. One is used for normal operation; the other is used for backup. Inside the central office (CO), an empty DS1 or E1 channel is used to carry the clocking information for the network elements (NEs) located in the CO. Therefore, each rack inside the CO is connected to the BITS source. Typical service-provider equipment has a connection point for BITS input. It can use any physical interface defined for E1, such as coax or wire-wrap pins to connect to BITS.

Several rules are related to the BITS synchronization method.

A typical carrier NE is normally equipped with a Stratum 3 clock. This clock is used when the NE is free running, without synchronization, toward an external clock source. In case of network synchronization, this NE runs at the same rate as the Stratum 1 reference clock.

The maximum slip rate allowed for a free-running SONET NE is 20 ppm. This is between the accuracy of a Stratum 3 and Stratum 4 clock.

A Stratum 3 clock source is more expensive than a Stratum 4 clock source. Packet-switching devices use buffering, so they do not need precise clock synchronization. Therefore, a Stratum 4 clock is reasonable for this use. Cisco routers are typically equipped with a Stratum 4 clock.

Service-provider NEs have a higher price tag and higher need for precise transmit clocks. Thus, a Stratum 3 clock is the correct choice because Stratum 4 is not good enough for DS-3.

In Figure 2-26, you can see a CO location and three other locations. If you had a BITS in all four locations, you would option all four nodes to be externally timed from each respective BITS source. This would ensure that your network would have the best synchronization plan available. So what do you do when the three locations don't have a BITS available for direct connection? You must use an alternative method to synchronize an MSPP network element.

Figure 2-26 *BITS*

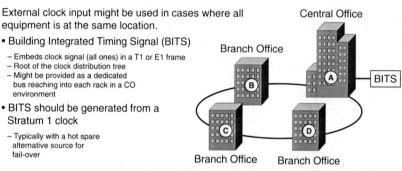

External clock input might be used in cases where all equipment is at the same location.

• Building Integrated Timing Signal (BITS)

   – Embeds clock signal (all ones) in a T1 or E1 frame
   – Root of the clock distribution tree
   – Might be provided as a dedicated
     bus reaching into each rack in a CO
     environment

• BITS should be generated from a
  Stratum 1 clock

   – Typically with a hot spare
    alternative source for
    fail-over

NODE A, in the Central Office, reveives it's clock from BITS.

Some alternative ways to synchronize an MSPP network element where there is no external timing source include line timing, loop timing, and through timing.

The first method of performing timing is to use an external clock source, such as the BITS.

The BITS supplies your NE with an external clock source that the NE uses to time the outgoing transmission lines. Normally, the BITS is equipped with a primary clock generator and a secondary clock generator for backup. Figure 2-27 shows an example of external timing.

Figure 2-27 *External Timing*

All signals transmitted from a specific node are synchronized to an external source received by that node; e.g., a BITS timing source

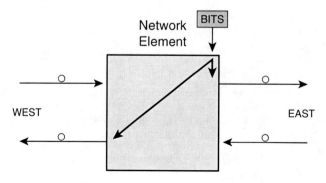

If no external timing source is available, you can use one of the incoming lines to regenerate the clocking signal. The incoming signal that has the shortest path to a primary reference source should be selected. In line timing, the clock is recovered from the received signal and is used to time all outgoing transmission lines. If there is more than one incoming line, such as at a cross-connect, and the sources of the different line signals are not synchronized, slips can occur. Figure 2-28 shows an example of line timing.

Figure 2-28 *Line Timing*

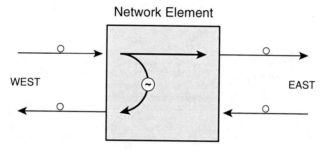

All transmitted signals from a specific node are synchronized to one received signal

Two other methods of synchronization are loop timing and through timing. Figures 2-29 and 2-30 shed some light on loop timing and through timing.

Figure 2-29 *Loop Timing*

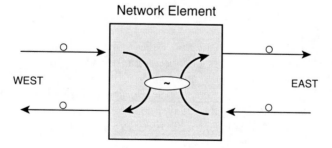

The transmit signal in an optical link, going east or west, is synchronized to the received signal from that same optical link

In loop timing, you use the incoming line signal as the timing source for the signal that is transmitted on the same interface. Loop timing is normally not used in ADMs or cross-connects that have multiple external interfaces because if data is sent from the incoming West channel to the outgoing East channel and both clocks are not accurate, slips can occur.

Through timing is normally used in ring configurations. The transmit signal in one direction of transmission is synchronized with the received signal from the same direction of transmission.

Through timing is also typically used by SONET signal regenerators, which normally only pass the data through the machinery, without cross-connect functionality.

Figure 2-30 *Through Timing*

The transmit signal going in one direction of transmission around the ring is synchronized to the received signal from that same direction of transmission

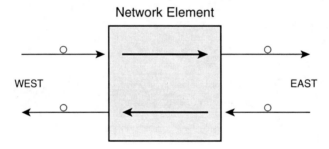

Network Element

WEST   EAST

Summary

This chapter introduced you to multiple topics that are foundational for understanding the use and design of MSPP networks. It presented an overview of optical transmission through glass fiber, including the components of an optical fiber, and the way in which these components interact to confine and reflect light signals to enable optically based communication. Additionally, this chapter discussed the primary technologies used in current MSPP networks, SONET and SDH. SONET is primarily deployed in North America; SDH is deployed around the world.

This chapter described the operation of SONET/SDH in detail, including the methods in which it is used to support the legacy digital signal hierarchies. It also explored the various types of SONET/SDH network elements, how they are used in the network, and how the various network-overhead levels enable reliable transmission. Finally, the chapter introduced the concepts of synchronization and timing in an MSPP network. Understanding these various concepts will provide a solid foundation for engineering and deploying MSPP networks.

This chapter covers the following topics:

- Storage
- Dense Wavelength-Division Multiplexing
- Ethernet

CHAPTER 3

Advanced Technologies over Multiservice Provisioning Platforms

Several technologies being deployed from Multiservice Provisioning Platform (MSPP) legacy platforms go beyond the traditional time-division multiplexing (TDM) services (DS1 and DS3 circuits) and optical services (OC-3, OC-12, OC-48, or even OC-192).

As we consider the traditional technologies over a Synchronous Optical Network (SONET) facility, it is important to note that these technologies are all encapsulated within a SONET frame and carried in the SONET payload. As we venture into the next-generation MSPP technologies, some of these advanced services are encapsulated within a SONET frame as the payload (such as Ethernet private line, shown in Figure 3-1), and some are deployed on the SONET platform but do not ride the SONET frame (such as dense wavelength-division multiplexing [DWDM], shown in Figure 3-2).

Figure 3-1 *Ethernet Traffic Carried within a SONET Payload*

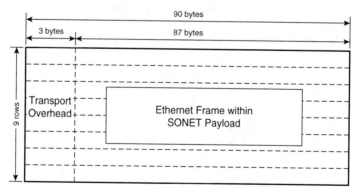

Sonet STS-1 Frame Format

Therefore, when the terminology "advanced technologies over MSPP" is used, you might have to reprogram your thinking to go beyond simply SONET encapsulation of services to that of any service that can be launched from the MSPP platform.

Figure 3-2 *DWDM from an MSPP*

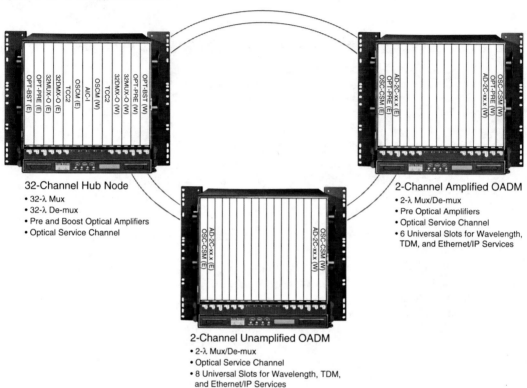

32-Channel Hub Node
- 32-λ Mux
- 32-λ De-mux
- Pre and Boost Optical Amplifiers
- Optical Service Channel

2-Channel Amplified OADM
- 2-λ Mux/De-mux
- Pre Optical Amplifiers
- Optical Service Channel
- 6 Universal Slots for Wavelength, TDM, and Ethernet/IP Services

2-Channel Unamplified OADM
- 2-λ Mux/De-mux
- Optical Service Channel
- 8 Universal Slots for Wavelength, TDM, and Ethernet/IP Services

This chapter covers three major advanced technologies:

- Storage-area networking (SAN)
- DWDM
- Ethernet

For each, you will look at a brief history of the evolution of the service in general and then focus on their integration into the MSPP platform.

Storage

IT organizations have been wrestling over whether the advantages of implementing a SAN solution justify the associated costs. Other organizations are exploring new storage options and whether SAN really has advantages over traditional storage options, such as Network Attached Storage (NAS). In this brief historical overview, you will be introduced to the basic purpose and function of a SAN and will examine its role in modern network environments. You will also see how SANs meet the network storage needs of today's organizations.

When the layers of even the most complex technologies are stripped back, you will likely find that they are rooted in common rudimentary principles. This is certainly true of storage-area networks (SANs). Behind the acronyms and fancy terminology lies a technology designed to provide a way of offering one of the oldest network services of providing data to users who are requesting it.

In very basic terms, a SAN can be anything from a pair of servers on a network that access a central pool of storage devices, as shown in Figure 3-3, to more than a thousand servers accessing multimillions of megabytes of storage. Theoretically, a SAN can be thought of as a separate network of storage devices that are physically removed from but still connected to the network, as shown in Figure 3-4. SANs evolved from the concept of taking storage devices—and, therefore, storage traffic—from the local-area network (LAN) and creating a separate back-end network designed specifically for data.

Figure 3-3   *Servers Accessing a Central Pool of Storage Devices*

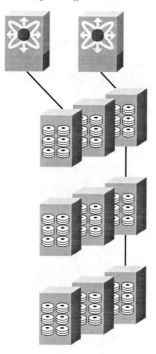

A Brief History of Storage

SANs represent the latest of an emerging sequence of phases in data storage technology. In this section, you will take a look at the evolution of Direct Attached Storage, NAS, and SAN. Just keep in mind that, regardless of the complexity, one basic phenomenon is occurring: clients acquiring data from a central repository. This evolution has been driven partly by the changing ways in which users use technology, and partly by the exponential increase in the volume of data that users need to store. It has also been driven by new technologies that enable users to store and manage data in a more effective manner.

Figure 3-4 *SAN: A Physically Separate Network Attached to a LAN*

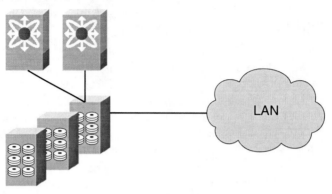

When mainframes were the dominant computing technology, data was stored physically separate from the actual processing unit but was still accessible only through the processing units. As personal computing-based servers proliferated, storage devices migrated to the interior of the devices or in external boxes that were connected directly to the system. Each of these approaches was valid in its time, but with users' growing need to store increasing volumes of data and make that data more accessible, other alternatives were needed. Enter network storage.

Network storage is a generic term used to describe network-based data storage, but many technologies within it make the science happen. The next section covers the evolution of network storage.

Direct Attached Storage

Traditionally, on client/server systems, data has been stored on devices that are either inside or directly attached to the server. Simply stated, Direct Attached Storage (DAS) refers to storage devices connected to a server. All information coming into or going out of DAS must go through the server, so heavy access to DAS can cause servers to slow down, as shown in Figure 3-5.

Figure 3-5 *Direct Attached Storage Example*

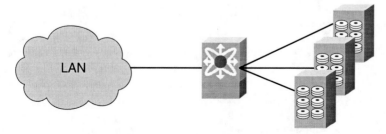

In DAS, the server acts as a gateway to the stored data. Next in the evolutionary chain came NAS, which removed the storage devices from behind the server and connected them directly to the network.

Network Attached Storage

Network Attached Storage (NAS) is a data-storage mechanism that uses special devices connected directly to the network media. These devices are assigned an Internet Protocol (IP) address and can then be accessed by clients using a server that acts as a gateway to the data or, in some cases, allows the device to be accessed directly by the clients without an intermediary, as shown in Figure 3-6.

Figure 3-6   *NAS*

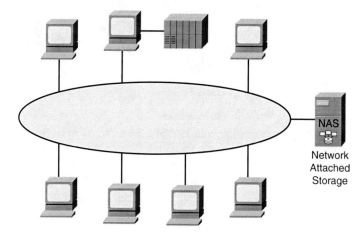

The benefit of the NAS structure is that, in an environment with many servers running different operating systems, storage of data can be centralized, as can the security, management, and backup of the data. An increasing number of businesses are already using NAS technology, if only with devices such as CD-ROM towers (standalone boxes that contain multiple CD-ROM drives) that are connected directly to the network.

Some of the advantages of NAS include scalability and fault tolerance. In a DAS environment, when a server goes down, the data that the server holds is no longer available. With NAS, the data is still available on the network and is accessible by clients.

A primary means of providing fault-tolerant technology is Redundant Array of Independent (or Inexpensive) Disks (RAID), which uses two or more drives working together. RAID disk drives are often used for servers; however, their use in personal computers (PCs) is limited. RAID can also be used to ensure that the NAS device does not become a single point of failure.

Storage-Area Networking

Storage-area networking (SAN) takes the principle one step further by allowing storage devices to exist on their own separate network and communicate directly with each other over very fast media. Users can gain access to these storage devices through server systems, which are connected to both the local-area network (LAN) and the SAN, as shown in Figure 3-7.

Figure 3-7 *A SAN with Interconnected Switches*

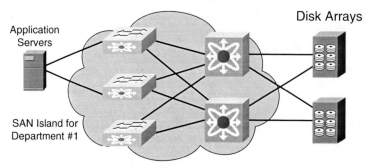

This is in contrast to the use of a traditional LAN for providing a connection for server-based storage, a strategy that limits overall network bandwidth. SANs address the bandwidth bottlenecks associated with LAN-based server storage and the scalability limitations found with Small Computer Systems Interface (SCSI) bus-based implementations. SANs provide modular scalability, high availability, increased fault tolerance, and centralized storage management. These advantages have led to an increase in the popularity of SANs because they are better suited to address the data-storage needs of today's data-intensive network environments.

Business Drivers Creating a Demand for SAN

Several business drivers are creating the demand and popularity for SANs:

- **Regulations**—Recent national disasters have driven regulatory authorities to mandate new standards for disaster recovery and business continuance across many sectors, including financial and banking, insurance, health care, and government entities. As an example, the Federal Reserve and the Securities and Exchange Commission (SEC) recently released a document titled *Interagency Paper on Sound Practices to Strengthen the Resilience of the U.S. Financial System,* which outlines objectives for rapid recovery and timely resumption of critical operations after a disaster. Similar regulations addressing specific requirements for health care, life sciences, and government have been issued or are under consideration.

- **Cost**—Factors include the cost of downtime (millions of dollars per hour for some institutions), more efficient use of storage resources, and reduced operational expenses.

- **Competition**—With competitive pressures created by industry deregulation and globalization, many businesses are now being judged on their business continuance plans more closely than ever. Many customers being courted are requesting documentation detailing disaster-recovery plans before they select providers or even business partners. Being in a position to recover quickly from an unplanned outage or from data corruption can be a vital competitive differentiator in today's marketplace. This rapid recovery capability will also help maintain customer and partner relationships if such an event does occur.

The advantages of SANs are numerous, but perhaps one of the best examples is that of the serverless backup (also commonly referred to as third-party copying). This system allows a disk storage device to copy data directly to a backup device across the high-speed links of the SAN without any intervention from a server. Data is kept on the SAN, which means that the transfer does not pollute the LAN, and the server-processing resources are still available to client systems.

SANs are most commonly implemented using a technology called Fibre Channel (FC). FC is a set of communication standards developed by the American National Standards Institute (ANSI). These standards define a high-performance data-communications technology that supports very fast data rates of more than 2 Gbps. FC can be used in a point-to-point configuration between two devices, in a ring type of model known as an arbitrated loop, and in a fabric model.

Devices on the SAN are normally connected through a special kind of switch called an FC switch, which performs basically the same function as a switch on an Ethernet network: It acts as a connectivity point for the devices. Because FC is a switched technology, it is capable of providing a dedicated path between the devices in the fabric so that they can use the entire bandwidth for the duration of the communication.

Regardless of whether the network-storage mechanism is DAS, NAS, or SAN, certain technologies are common. Examples of these technologies include SCSI and RAID.

For years, SCSI has been providing a high-speed, reliable method of data storage. Over the years, SCSI has evolved through many standards to the point that it is now the storage technology of choice. Related to but not reliant on SCSI is RAID. RAID is a series of standards that provide improved performance and fault tolerance for disk failures. Such protection is necessary because disks account for about 50 percent of all hardware device failures on server systems. As with SCSI, the technologies such as RAID used to implement data storage have evolved, developed, and matured over the years.

The storage devices are connected to the FC switch using either multimode or single-mode fiber-optic cable. Multimode cable is used for short distances (up to 2 km), and single-mode cable is used for longer distances. In the storage devices themselves, special FC interfaces provide the connectivity points. These interfaces can take the form of built-in adapters, which are commonly found in storage subsystems designed for SANs, or can be interface cards much like a network card, which are installed into server systems.

So how do you determine whether you should be moving toward a SAN? If you need to centralize or streamline your data storage, a SAN might be right for you. Of course, there is one barrier between you and storage heaven: money. SANs remain the domain of big business because the price tag of SAN equipment is likely to remain at a level outside the reach of small or even medium-size businesses. However; if prices fall significantly, SANs will find their way into organizations of smaller sizes.

Evolution of SAN

The evolution of SAN is best described in three phases, each of which has its own features and benefits of configuring, consolidating, and evolution:

- **Phase I**—Configures SANs into homogeneous islands, as shown in Figure 3-8. Each of the storage networks is segmented based on some given criteria, such as workgroup, geography, or product.

Figure 3-8 *Isolated Islands of Storage Whose Segmentation Is Based on Organization*

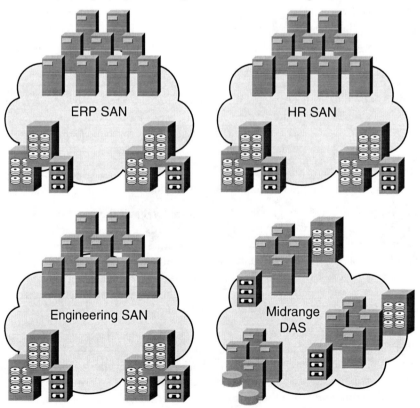

- **Phase II**—Consolidates these storage networks and virtualizes the storage so that storage is shared or pooled among the various work groups. Technologies such as virtual SANs (similar to virtual LANS [VLANs]) are used to provide security and scalability, while reducing total cost of capital. This is often called a multilayer SAN (see Figure 3-9).

Figure 3-9   *Multilayer SAN*

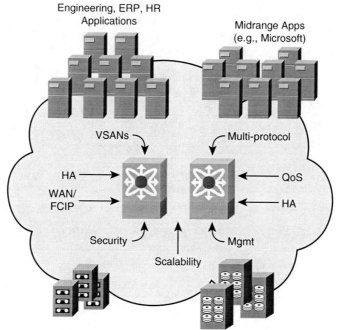

Pooled Disk and Tape

- **Phase III**—Involves adding features such as dynamic provisioning, LAN free backup, and data mobility to the SAN. This avoids having to deploy a separate infrastructure per application environment or department, creating one physical infrastructure with many logical infrastructures. Thus, there is improved use of resources. On-demand provisioning allows networking, storage, and server components to be allocated quickly and seamlessly. This also results in facilities improvements because of improved density and lower power and cabling requirements.

Phase III is often referred to as a multilayer storage utility because this network is more seamless and totally integrated so that it appears as one entity or utility to which you can just plug in. This is analogous to receiving power to your home. Many components are involved in getting power to your home, including transformers, generators, Automatic Transfer Switches, and so on. However, from the enterprise's perspective, only the utility handles the power. Everything behind the outlets is handled seamlessly by the utility provider. Be it water, cable television, power, or "storage," the enterprise sees each as a utility, even though many components are involved in delivering them. Figure 3-10 shows a multilayer storage utility.

Figure 3-10 *Multilayer Storage Utility: One Seamless, Integrated System*

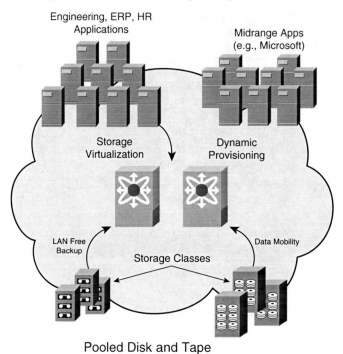

The three major SAN protocols include FC, ESCON, and FICON, and are covered in the following section.

Fibre Channel

FC is a layered network protocol suite developed by ANSI and typically used for networking between host servers and storage devices, and between storage devices. Transfer speeds come in three rates: 1.0625 Gbps, 2.125 Gbps, and 4 Gbps. With single-mode fiber connections, FC has a maximum distance of about 10 km (6.2 miles).

The primary problem with transparently extending FC over long distances stems from its flow-control mechanism and its potential effect on an application's effective input/output (IO) performance. To ensure that input buffers do not get overrun and start dropping FC frames, a system of buffer-to-buffer credits provides a throttling mechanism to the transmitting storage or host devices to slow the flow of frames. The general principle is that one buffer-to-buffer credit is required for every 2 km (1.2 miles) to sustain 1 Gbps of bandwidth, and one buffer-to-buffer credit is required for every 1 km (0.6 miles) between two interfaces on a link for 2 Gbps. These numbers are derived using full-size FC frames (2148 bytes); if using smaller frames, the number of buffer credits required significantly increases. Without SAN extension methods in place, a typical FC fabric cannot exceed 10 km (6.2 miles). To achieve greater distances with FC SAN extensions, SAN switches are used to provide additional inline buffer credits. These credits are required because most storage devices support very few credits (less than 10) of their own, thereby limiting the capability to directly extend a storage array.

Enterprise Systems Connection

Enterprise Systems Connection (ESCON) is a 200-Mbps unidirectional serial bit transmission protocol used to dynamically connect IBM or IBM-compatible mainframes with their various control units. ESCON provides nonblocking access through either point-to-point connections or high-speed switches called ESCON directors. ESCON performance is seriously affected if the distance spanned is greater than 8 km (5 miles).

Fiber Connection

Fiber Connection (FICON) is the next-generation bidirectional channel protocol used to connect mainframes directly with control units or ESCON aggregation switches, such as ESCON directors with a bridge card. FICON runs over FC at a data rate of 1.062 Gbps by using its multiplexing capabilities. One of the main advantages of FICON is its performance stability over distances. FICON can reach a distance of 100 km (62 miles) before experiencing any significant drop in data throughput.

- **Present and future demand**—The type and quantity (density) of SAN extension protocols to be transported, as well as specific traffic patterns and restoration techniques, need to be considered. The type and density requirements help determine the technology options and specific products that should be implemented. Growth should be factored into the initial design to ensure a cost-effective upgrade path.

- **Distances**—Because of the strict latency requirements of SAN applications, especially those found in synchronous environments, performance could be severely affected by the type of SAN extension technology implemented. Table 3-2 provides some guidance on distance restrictions and other considerations for each technology option.

- **Recovery objectives**—A business-continuity strategy can be implemented to reduce an organization's annual downtime and to reduce the potential costs and intangible issues associated with downtime. Recovery with local or remote tape backup could require days to implement, whereas geographically dispersed clusters with synchronous mirroring can result in recovery times measured in minutes. Ultimately, the business risks and costs of each solution have to be weighed to determine the appropriate recovery objective for each enterprise.

- **Original storage manufacturer certifications**—Manufacturers such as IBM, EMC, HP, and Hitachi Data Systems require rigorous testing and associated certifications for SAN extension technologies and for specific vendor products. Implementing a network containing elements without the proper certification can result in limited support from the manufacturer in the event of network problems.

Unlike ESCON, FICON supports data transfers and enables greater rates over longer distances. FICON uses a layer that is based on technology developed for FC and multiplexing technology, which allows small data transfers to be transmitted at the same time as larger ones. IBM first introduced the technology in 1998 on its G5 servers.

FICON can support multiple concurrent data transfers (up to 16 concurrent operations), as well as full-duplex channel operations (multiple simultaneous reads and writes), compared to the half-duplex operation of ESCON.

Table 3-1 *FC Levels*

Level	Functionality
FC-4	Mapping ATM, SCSI-3, IPI-3, HIPPI, SBCCS, FICON, and LE
FC-3	Common services
FC-2	Framing protocol
FC-1	Encode/decode (8B/10B)
FC-0	Physical

Table 3-2 *SAN Extension Options*

	FC over Dark Fiber	FC over CWDM	DWDM	SONET/SDH	FCIP
SAN protocols supported	FC	FC	FC, FICON, ESCON, IBM Sysplex Timer, IBM Coupling Facility	FC, FICON	FC over IP
SAN distances supported (FC/ FCIP only for comparative purposes)	90 km (56 miles)*	60 to 66 km (37 to 41 miles)**	Up to 200 km (124 miles)***	2800 km (1740 miles) with buffer credit support	Distance limitation dependent on the latency tolerance of the end application

**Longest tested distance is 5800 km (3604 miles) |

Table 3-2   *SAN Extension Options (Continued)*

	FC over Dark Fiber	FC over CWDM	DWDM	SONET/SDH	FCIP
SAN bandwidth options (per fiber pair)	1-Gbps FC (1.0625 Gbps), 2-Gbps FC (2.125 Gbps), 4-Gbps FC	1-Gbps FC (1.0625 Gbps), 2-Gbps FC (2.125 Gbps), up to 8 channels	Up to 256 FC/FICON channels, up to 1280 ESCON channels, up to 32 channels at 10 Gbps	1-Gbps FC (1.0625 Gbps), up to 32 channels with subrating, 2-Gbps FC (2.125 Gbps), up to 16 channels with subrating	1-Gbps FC (1.0625 Gbps)
Network-protection options	FSPF, PortChannel, isolation with VSANs	FSPF, PortChannel, isolation with VSANs	Client, 1+1, y-cable, switch fabric protected, switch fabric protected trunk, and protection switch module, unprotected	UPSR/SNCP, 2F and 4F BLSR/ MS-SPR, PPMN, 1+1 APS/MSP, unprotected	VRRP, redundant FCIP tunnels, FSPF, PortChannel, isolation with VSANs
Other protocols supported	—	CWDM filters also support GigE	OC-3/12/48/ 192, STM-1/4/ 16/64, GigE, 10-Gigabit Fast Ethernet, D1 Video	DS-1, DS-3, OC-3/12/48/192, E-1, E-3, E-4, STM-1E, STM-1/4/16/64, 10/100-Mbps Ethernet, GigE	—

*Assumes the use of CWDM SFPs, no filters

**Assumes point-to-point configuration

***Actual distances depend on the characteristics of fiber used.

FICON is mapped over the FC-2 protocol layer (refer back to Table 3-1) in the FC protocol stack, in both 1-Gbps and 2-Gbps implementations. The FC standard uses the term *Level* instead of *Layer* because there is no direct relationship between the Open Systems Interconnection (OSI) layers of a protocol stack and the levels in the FC standard.

Within the FC standard, FICON is defined as a Level 4 protocol called SB-2, which is the generic terminology for the IBM single-byte command architecture for attached I/O devices. FICON and SB-2 are interchangeable terms; they are connectionless point-to-point or switched point-to-point FC topology.

FCIP

Finally, before delving into the SAN over MSPP, it is important to note that FC can be tunneled over an IP network known as FCIP, as shown in Figure 3-11. FC over IP (FCIP) is a protocol specification developed by the Internet Engineering Task Force (IETF) that allows a device to transparently tunnel FC frames over an IP network. An FCIP gateway or edge device attaches to an FC switch and provides an interface to the IP network. At the remote SAN island, another FCIP device receives incoming FCIP traffic and places FC frames back onto the SAN. FCIP devices provide FC expansion port connectivity, creating a single FC fabric.

FCIP moves encapsulated FC data through a "dumb" tunnel, essentially creating an extended routing system of FC switches. This protocol is best used in point-to-point connections between SANs because it cannot take advantage of routing or other IP management features. And because FCIP creates a single fabric, traffic flows could be disrupted if a storage switch goes down.

One of the primary advantages of FCIP for remote connectivity is its capability to extend distances using the Transmission Control Protocol/Internet Protocol (TCP/IP). However, distance achieved at the expense of performance is an unacceptable trade-off for IT organizations that demand full utilization of expensive wide-area network (WAN) bandwidth. IETF RFC 1323 adds Transmission Control Protocol (TCP) options for performance, including the capability to scale the standard TCP window size up to 1 GB. As the TCP window size widens, the sustained bandwidth rate across a long-haul (more latency) TCP connection increases. From early field trials, distances spanning more than 5806 km (3600 miles) were feasible for disk replication in asynchronous mode. Even greater transport distances are achievable. Theoretically, a 32-MB TCP window with a 1-Gbps bandwidth can be extended over 50,000 km (31,069 miles) with 256 ms of latency.

Another advantage of FCIP is the capability to use existing infrastructures that provide IP services. For IT organizations that are deploying routers for IP transport between their primary data centers and their disaster-recovery sites, and with quality of service (QoS) enabled, FCIP can be used for SAN extension applications. For larger IT organizations that have already invested in or are leasing SONET/Synchronous Digital Hierarchy (SDH) infrastructures, FCIP can provide the most flexibility in adding SAN extension services because no additional hardware is required.

For enterprises that are required to deploy SAN extensions across various remote offices with the central office (CO), a hub-and-spoke configuration of FCIP connections is also possible. In this manner, applications such as disk replication can be used between the disk arrays of each individual office and the CO's disk array, but not necessarily between the individual offices' disk arrays themselves. With this scenario, the most cost-effective method of deployment is to use FCIP along routers.

Figure 3-11 *FCIP: FC Tunneled over IP*

- Transparently joining SAN islands over WAN
- Transparent bridging of FC over TCP/IP
- Extended distance (>2000 km)

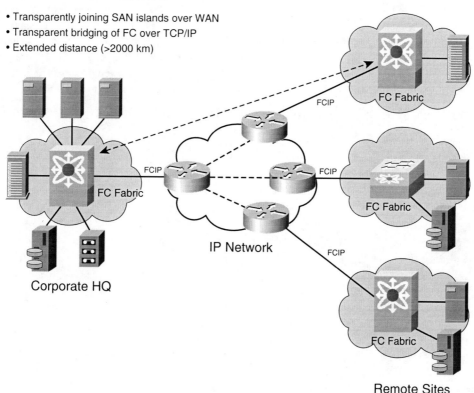

SAN over MSPP

FC technology has become the protocol of choice for the SAN environment. It has also become common as a service interface in metro DWDM networks, and it is considered one of the primary drivers in the DWDM market segment. However, the lack of dark fiber available for lease in the access portion of the network has left SAN managers searching for an affordable and realizable solution to their storage transport needs. Thus, service providers have an opportunity to generate revenue by efficiently connecting and transporting the user's data traffic via FC handoffs. Service providers must deploy metro transport equipment that will enable them to deliver these services cost-effectively and with the reliability required by their service-level agreements (SLAs). This growth mirrors the growth in Ethernet-based services and is expected to follow a similar path to adoption— that is, a transport evolution in which TDM, Ethernet, and now FC move across the same infrastructure, meeting the needs of the enterprise end user without requiring a complete hardware upgrade of a service provider's existing infrastructure.

Consider a couple of the traditional FCIP over SONET configurations. Figure 3-12 shows a basic configuration, in which the Gigabit Ethernet (GigE) port of the IP Storage Services Module is connected directly to the GigE port of an MSPP. This scenario assumes that a dedicated GigE port is available on the MSPP. Another possible configuration is to include routers between the IP Storage Services Module and the MSPP, as shown in Figure 3-13. In this case, the MSPP might not necessarily have a GigE card, so a router is required to connect the GigE connection of the IP Storage Services Module to the MSPP.

Figure 3-12 *IP Storage Services Module Connected Directly to an MSPP*

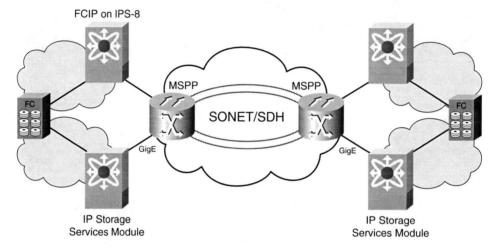

Figure 3-13 *IP Storage Services Module Connected Routers Interfaced to MSPP*

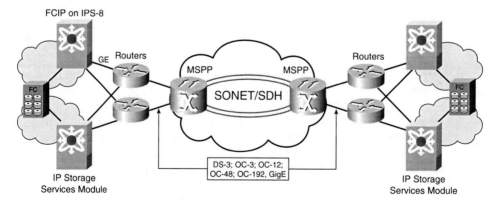

MSPP with Integrated Storage Card

The storage card, such as is found in the Cisco ONS 15454 MSPP, is a single-slot card with multiple client ports, each supporting 1.0625- or 2.125-Gbps FC/FICON. It uses pluggable gigabit interface converter (GBIC) optical modules for the client interfaces, enabling greater user flexibility. The payload from a client interface is mapped directly to SONET/SDH payload through transparent generic framing procedure (GFP-T) encapsulation. This payload is then cross-connected to the system's optical trunk interfaces (up to OC-192) for transport, along with other services, to other network elements.

The new card fills the FC over SONET gaps in the transport category of the application. This allows MSPP manufacturers to provide 100 percent availability of the FC need, while also providing end-to-end coverage of data center and enterprise storage networking solutions across the metropolitan, regional, and wide area networks, as shown in Figure 3-14.

Figure 3-14 *Integrated Storage Card within an MSPP*

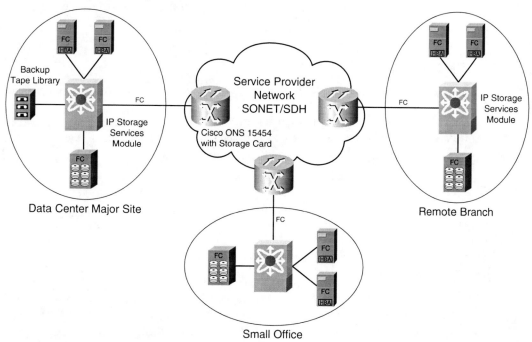

The storage interface card plugs into the existing MSPP chassis and is managed through the existing management system. Its introduction does not require a major investment in capital expenditures (CapEx) or operational expenditures (OpEx), but rather, an evolutionary extension of services. For the service provider, this creates an opportunity to further capture market and revenues from existing and often extensive MSPP installations. For the enterprise, this equals access to new storage over SONET/SDH services, enabling it to deploy needed SAN extensions and meet business-continuance objectives.

Storage Card Highlights

Consider the storage features of Cisco ONS 15454 MSPP:

- It supports 1-Gbps and also 2-Gbps FC with low-latency GFP-T mapping, allowing customers to grow beyond 1-Gbps FC.

- It supports FC over protected SONET/SDH transport networks in a single network element: 16 line-rate FC on a single shelf over a fully protected transport network, such as 4F bidirectional line-switched ring (BLSR) OC-192 and dual 2F-BLSR/ unidirectional path-switched ring (UPSR) OC-192.

- It lowers CapEx and OpEx costs by using existing infrastructure and management tools.

- It increases the service-offering capabilities.

- It does not require upgrade of costly components of the MSPP, such as the switch matrix of the network element.

SAN Management

Storage networking over the MSPP continues the simple, fast, easy approach introduced in implementing traditional services in the MSPP. The GUI applications greatly increase the speed of provisioning, testing, turn-up, and even troubleshooting aspects of storage over MSPP, and they reduce the need for an additional OSS to implement this service.

DWDM

As a means of introduction, next you will look at a brief history of DWDM and how DWDM is delivered from MSPPs.

History of DWDM

In the mid-1980s, the U.S. government deregulated telephone service, allowing small telephone companies to compete with the giant AT&T. Companies such as MCI and Sprint quickly went to work installing regional fiber-optic telecommunications networks throughout the world. Taking advantage of railroad lines, gas pipes, and other rights of way, these companies laid miles of fiber-optic cable, allowing the deployment of these networks to continue throughout the 1980s. However, this created the need to expand fiber's transmission capabilities.

In 1990, Bell Labs transmitted a 2.5-Gbps signal over 7500 km without regeneration. The system used a soliton laser and an erbium-doped fiber amplifier (EDFA) that allowed the light wave to maintain its shape and density. In 1998, they went one step further as researchers transmitted 100 simultaneous optical signals, each at a data rate of 10 Gbps, for a distance of nearly 250 miles. In this experiment, DWDM (technology that allows multiple wavelengths to be combined into one optical signal) increased the total data rate on one fiber to 1 terabit per second (Tbps, or 10^{12} bits per second).

Today DWDM technology continues to develop. As the demand for data bandwidth increases (driven by the phenomenal growth of the Internet and other LAN-based applications such as storage, voice, and video), the move to optical networking is the focus of new technologies. At the time of this writing, more than one billion people have Internet access and use it regularly. More than 50 million households are "wired." The World Wide Web already hosts billions of web pages, and, according to estimates, people upload more than 3.5 million new web pages every day.

The important factor in these developments is the increase in a fiber's capacity to transmit data, which has grown by a factor of 200 in the last decade. Because of fiber-optic technology's immense potential bandwidth (50 terahertz [THz] or greater), there are extraordinary possibilities for future fiber-optic applications. Alas, as of this writing, carriers are planning and launching broadband services, including data, audio, and especially video, into the home.

DWDM technology multiplies the total bandwidth of a single fiber by carrying separate signals on different wavelengths of light. DWDM has been a core technology in long-haul core networks, where it has been used successfully for many years to vastly increase the capacity of long-haul fibers. Metro networks are now realizing the same benefit: With DWDM, a fiber that now carries a single OC-48 can carry a dozen or more equivalent colors of light. This transforms what was a single-lane road to a multilane freeway.

One distinction to note is that metro DWDM must be designed differently than long-haul DWDM. In long-haul applications, DWDM systems must handle the attenuation and signal loss that are resident in connections over hundreds of miles. In metro applications, however, where distances are measured in tens of miles, attenuation is not the primary challenge. Unlike point-to-point long-haul connections, metro rings have many add and drop access points to the network. Therefore, metro DWDM systems must be designed to provide the flexibility for adding and dropping individual wavelengths at different access points in the network, while passing along the wavelengths that are not needed at those points.

As you already know, DWDM is an MSPP service that does not ride within a SONET frame. On the contrary, as Figure 3-15 and Figure 3-16 show, the SONET frames can be carried within the wavelength.

Adding wavelengths (or lambdas, as they are often called, after the Greek letter that is the physical symbol for wavelength) enlarges the pipe, but it is important to use the extra capacity efficiently as well. Next-generation metro transport platforms provide the versatility needed to groom smaller services efficiently onto each wavelength. As you have seen already, next-generation platforms can handle traditional DS1 and DS3 services, 10-/100-Mbps, GigE services, and interfaces to TDM transport that scale from DS1 (1.5 Mbps) to OC-192 (10 Gbps). Used in combination with metro DWDM technology integrated into the platform, these systems give carriers the capability of scaling their service offerings, including these:

- Wavelength services for large customers who need storage-area networking, for example, which requires a large amount of bandwidth for frequent backups. In this application, an entire wavelength can consist of a line-rate GigE signal or one of several other interfaces, including FC, ESCON, FICON, and D1 video.

- Subwavelength services that guarantee full line rate for a portion of a wavelength at any time. Next-generation systems support transport for any data, TDM, or combination of services on any wavelength, eliminating the wasted bandwidth inherent in legacy systems.

- Subwavelength services that are statistically multiplexed, allowing the provider to oversubscribe the connection to maximize use based upon demand, as well as time-of-day traffic requirements. For example, a service might be used heavily for business traffic during the day and for residential traffic during the evening, although it could not handle the full level of both types of traffic at the same time.

Figure 3-15 shows the aggregation of these various types of services on a DWDM fiber. Figure 3-16 shows the varied layering of transport and service protocols that is possible over DWDM wavelengths.

Figure 3-15 *Metro DWDM Bandwidth Management*

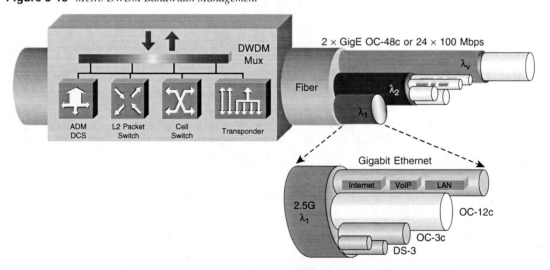

By enabling such a wide range of service offerings, DWDM plus next-generation metro transport addresses the overriding concerns of service density and service velocity: the amount of time it takes to deploy a service to a customer. Greater capacity and efficiency enable carriers to approach the maximum service density—that is, the greatest total number of services that can be provided on each wavelength of the fiber.

Figure 3-16 *Metro DWDM as Enabler*

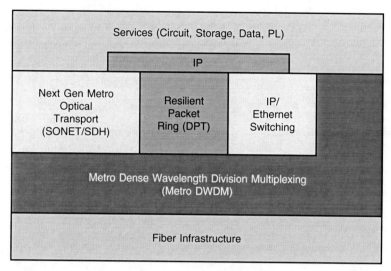

High service density means that the greatest number of customers are served by a given capital investment in network infrastructure. At the same time, carriers can achieve higher service velocity in a competitive marketplace. With maximum bandwidth available and flexible bandwidth management, carriers can introduce new services and modify existing ones more quickly, and then make service alternatives available faster to a wider range of customers.

Fiber-Optic Cable

Fiber-optic cable comes in various sizes and shapes. As with coaxial cable, its actual construction is a function of its intended application. It also has a similar "feel" and appearance. Figure 3-17 is a sketch of a typical fiber-optic cable.

Figure 3-17 *Fiber-Optic Cable Construction*

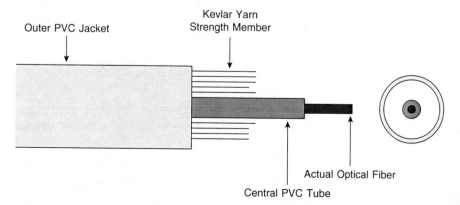

The basic optical fiber is provided with a buffer coating, which is mainly used for protection during the manufacturing process. This fiber is then enclosed in a central PVC loose tube, which allows the fiber to flex and bend, particularly when going around corners or when being pulled through conduits. Around the loose tube is a braided Kevlar yarn strength member, which absorbs most of the strain put on the fiber during installation.

Acceptance of Light onto Fiber

Light is directed into a fiber in two ways. One is by pigtailing; the other is by placing the fiber's tip close to a light emitting diode (LED). When the proximity type of coupling is employed, the amount of light that will enter the fiber is a function of one of four factors:

- **Intensity of the LED**—The intensity of the LED is a function of its design and is usually specified in terms of total power output at a particular drive current. Sometimes this figure is given as actual power that is delivered into a particular type of fiber. All other factors being equal, more power provided by the LED translates to more power "launched" into the fiber.

- **Area of the light-emitting surface**—The amount of light "launched" into a fiber is a function of the area of the light-emitting surface compared to the area of the light-accepting core of the fiber. The smaller this ratio is, the more light is launched into the fiber.

- **Acceptance angle of the fiber**—The acceptance angle of a fiber is expressed in terms of numeric aperture. The numerical aperture (NA) is defined as the sine of the acceptance angle of the fiber. Typical NA values are 0.1 to 0.4, which correspond to acceptance angles of 5.7 degrees to 23.6 degrees. Optical fibers transmit only light that enters at an angle that is equal to or less than the acceptance angle for the particular fiber.

- **Other losses from reflections and scattering**—Other than opaque obstructions on the surface of a fiber, there is always a loss that results from reflection from the entrance and exit surface of any fiber. This loss is called the Fresnell Loss and is equal to about 4 percent for each transition between air and glass. Special coupling gels can be applied between glass surfaces to reduce this loss, when necessary.

Wavelength-Division Multiplexing: Course Wavelength-Division Multiplexing versus DWDM

Before we go any further, let's distinguish among wavelength-division multiplexing (WDM), course wavelength-division multiplexing, and DWDM, even though the focus remains on DWDM.

WDM technology was developed to get more capacity from the existing fiber-optic cable plant by using channels (wavelengths) to carry multiple signals on a single fiber.

Two major categories of WDM are used: CWDM and DWDM.

A major difference between the two, as their names imply, is the channel spacing within the window of the optical spectrum. The wide pass-band (20 ± 6–7 nm) of CWDM channels allows for the use of less expensive components, such as uncooled lasers and thin-film filter technology. CWDM systems provide cost advantages over DWDM in the same application. As such, many people push CWDM as a more appropriate platform for the shorter distances typically found in metro access networks.

Sometimes metro networks require longer distances and more wavelengths than CWDM can provide, however. Today CWDM does not practically support more than the 18 channels between 1271 and 1611 nm, standardized by the ITU Telecommunication Standardization Sector (ITU-T) G.694.2 wavelength grid. Relatively low-cost "metro" DWDM can pack many 2.5- and 10-Gbps wavelengths (up to 40), onto a single fiber. However, to do so, precision filters, cooled lasers, and more space are needed, which can make DWDM too expensive for some edge networks.

What is the best solution for a given application? Depending on cost, distance, and the number of channels, metro-area networks might benefit from a mixture of both coarse and dense WDM technologies.

Early WDM deployments emerged in the form of "doublers," which used 1310-nm and 1550-nm lasers through passive filters to transport two signals on a single fiber. This simple approach was reliable, low in cost, and easy to operate, which made it suitable for carrier networks.

Other WDM techniques were customized for specific applications. CWDM was used mostly in LANs because of the reach limitations imposed by operating in the 850-nm range. By the early 1990s, WDM development was focused on solving capacity shortages experienced by interexchange carriers. Their national backbone networks presented a different set of parameters and cost structure. This enabled the use of more complex and expensive components, which made high-capacity transport over long distances possible. With its narrow channel spacing (1.6 and 0.8 nm), DWDM allowed many wavelengths in a small window of optical spectrum. Narrow channel spacing is important for long distances because fiber attenuation is lowest in the C-band.

DWDM Integrated in MSPP

Traditionally, SONET platforms have been dedicated to services that could be encapsulated within SONET frames. Today vendors not only can deliver SONET services from MSPPs, but they also can hand off these services in a DWDM wavelength service.

Figure 3-18 shows a DWDM networked environment that uses the MSPP architecture.

Five aspects will be discussed with regard to integrating DWDM into MSPPs; active vs. passive DWDM, EDFAs, DWDM benefits, protection options, and market drivers for MSPP-based DWDM.

Figure 3-18 *Metro DWDM Integrated within the MSPP Architecture*

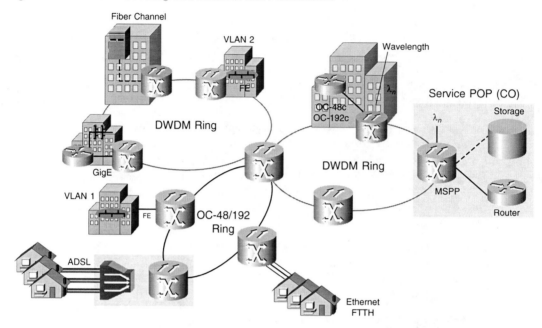

Active and Passive DWDM

DWDM can be implemented with an MSPP in two ways. Most often when you think about DWDM systems, you probably think of active DWDM systems. However, the multiplexing of multiple light sources is always a "passive" activity. Wavelength conversion and amplification is always the "active" DWDM activity.

Figure 3-19 shows an MSPP chassis with integrated DWDM optics in which the optics cards (in this case, OC-48s) use one of the ITU wavelengths and interfaces to an external filter. This filter multiplexes the wavelengths from various optics cards within multiple chasses and transports them over the fiber, where they are demultiplexed on the other end. This is known as passive DWDM with respect to the MSPP because the filter is a separate device.

This inefficient use of the rack and shelf space has led to the development of active DWDM from the MSPP. With active DWDM, the transponding of the ITU wavelength to a standard 1550-nm wavelength is performed by converting the MSPP shelf into various components required in a DWDM system. This conversion has greatly increased the density of wavelengths within a given footprint. Under the passive example shown in Figure 3-20, only 16 wavelengths could be configured within a bay, 4 per chassis. With today's multiport, multirate optical cards, this density can be doubled to 8 wavelengths per shelf and 32 per rack.

Figure 3-19 *MSPP Chassis with Integrated ITU Optics Card Connected to East and West DWDM Filters*

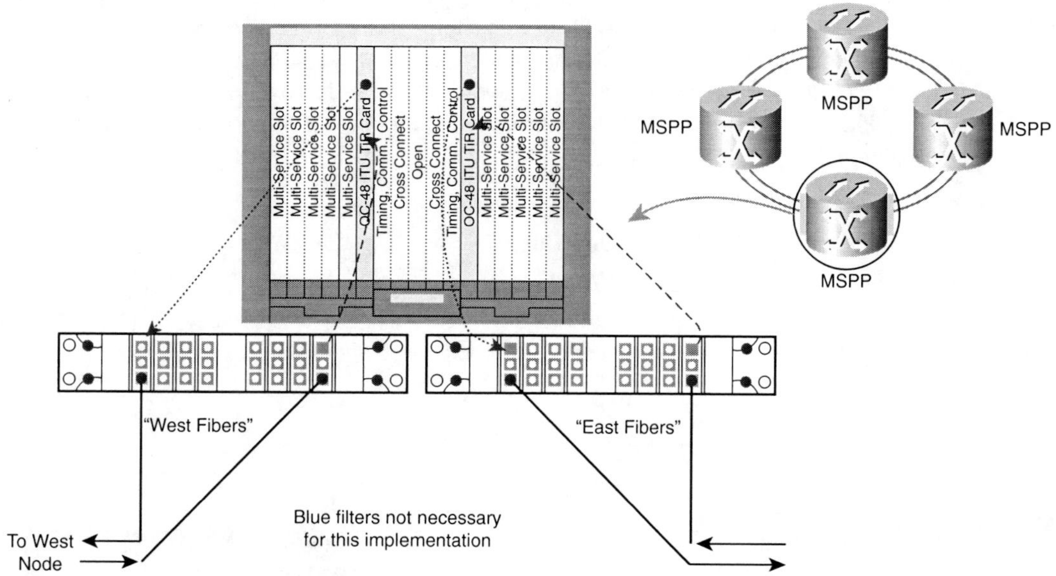

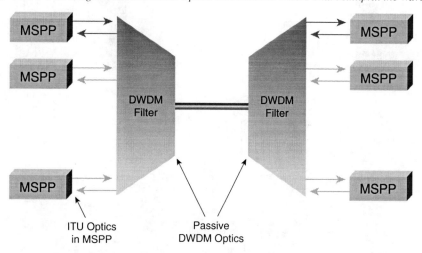

Figure 3-20 *MSPP with Integrated ITU DWDM Optics Connected to Filters That Multiplex the Wavelengths*

With the integrated active DWDM solution of Figure 3-20, one MSPP chassis can be converted into a 32-channel multiplexer/demultiplexer using reconfigurable optical add/drop multiplexing (ROADM) technology. Other chassis can be converted into a multichannel optical add/drop multiplexer (OADM), which can receive and distribute multiple wavelengths per shelf. The implication of this is that up to 32 wavelengths can be terminated within a bay or rack, a factor of eight times the density of even early MSPPs using a passive external filter. The traffic from within each wavelength dropped into an MSPP shelf from the ROADM hub shelf can be groomed or extracted from the wavelengths carrying it, as needed, and dropped out of the OADM shelves, as shown in Figure 3-21. ROADM is an option that can be deployed in place of fixed-wavelength OADMs. Cisco Systems ROADM technology, for example, consists of two modules:

- 32-channel reconfigurable multiplexer (two-slot module)
- 32-channel reconfigurable demultiplexer (one-slot module)

It uses two sets of these modules for East and West connectivity, allowing for 32-channel add/drop/pass-through support with no hardware changes. At each node, software controls which wavelengths are added, dropped, or passed through, on a per-wavelength basis.

Figure 3-21 *32 ITU MSPP Wavelengths Connected to OADM Shelves*

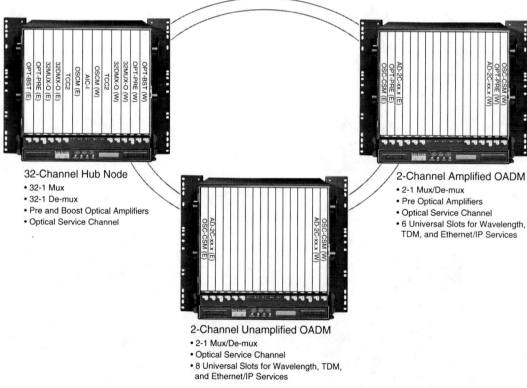

32-Channel Hub Node
- 32-1 Mux
- 32-1 De-mux
- Pre and Boost Optical Amplifiers
- Optical Service Channel

2-Channel Amplified OADM
- 2-1 Mux/De-mux
- Pre Optical Amplifiers
- Optical Service Channel
- 6 Universal Slots for Wavelength, TDM, and Ethernet/IP Services

2-Channel Unamplified OADM
- 2-1 Mux/De-mux
- Optical Service Channel
- 8 Universal Slots for Wavelength, TDM, and Ethernet/IP Services

Erbium-Doped Fiber Amplifiers

Erbium-doped fiber amplifiers (EDFAs) can be integrated within the DWDM MSPP shelf as well as optical service channel cards for management. These amplifiers extend the distance of the signal by amplifying it. Features of EDFAs include these:

- **Constant flat gain**—Constant gain and noise control simplify network design.
- **Metro optimized automatic gain control**—Highly precise, rapid automatic gain control (AGC) capabilities allow the EDFAs to be used as a booster or inline amplifier.
- **Variable gain**—The variable gain capabilities of EDFAs are critical to network designs in which amplifier spacing must be flexible.

  Variable gain allows for the addition or elimination of optical elements, such as OADM, without drastic network redesigns or costly equipment changes.

  The adjustable gain of the EDFAs can be used to reset a network to a better operating point after a change in span loss.

DWDM Advantages

Many benefits are tied to having a DWDM over MSPP, including scalability, rapid service velocity, and flexibility (to name a few). Some of these benefits are shown as follows:

- Scalability, up to 64 wavelengths in a single network for superior capital investment protection
- Transport of 150-Mbps to 10-Gbps wavelength services, as well as aggregated TDM and data services, providing maximum service flexibility.
- Transmission capability from tens to hundreds of kilometers (up to 1000 km) through the use of advanced amplification, dispersion compensation, and forward error correction (FEC) technologies.
- "Plug-and-play" card architecture that provides complete flexibility in configuring DWDM network elements such as terminal nodes, optical add/drop nodes, line amplifiers, and dispersion compensation.
- Vastly improved shelf density for high-bandwidth (10-Gbps) wavelength services.
- Flexible 1- to 64-channel OADM granularity, supporting both band and channel OADMs, for reduced complexity in network planning and service forecasting.
- Seamless use of pre- and post-amplification.
- Use of software-provisionable, small form-factor pluggable (SFP) client connectors, and wavelength tunability for reduced card inventory requirements.
- Multilevel service monitoring: SONET/SDH, G.709 digital wrapper, and optical service channel for unparalleled service reliability.

Figure 3-22 shows an active transponder-based DWDM.

Figure 3-22 *Active Transponder-Based DWDM with EDFAs Integrated into the Shelf*

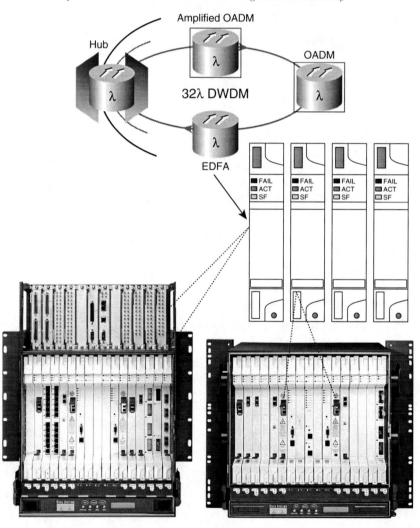

Protection Options

Several ways exist for protecting an MSPP-based DWDM system in the event of a fiber cut or signal degradation. Such protection options include client protection, Y-cable protection, and wavelength splitting.

Figure 3-23 shows these protection types. They are described as follows:

- **Client protection**—Client protection is an option in which signaling is the function of the client equipment. Clients can use linear 1+1, SONET/SDH, or other protection mechanisms; MSPP provides the diverse routes through the intelligent optical network.

- **Y-cable protection**—Y-cable protection is an option in which the client signal is split and diversely routed through the network. Protection is provided by the MSPP.

- **Wavelength splitting**—In wavelength splitting, the DWDM signal is split and diversely routed. In this case, protection is provided by the MSPP. Protection options depend on the service agreement and can be combined for maximum reliability.

Reliability for these options varies, depending on the client network architectures and service-level agreements (SLA) provided to the client. Thus, there is no "one size fits all" approach to protection.

Figure 3-23 *Client, Y-Cable, and Fiber Protection Options*

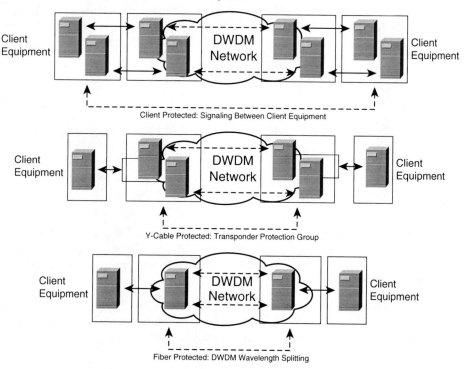

Market Drivers for MSPP-Based DWDM

One of the main obstacles to the adoption of DWDM technology in metro networks is the inflexibility associated with first-cost and network growth. The MSPP-based DWDM has been architected for networking flexibility and growth:

- **Added wavelengths**—Wavelengths can be added as needed without impacting other wavelengths in the network, and without having to adjust multiple optical parameters across the network.

- **Migration from 2.5-Gbps to 10-Gbps wavelengths**—Most wavelengths today have 2.5 Gbps bandwidth capacity, but it is clear that 10-Gbps wavelengths will be needed in the future. MSPP DWDM has been designed for wavelength growth. For example, specially designed amplifiers allow dispersion to be managed without adversely affecting link budgets when wavelength capacity is upgraded.

- **Wavelength add/drop flexibility**—A flexible OADM architecture allows wavelengths to be added and dropped or passed through, allowing configurations that can be changed on a per-wavelength basis without affecting other wavelengths.

Ethernet

As you have seen for storage and DWDM, Ethernet is an advanced technology. Before launching into a discussion of its use over MSPP, let's take a look at its brief history. One point to clarify at the onset is that Ethernet has been around for several decades. Ethernet itself is not a "new" technology, but its use over MSPP is an emerging technique in delivering Ethernet transport.

A Brief History of Ethernet

Personal computers hadn't proliferated in any significant way when researchers and developers started trialing what would later turn out to be the next phase of the PC revolution: connecting these devices to a network. The year 1977 is widely recognized as the PC's big arrival; however, Ethernet—the technology that today attaches millions of PCs to LANs—was invented four years earlier, in the spring of 1973.

The source of this forethought was Xerox Corporation's Palo Alto Research Center (PARC). In 1972, PARC researchers were working on both a prototype of the Alto computer—a personal workstation with a graphical user interface—and a page-per-second laser printer. The plan was for all PARC employees to have computers and to tie all the computers to the laser printer.

The task of creating the network fell to Bob Metcalfe, an MIT graduate who had joined Xerox that year. As Metcalfe says, the two novel requirements of this network were that it had to be very fast to accommodate the laser printer and that it had to connect hundreds of computers.

By the end of 1972, Metcalfe and other PARC experts had completed an experimental 3-Mbps PC LAN. The following year, Metcalfe defined the general principles of what became the first PC LAN backbone. Additionally, this team developed the first PC LAN board that can be installed inside a PC to create a network.

Metcalfe eventually named this PC LAN backbone as Ethernet, based on the idea of "lumeniferous ether," which is the medium that scientists once thought carried electromagnetic waves through space.

Ethernet defines the wire and chip specifications of PC networking, along with the software specifications regarding how data is transmitted. One of the pillars is its system of collision detection and recovery capability, called carrier sense multiple access collision detect (CSMA/CD), which we discuss later in this chapter.

Metcalfe worked feverishly to get Intel Corp., Digital, and Xerox to agree to work on using Ethernet as the standard way of sending packets in a PC network. Thus, 3Com (three companies) Corporation was born. 3Com introduced its first product, EtherLink (the first PC Ethernet network interface card), in 1982. Early 3Com customers included Trans-America Corp. and the White House.

Ethernet gained popularity in 1983 and was soon named an international standard by the Institute of Electrical and Electronics Engineers, Inc. (IEEE). However, one major computer force did not get on board: IBM, which developed a very different LAN mechanism called Token Ring. Despite IBM's resistance, Ethernet went on to become the most widely installed technology for creating LANs. Today it is common to have Fast Ethernet, which runs at 100 Mbps, and GigE, which operates at 1 Gbps. Most desktop PCs in large corporations are running at 10/100 Mbps. The network senses the speed of the PC card and automatically adjusts to it, which is known as autosensing.

Fast Ethernet

The Fast Ethernet (FE) standard was officially ratified in the summer of 1995. FE is ten times the speed of 10BaseT Ethernet. Fast Ethernet (also known as 100BaseT) uses the same CSMA/CD protocol and Category 5 cabling support as its predecessor, while offering new features, such as full-duplex operation and autonegotiation. FE calls for three types of transmissions over various physical media:

- **100BaseTX**—Is the most common application whose cabling is similar to 10BaseT. This uses Category 5–rated twisted-pair copper cable to connect various data-networking elements, using an RJ-45 jack.

- **100BaseFX**—Used predominately to connect switches either between wiring closets or between buildings using multimode fiber-optic cable.

- **100BaseT4**—Uses two more pairs of wiring, which enables Fast Ethernet to operate over Category 3–rated cables or above.

GigE

The next evolutionary leap for Ethernet was driven by the Gigabit Ethernet Alliance, which was formed in 1996 and ratified in the summer of 1999. It specified a physical layer using a mixture of established technologies from the original Ethernet specification and the ANSI X3T11 FC specification:

- **1000BaseX**—A standard based on the FC physical layer. It specifies the technology for connecting workstations, supercomputers, storage devices, and other devices with fiber-optic and copper shielded twisted-pair (STP) media based on the cable distance.

- **1000BaseT**—A GigE standard for long-haul copper unshielded-twisted pair (UTP) media.

Because it is similar to 10-Mbps and 100-Mbps Ethernet, GigE offers an easy, incremental migration path for bandwidth requirements. IEEE 802.3 framing and CSMA/CD are common among all three standards.

The common framing and packet size (64- to 1518-byte packets) is key to the ubiquitous connectivity that 10-/100-/1000-Mbps Ethernet offers through LAN switches and routers in the WAN. Figure 3-24 shows the GigE frame format.

Figure 3-24 *GigE Frame Format*

Bytes	8	6	6	2	0–1500
	Preamble	Destination address	Source address	Length of data field	Protocol header, data and padding

Ethernet Emerges

Why did Ethernet emerge as the victor alongside competitors such as Token Ring?

Since its infancy, Ethernet has thrived primarily because of its flexibility and ease of implementation. To say "LAN" or "network card" is understood to mean "Ethernet." With the capability to use existing UTP telephone wire for 10 Mbps Ethernet, the path into the home and small office was paved for its long-term proliferation.

The CSMA/CD Media Access Control (MAC) protocol defines the rules and conventions for access in a shared network. The name itself implies how the traffic is controlled.

1 First, devices attached to the network check, or sense, the carrier (wire) before transmitting.

2 The device waits before transmitting if the media is in use. ("Multiple access" refers to many devices sharing the same network medium.)

If two devices transmit at the same time, a collision occurs. A collision-detection mechanism retransmits after a random timer "times out" for each device.

With switched Ethernet, each sender and receiver pair provides the full bandwidth. Interface cards or internal circuitry are used to deliver the switched Ethernet signaling and cabling conventions specify the use of a transceiver to attach a cable to the physical network medium. Transceivers in the network cards or internal circuitry perform many of the physical layer functions, including carrier sensing and collision detection.

Let's take a look at the growth of Ethernet beyond the LAN and into the metropolitan-area network (MAN), which is enabled by MSPPs.

Ethernet over MSPP

Today's evolving networks are driven by the demand for a wide variety of high-bandwidth data services. Enterprises must scale up and centralize their information technology to stay competitive. Service providers must increase capacity and service offerings to meet customer requirements while maintaining their own profitability. Both enterprises and service providers need to lower capital and operating expenditures as they evolve their networks to a simplified architecture. Additionally, service providers must accelerate time to market for the delivery of value-added services, and enterprises must accelerate and simplify the process of adding new users. Increasingly, service providers and enterprises are looking to Ethernet as an option because of its bandwidth capabilities, perceived cost advantages, and ubiquity in the enterprise.

The vast fiber build-out over the last few years has caused an emergence of next-generation services in the metropolitan market, including wavelength services and Ethernet services. As discussed, Ethernet providers can deploy a single interface type and then remotely change the end user's bandwidth profile without the complexity or cost associated with Asynchronous Transfer Mode (ATM) and at higher speeds than Frame Relay. ATM requires complex network protocols, including Private Network Node Interface (PNNI), to disseminate address information; LAN Emulation (LANE), which does not scale at all in the WAN; and RFC 1483 ATM Bridged Encapsulation, which works only for point-to-point circuits. On the other hand, whereas Frame Relay is simple to operate, its maximum speed is about 50 Mbps. Ethernet scales from 1 Mbps to 10 Gbps in small increments.

Because of Ethernet's cost advantages, relative simplicity, and scalability, service providers have become very interested in offering it. Service providers use it for hand-off between their network and the enterprise customer, and for transporting those Ethernet frames through the service provider network.

Many, if not most, service providers today use a transport layer made up of SONET or SDH. Therefore, any discussion of Ethernet service offerings must include a means of using the installed infrastructure. An MSPP is a platform that can transport traditional TDM traffic, such as voice, and also provide the foundational infrastructure for data traffic, for which Ethernet is optimized. The capability to integrate these capabilities allows the service provider to deploy a cost-effective, flexible architecture that can support a variety of different services—hence, the emergence of Ethernet over MSPP.

Why Ethernet over MSPP?

Ethernet over MSPP solutions enable service providers and enterprises to take advantage of fiber-optic capabilities to provide much higher levels of service density. This, in turn, lowers the cost per bit delivered throughout the network. MSPP solutions deliver profitability for carriers and cost reduction for enterprises through the following:

- Backward compatibility with legacy optical systems, supporting all restoration techniques, topologies, and transmission criteria used in legacy TDM and optical networks

- Eliminated need for overlay networks while providing support at the network edge for all optical and data interfaces, thus maximizing the types of services offered at the network edge

- Use of a single end-to-end provisioning and management system to reduce management overhead and expenditure

- Rapid service deployment

A significant advantage of Ethernet over MSPP is that it eliminates the need for parallel and overlay networks. In the past, services such as DS1s and DS3s, Frame Relay, and ATM required multiple access network elements and, in many cases, separate networks. These services were overlaid onto TDM networks or were built as completely separate networks. Multiple overlay networks pose many challenges:

- Separate fiber/copper physical layer

- Separate element and network management

- Separate provisioning schemes

- Training for all of the above

- An overlay workforce

All of these come at a significant cost so that, even if a new service's network elements are less expensive than additional TDM network elements, the operational expenses far outweigh the capital expenses saved by buying less expensive network elements.

Therefore, if you want to provide new Ethernet data services, you have to build an overlay network, as shown in Figure 3-25.

MSPP allows for one simple integrated network, as shown in Figure 3-26.

Another important feature of Ethernet over MSPPs is that MSPPs support existing management systems. There are virtually as many management systems as there are carriers. These systems can include one or more of the following: network element vendor systems, internally developed systems, and third-party systems. The key is flexibility. The MSPP must support all the legacy and new management protocols, including Translation Language-1 (TL-1), Simple Network Management Protocol (SNMP), and Common Object Request Broker Architecture (CORBA). SNMP and CORBA are present in many of today's non-MSPP network elements, but TL-1 is not. TL-1, which was developed for TDM networks, is the dominant legacy protocol and is a must-have for large service providers delivering a variety of services.

Figure 3-25 *An Additional Network Built Out to Accommodate New Services, Sometimes
Called an "Overbuild"*

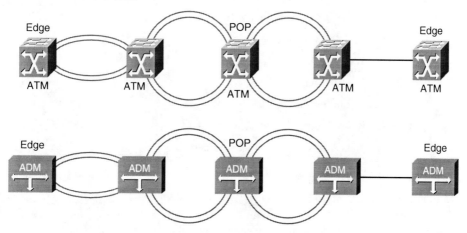

Figure 3-26 *An Integrated Network Built Out over MSPP*

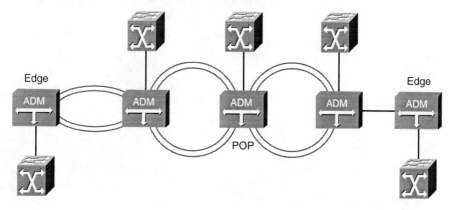

The final key advantage of Ethernet over MSPPs is that carriers can offer rapid service
deployment in two ways. The first is the time it takes to get the Ethernet service network in
place. Most service providers already have a physical presence near their customers.
However, if their existing network elements are not MSPPs, they have to build the new
overlay network before they can turn up service. This can take months. With an MSPP,
adding a new service is as simple as adding a new card to the MSPP, so network deployment
can go from months to virtually on-demand. Furthermore, because many MSPPs support
DWDM, as the number of customers grows, the bandwidth back to the central office can
be scaled gracefully by adding cards instead of pulling new fiber or adding an overlay
DWDM system.

Metro Ethernet Services

As discussed in Chapter 1, "Market Drivers for Multiservice Provisioning Platforms," several major deployment models exist for Ethernet services:

- Ethernet Private Line Service
- Ethernet Wire Service
- Ethernet Relay Service
- Ethernet Multipoint Service
- Ethernet Relay Multipoint Service

Here is a brief review of these Ethernet services.

Ethernet Private Line Service

Ethernet Private Line (EPL) Service, shown in Figure 3-27, is a dedicated, point-to-point, fixed-bandwidth, nonswitched link between two customer locations, with guaranteed bandwidth and payload transparency end to end. The EPL service is ideal for transparent LAN interconnection and data center integration, for which wire-speed performance and VLAN transparency are important. Although TDM and OC-N based facilities have been the traditional means of providing Private Line Service, the EPL service is Ethernet over SONET.

Figure 3-27 *Ethernet Private Line Service Using MSPP over DWDM*

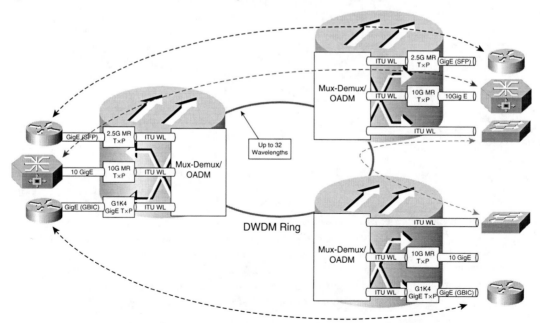

Traditionally, Private Line Services (PLSs) have been used for TDM applications such as voice or data, and they do not require the service provider to offer any added value, such as

Layer 3 (network) or Layer 2 addressing. An Ethernet PLS is a point-to-point Ethernet connection between two subscriber locations. It is symmetrical, providing the same bandwidth performance for sending or receiving. Ethernet PLS is equivalent to a Frame Relay permanent virtual circuit (PVC), but with a greater range of bandwidth, the capability to provision bandwidth in increments, and more service options. Additionally, it is less expensive and easier to manage than a Frame Relay PVC because the customer premises equipment (CPE) costs are lower for subscribers, and subscribers do not need to purchase and manage a Frame Relay switch or a WAN router with a Frame Relay interface.

Ethernet Wire Service

Like the EPL Service, the Ethernet Wire Service (EWS), depicted in Figure 3-28, is a point-to-point connection between a pair of sites, sometimes called an Ethernet virtual circuit (EVC). EWS differs from EPL in that it is typically provided over a shared, switched infrastructure within the service-provider network and can be shared between one or more other customers. The benefit of EWS to the customer is that it typically is offered with a wider choice of committed bandwidth levels up to wire speed. To help ensure privacy, the service provider segregates each subscriber's traffic by applying VLAN tags on each EVC.

EWS is considered a port-based service. All customer packets are transmitted to the destination port transparently, and the customers' VLAN tags are preserved from the customer equipment through the service-provider network. This capability is called all-to-one bundling.

Figure 3-28 shows EWS over MSPP.

Figure 3-28 *Ethernet Wire Service with Multiple VLANs over SONET*

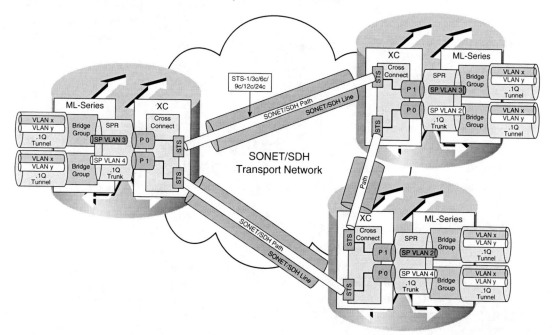

Ethernet Relay Service

Ethernet Relay Service (ERS), shown in Figure 3-29, enables multiple instances of service to be multiplexed onto a single customer User-Network Interface (UNI) so that the UNI can belong to multiple ERS. The resulting "multiplexed UNI" supports point-to-multipoint connections between two or more customer-specified sites, similar to Frame Relay service. ERS also provides Ethernet access to other Layer 2 services (Frame Relay and ATM) so that the service provider's customers can begin using Ethernet services without replacing their existing legacy systems.

ERS is ideal for interconnecting routers in an enterprise network, and for connecting to Internet service providers (ISPs) and other service providers for dedicated Internet access (DIA), virtual private network (VPN) services, and other value-added services. Service providers can multiplex connections from many end customers onto a single Ethernet port at the point of presence (POP), for efficiency and ease of management. The connection identifier in ERS is a VLAN tag. Each customer VLAN tag is mapped to a specific Ethernet virtual connection.

Figure 3-29 *Ethernet Relay Service, XC*

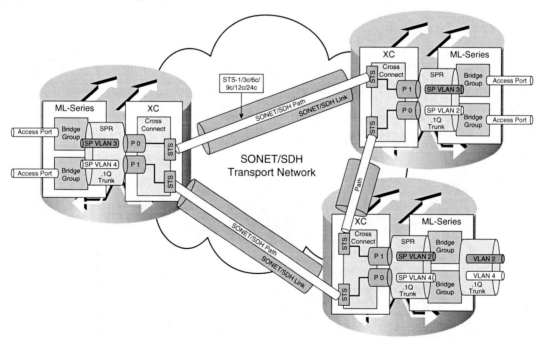

Ethernet Multipoint Service

A multipoint-to-multipoint version of EWS, Ethernet Multipoint Service (EMS), shown in Figure 3-30, shares the same technical access requirements and characteristics. The service-provider network acts as a virtual switch for the customer, providing the capability to connect multiple customer sites and allow for any-to-any communication. The enabling technology is virtual private LAN service (VPLS), implemented at the network-provider edge (N-PE).

Figure 3-30   *Ethernet Multipoint Service with Multiple VLANS over SONET*

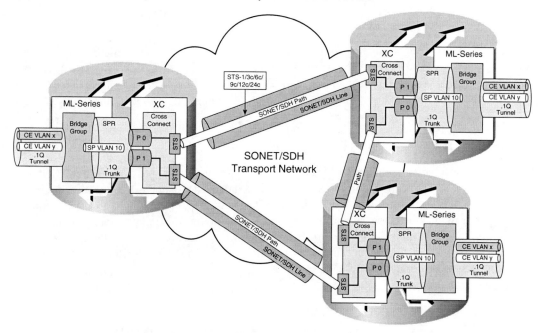

Ethernet Relay Multipoint Service

The Ethernet Relay Multipoint Service (ERMS) is a hybrid of EMS and ERS. It offers the any-to-any connectivity characteristics of EMS, as well as the service multiplexing of ERS. This combination enables a single UNI to support a customer's intranet connection, and one or more additional EVCs for connection to outside networks, ISPs, or content providers.

Table 3-3 summarizes the characteristics of metro Ethernet access solutions.

Table 3-3   *Summary of Metro Ethernet Access Services*

Service	EVC Type	CPE	Characteristics
EPL	P-to-P	Router	VLAN transparency, bundling
EWS	P-to-P	Router	VLAN transparency, bundling, Layer 2 Tunneling Protocol
ERS	P-to-P	Router	Service multiplexing
EMS	MP-to-MP	Router	VLAN transparency, bundling, Layer 2 Tunneling Protocol
ERMS	MP-to-MP	Router	Service multiplexing, VLAN transparency, bundling, Layer 2 Tunneling Protocol

The aforementioned services describe the way in which a service provider markets its Ethernet service, or even how an enterprise might deploy its own private service, but they do not provide any specification for the underlying infrastructure. Even though it is not necessary that the Ethernet services be deployed over an MSPP architecture, Figure 3-27 through Figure 3-30 showed these Ethernet services deployed over MSPPs that use either native SONET or DWDM for transport.

Two of the major Ethernet-over-SONET infrastructure architectures supported by Ethernet over MSPP are point to point, or SONET mapping, and resilient packet ring (RPR) (which is a type of multilayer switched Ethernet). Each configuration can be implemented in a BLSR, UPSR, or linear automatic protection switching (APS) network topology.

Point-to-Point Ethernet over MSPP

Point-to-point configurations over a BLSR or a linear APS are provided with full SONET switching protection. Point-to-point circuits do not need a spanning tree because the circuit has only two termination points. Therefore, the point-to-point configuration allows a simple circuit creation between two Ethernet termination points, making it a viable option for network operators looking to provide 10-/100-Mbps access drops for high-capacity customer LAN interconnects, Internet traffic, and cable modem traffic aggregation. This service is commonly referred to as EPL.

SONET Mapping

Mapping involves encapsulating the Ethernet data directly into the STS bandwidth of SONET and transporting the Ethernet within the SONET payload around the ring from one MSPP to another, where either it is either dropped or it continues to the next MSPP node. In this application, for example, a 10-Mb Ethernet circuit could be mapped directly into an STS-1, a 100-Mb circuit could be mapped into an STS-3c, and a GigE circuit could be mapped into 24 STSs.

However, STS bandwidth scaling does allow for rudimentary statistical multiplexing and bandwidth oversubscription. This involves mapping two or more Ethernet circuits into a given STS-Nc payload. For example, assume that two customers, a school district and a local cable provider, deliver cable modem–based residential subscriber services. Both customers are provided a 100-Mbps interface to their backbone switch and cable modem terminating device, respectively. Because of time-of-day demand fluctuations, neither customer is using the full provided bandwidth simultaneously. As such, the service provider might choose to place traffic from both customers onto a single STS-3c circuit across the SONET backbone. (Note that traffic is logically separated with IEEE 802.1Q tags placed at port ingress.)

Previously, each 100-Mbps customer circuit consumed a full OC-3c (155 Mbps) of bandwidth across the network. Through STS bandwidth scaling, however, one OC-3c pipe has been preserved. This enhances service-provider profitability by allowing the service provider to generate additional revenue by delivering additional data and TDM services with no effect on CapEx.

Limitations of Point-to-Point Ethernet over SONET

Ring topology is a natural match for SONET-based TDM networks that constitute the bulk of existing metro-network infrastructure. However, there are well-known disadvantages to using SONET for transporting data traffic (or point-to-point SONET data solutions, such as Ethernet over SONET).

SONET was designed for point-to-point, circuit-switched applications (such as voice traffic), and most of its limitations stem from these origins. These are some of the disadvantages of using SONET rings for data transport:

- **Fixed circuits**—SONET provisions point-to-point circuits between ring nodes. Each circuit is allocated a fixed amount of bandwidth that is wasted when not used. For the SONET network that is used for access, each node on the ring is allocated only one quarter of the ring's total bandwidth (say, OC-3 each on an OC-12 ring). That fixed allocation puts a limit on the maximum burst traffic data-transfer rate between endpoints. This is a disadvantage for data traffic, which is inherently bursty.

- **Waste of bandwidth for meshing**—If the network design calls for a logical mesh, the network designer must divide the OC-12 of ring bandwidth into $n(n-1)/2$ circuits, where n is the number of nodes provisioned. Provisioning the circuits necessary to create a logical mesh over a SONET ring not only is difficult, but it also results in extremely inefficient use of ring bandwidth. Because the amount of data traffic that stays within metro networks is increasing, a fully meshed network that is easy to deploy, maintain, and upgrade is becoming an important requirement.

- **Multicast traffic**—On a SONET ring, multicast traffic requires each source to allocate a separate circuit for each destination. A separate copy of the packet is sent to each destination. The result is multiple copies of multicast packets traveling around the ring, wasting bandwidth.

- **Wasted protection bandwidth**—Typically, 50 percent of ring bandwidth is reserved for protection. Although protection is obviously important, SONET does not achieve this goal in an efficient manner that gives the provider the choice of how much bandwidth to reserve for protection.

Ethernet over a Ring?

Will Ethernet over a ring improve upon point-to-point Ethernet over SONET? Ethernet does make efficient use of available bandwidth for data traffic and offers a far simpler and inexpensive solution for data traffic. However, because Ethernet is optimized for point-to-point or meshed topologies, it does not make the most of the ring topology.

Unlike SONET, Ethernet does not take advantage of a ring topology to implement a fast protection mechanism. Ethernet generally relies on the Spanning Tree Protocol to eliminate all loops from a switched network, which is notoriously slow. Even though the Spanning Tree Protocol can be used to achieve path redundancy, its comparatively slow recovery mechanism requires the failure condition to be propagated serially to each upstream node after a fiber cut. Link aggregation (802.1ad) can provide a link-level resiliency solution, but it is comparatively slow (about 500 ms vs. 50 ms) and is not appropriate for providing path-level protection.

Ethernet is also not good at creating an environment for equitable sharing of ring bandwidth. Ethernet switches can provide link-level fairness, but this does not necessarily or easily translate into overall fairness in bandwidth allocation. A simpler and more efficient method comes from taking advantage of the ring topology to create a universal equity plan for bandwidth allocation.

As we've discussed, neither SONET nor Ethernet is ideal for handling data traffic on a ring network. SONET does take advantage of the ring topology, but it does not handle data traffic efficiently and wastes ring bandwidth. Although Ethernet is a natural fit for data traffic, it is actually difficult to implement on a ring and does not make the most of the ring's capabilities.

One final note before we venture into our next topic of RPR: The Rapid Spanning Tree Protocol (RSTP) (802.1w) is another step in the evolution of the Ethernet over SONET that evolved from the Spanning Tree Protocol (802.1d standard) and provides for faster spanning-tree convergence after a topology change. The terminology of STP (and the parameters) remains the same in RSTP. This was used as a means of ring convergence before the development of RPR, which we discuss next.

Resilient Packet Ring

Resilient packet ring is an emerging network architecture designed to meet the requirements of a packet-based metropolitan-area network. Unlike incumbent architectures based on Ethernet switches or SONET add/drop muxes (ADMs), RPR approaches the metro bandwidth limitation problem differently. RPR provides more than just mere SONET mapping of Ethernet over a self-healing, "resilient" ring.

This problem of effectively managing a shared resource, the fiber ring, which needs to be shared across thousands of subscribers in a metro area, is most efficiently solved at the MAC layer of the protocol stack.

By creating a MAC protocol for ring networks, RPR attempts to find a fundamental solution to the metro bottleneck problem. Other solutions attempt to make incremental changes to existing products but do not address the fundamental problem and, hence, are inefficient. Neither SONET nor Ethernet switches address the need for a MAC layer designed for the MAN. SONET employs Layer 1 techniques (point-to-point connections) to manage capacity on a ring. Ethernet switches rely on Ethernet bridging or IP routing for bandwidth management. Consequently, the network is either underutilized, in the case of SONET, or nondeterministic, in the case of Ethernet switches.

Instead of being a total replacement of SONET and Ethernet, RPR is complementary to both. Both SONET and Ethernet are excellent Layer 1 technologies. Whereas SONET was designed as a Layer 1 technology, Ethernet has evolved into one. Through its various evolutions, Ethernet has transformed from the CSMA/CD shared-media network architecture to a full-duplex, point-to-point switched network architecture.

Most of the development in Ethernet has been focused on its physical layer, or Layer 1, increasing the speed at which it operates. The MAC layer has been largely unchanged. The portion of the MAC layer that continues to thrive is the MAC frame format. RPR is a MAC protocol and operates at Layer 2 of the OSI protocol stack. By design, RPR is Layer 1 agnostic, which means that RPR can run over either SONET or Ethernet. RPR enables carriers and enterprises to build more scalable and efficient metro networks using SONET or Ethernet as physical layers.

RPR Characteristics

RPR has several unique attributes that make it an ideal platform for delivery of data services in metro networks.

Resiliency   The Ethernet traffic is sent in both directions of a dual counter-rotating ring to achieve the maximum bandwidth utilization on the SONET/SDH ring. Ring failover is often described as "self-healing" or "automatic recovery." SONET rings can recover in less than 50 ms.

Sharing Bandwidth Equitably   SONET rings also have an innate advantage for implementing algorithms to control bandwidth use. Ring bandwidth is a public resource and is susceptible to being dominated by individual users or nodes. An algorithm that allocates the bandwidth in a just manner is a means of providing every customer on the ring with an equitable amount of the ring bandwidth, ideally without the burden of numerous provisioned circuits. A ring-level fairness algorithm can and should allocate ring bandwidth as a single resource. Bandwidth policies that can allow maximum ring bandwidth to be used between any two nodes when there is no congestion can be implemented without the inflexibility of a fixed circuit-based system such as SONET, but with greater effectiveness than point-to-point Ethernet.

Easier Activation of Services   A common challenge often experienced in data service customers is the time it takes for carriers to provision services. Installation, testing, and provisioning times can take anywhere from 6 weeks to 6 months for DS1 and DS3 services; services at OC-N rates can take even more time.

A significant portion of this delay in service lead times can be attributed to the underlying SONET infrastructure and its circuit-based provisioning model. Traditionally, the creation of an end-to-end circuit took numerous steps, especially before MSPP.

Initially the network technician identifies the circuit's physical endpoints to the operational support system. The technician must then configure each node within the ring for all the required circuits that will either pass through a node or continue around the ring. This provisioning operation can be time and labor intensive. MSPPs automate some of the circuit-provisioning steps. But the technician still needs to conduct traffic engineering manually to optimize bandwidth utilization on the ring. The technician must be aware of the network topology, the traffic distribution on the ring, and the available bandwidth on every span traversed by the circuit. Service provisioning on a network of Ethernet switches is improved because provisioning of circuits is not required through each node. However, circuit provisioning still occurs node by node. Additionally, if carriers want to deliver SLAs over the network, the network planner still needs to manually provision the network for the required traffic.

By comparison, an RPR system provides a very basic service model. In an RPR system, the ring functions as a shared medium. All the nodes on the ring share bandwidth on the packet ring. Each node has visibility into the capacity available on the ring. Therefore, circuit provisioning of a new service is much easier. There is no need for a node-by-node and link-by-link capacity planning, engineering, and provisioning exercise. The network operator simply identifies a traffic flow and specifies the QoS that each traffic type should get as it traverses the ring. Thus, there is no need for circuit provisioning because each node is aware of every other node on the ring, based on the MAC address.

Broadcast or Multicast Traffic Is Better Handled   RPRs are a natural fit for broadcast and multicast traffic. As already shown, for unicast traffic, or traffic from one entity to another, nodes on an RPR generally have the choice of stripping packets from the ring or forwarding them. However, for a multicast, the nodes can simply receive the packet and forward it, until the source node strips the packet. This means that multicasting or broadcasting a data packet requires that only one copy be sent around the ring, not n copies, where n is the number of nodes. This reduces the amount of bandwidth required by a factor of n.

Layer 1 Flexibility   The basic advantage of a packet ring is that each node can assume that a packet sent on the ring will eventually reach its destination node, regardless of which path around the ring has taken. Because the nodes identify themselves with the ring, only three basic packet-handling actions are needed: insertion (adding a packet into the ring), forwarding (sending the packet onward), and stripping (taking the packet off the ring). This decreases the magnitude of processing required for individual nodes to communicate with each other, especially as compared with a meshed network, in which each node has to decide which exit port to use for each packet as a part of the forwarding process.

RPR: A Multilayer Switched Ethernet Architecture over MSPP

The term *multilayer switched Ethernet* is used here because RPR goes beyond mere Layer 1 SONET payload mapping of Ethernet, a "mapper" approach, and uses Layer 2 and even Layer 3 features from the OSI reference model for data networking. This technology truly delivers on the promise of the MSPP and can be found on a single card. Cisco calls the card the ML Card for Multi Layer as a part of the ONS 15454 MSPP. This card supports multiple levels of priority of customer traffic that can be managed using existing operations support systems (OSS).

This multilayer switched design offers service providers and enterprises alike several key features and benefits:

- **The multilayer switched design brings packet processing to the SONET platform**—The benefit is that new services can be created around the notion of guaranteed and peak bandwidth, a feature that really enhances the service-provider business model.

- **The multilayer switched design offers the capability to create multipoint services**—This means that the provider can deploy the equivalent of a private-line service and a Frame Relay service out of the same transmission network infrastructure, thereby realizing significant cost savings.

- **The multilayer switched design delivers carrier-class services**—The key benefit is that the resiliency of the service is derived from the SONET/SDH 50-ms failover.

- **The multilayer switched design integrates into Transaction Language One (TL-1) and SNMP**—The key benefit is that these services can be created to a large extent within the existing service provider provisioning systems. Therefore, there is minimal disruption to existing business processes. Through an EMS multilayer switched design, Ethernet cards extend the data service capabilities of this technology, enabling service providers to evolve the data services available over their optical transport networks.

  The Cisco Systems multilayer switched design consists of two cards: a 12-port, 10/100BaseT module faceplate-mounted RJ-45 connectors, and a 2-port GigE module with two receptacle slots for field-installable, industry-standard SFP optical modules.

  Additionally, each service interface supports bandwidth guarantees down to 1 Mbps, enabling service providers to aggregate traffic from multiple customers onto shared network bandwidth, while still offering TDM or optical services from the same platform.

Q-in-Q

The multilayer switched design supports Q-in-Q, a technique that expands the VLAN space by retagging the tagged packets entering the service provider infrastructure. When a service provider's ingress interface receives an Ethernet frame from the end user, a second-level 802.1Q tag is placed in that frame, immediately preceding the original end-user 802.1Q tag. The service provider's network then uses this second tag as the frame transits the metro network. The multilayer switched card interface of the egress removes the second tag and hands off the original frame to the end customer. This builds a Layer 2 VPN in which traffic from different business customers is segregated inside the service provider network, yet the service provider can deliver a service that is completely transparent to the Layer 2 VLAN configuration of each enterprise customer.

Although Q-in-Q provides a solid solution for smaller networks, its VLAN ID limitations and reliance on the IEEE 802.1d spanning-tree algorithm make it difficult to scale to meet the demands of larger networks. Therefore, other innovations, such as Ethernet over MPLS (EoMPLS), must be introduced. As the name implies, EoMPLS encapsulates the Ethernet frames into an MPLS label switch path, which allows a Multiprotocol Label Switching (MPLS) core to provide transport of native Ethernet frames.

Several other important concepts related to Ethernet over SONET must be mentioned.

Virtual Concatenation VCAT

As synchronous transport signals and virtual containers (STSs/VCs) are provisioned, gaps can form in the overall flows. This is similar to a fragmented disk on a personal computer. However, unlike computer memory managers, TDM blocks of contiguous payload cannot be cut into fragments to fit into the unused TDM flow. For example, a concatenated STS-12c flow cannot be chopped up and mapped to 12 STs-1 flows. VCAT solves this shortfall by providing the capability to transmit and receive several noncontiguous STSs/VCs, fragments, as a single flow. This grouping of STSs/VCs is called a VCAT group (VCG).

VCAT drastically increases the utilization for Ethernet over TDM infrastructures. This enables carriers to accommodate more customers per metro area than without VCAT.

Carriers looking to reduce capital expenditures, while meeting the demands of data traffic growth and new service offerings, need to extract maximum value from their existing networks. Emerging mapper or framer technologies, such as VCAT and link capacity-adjustment scheme (LCAS), enable carriers to upgrade their existing SONET networks with minimal investment. These emerging technologies will help increase the bottom line of carriers by enabling new services through more rapid provisioning, increased scalability, and much higher bandwidth utilization when transporting Ethernet over SONET and packet over SONET data.

VCAT significantly improves the efficiency of data transport, along with the scalability of legacy SONET networks, by grouping the synchronous payload envelopes (SPEs) of SONET frames in a nonconsecutive manner to create VCAT groups. Traditionally, layer capacity formats were available only in contiguous concatenated groups of specific size. SPEs that belong to a virtual concatenated group are called members of that group. This VCAT method allows finer granularity for the provisioning of bandwidth services and is an extension of an existing concatenation method, contiguous concatenation, in which groups are presented in a consecutive manner and with gross granularity.

Different granularities of virtual concatenated groups are required for different parts of the network, such as the core or edge. VCAT applies to low-order (VT-1.5) and high-order (STS-1) paths. Low-order virtual concatenated groups are suitable at the edge, and the high-order VCAT groups are suitable for the core of the MAN.

VCAT allows for the efficient transport of GigE. Traditionally, GigE is transported over SONET networks using the nearest contiguous concatenation group size available, an OC-48c (2.488 Gbps), wasting approximately 60 percent of the connection's bandwidth. Some proprietary methods exist for mapping Ethernet over SONET, but they, too, are inefficient. With VCAT, 21 STS-1s of an OC-48 can be assigned for transporting one GigE. The remaining 27 STS-1s are still free to be assigned either to another GigE or to any other data client signal, such as ESCON, FICON, or FC.

VCAT improves bandwidth efficiency more than 100 percent when transporting clients such as GigE using standard mapping, or around 25 percent when compared to proprietary mapping mechanisms (for example, GigE over OC-24c). This suggests that carriers could significantly improve their existing networks' capacity by using VCAT. Furthermore, carriers gain scalability by increasing the use of the network in smaller incremental steps. In addition, the signals created by VCAT framers are still completely SONET, so a carrier needs to merely upgrade line cards at the access points of the network, not the elements in the core.

Whereas VCAT provides the capability to "right-size" SONET channels, LCAS increases the flexibility of VCAT by allowing dynamic reconfiguration of VCAT groups. Together the technologies allow for much more efficient use of existing infrastructure, giving service providers the capability to introduce new services with minimal investment.

LCAS

LCAS allows carriers to move away from the sluggish and inefficient provisioning process of traditional SONET networks and offers a means to incrementally enlarge or minimize the size of a SONET data circuit without impacting the transported data. LCAS uses a request/acknowledge mechanism that allows for the addition or deletion of STS-1s without affecting traffic. The LCAS protocol works unidirectionally, enabling carriers to provide asymmetric bandwidth. Thus, provisioning more bandwidth over a SONET link using

LCAS to add or remove members (STS-1s) of a VCAT group is simple and provides the benefit of not requiring a 50 ms service interruption.

The LCAS protocol uses the H4 control packet, which consists of the H4 byte of a 16-frame multiframe. The H4 control packet contains information of the member's sequence (sequence indicator, SQ#) and alignment (multiframe indicator, MFI) of a virtual concatenated group.

LCAS operates at the endpoints of the connection only, so it does not need to be implemented at the nodes where connections cross or in trunk line cards. This allows carriers to deploy LCAS in a simple manner, by installing new tributary cards. Likewise, they can scale LCAS implementations by adding more tributary cards without requiring hardware upgrades to add/drop multiplexers, for example, throughout the entire network.

One of the greatest benefits of LCAS for carriers is the capability to "reuse" bandwidth to generate more revenue and offer enhanced services that allow higher bandwidth transmission when needed. This will be a key reason for carriers to implement next-generation SONET gear, along with the potential extra revenue stream from such.

Generic Framing Procedure

Generic Framing Procedure (GFP) defines a standard encapsulation of both L2/L3 protocol data unit (PDU), client signals (GFP-F), and the mapping of block coded client signals (GFP-T). In addition, it performs multiplexing of multiple client signals into a single payload, even when they are not the same protocol. This allows MSPP users to use their TDM paths as one large pipe, in which all the protocols can take advantage of unused bandwidth. In the past, each protocol had to ride over, and had burst rates limited to, a small portion of the overall line rate, not the total line rate. Furthermore, the overbooking of large pipes is not only possible, but also manageable because GFP enables you to set traffic priority and discard eligibility.

GFP is comprised of common functions and payload-specific functions. Common functions are those shared by all payloads; payload-specific functions are different depending on the payload type. These are the two payload modes:

- **Transparent mode**—Uses block code–oriented adaptation to transport constant bit rate traffic and low-latency traffic
- **Frame mode**—Transports PDU payloads, including Ethernet and PPP

· GFP is a complete mapping protocol that can be used to map data packets as well as SAN block traffic. These are not just two sets of protocols—they are two different market segments. Deploying GFP will further a provider's capability to leverage the existing infrastructure.

QoS

A key feature of multilayer switched Ethernet is QoS. QoS is a means of prioritizing traffic based on its class, thereby allowing latency-sensitive data to take priority over non-latency-sensitive data (as in voice traffic over e-mail traffic), as shown in Figures 3-31 and 3-32.

Figure 3-31 *QoS Flow Process*

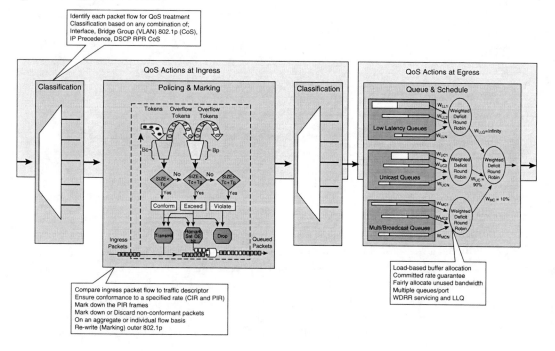

Figure 3-32 *QoS Process Showing an Ethernet Frame Flow Around a Resilient Packet Ring*

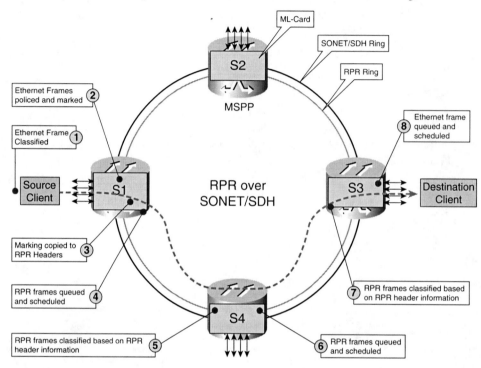

Summary

Ethernet over SONET is the wave of the future for customer access. As this book is being written, carriers are planning for Ethernet access to be the next Frame Relay access for customers. Ethernet over SONET enables end customers to connect and extend their LAN by simply plugging it into the WAN with an RJ-45 connector. RPR allows customers to use bandwidth on a SONET-based ring similar to the CSMA/CD protocol of a switched network, thanks to the MAC of RPR. QoS and other features allow for voice and multicast video to be deployed over the WAN simply, quickly, and with efficient use of bandwidth.

SAN demand, which is growing at an enormous rate year over year, is satisfied more easily through its integration into the MSPP platform. It allows direct connection of an FC port from a storage services server to an MSPP card port, and then a direct mapping to the SONET frame. This eliminates the need to map FC into GE pipes for increased distance, or add hardware such as a router, to transport it over the metropolitan area and beyond. This allows both service providers that offer managed storage services and enterprise users to capitalize on their existing infrastructures by simply inserting another card into the MSPP.

Traditional metro DWDM solutions have rigid network architectures and require considerable manual interaction to manage, particularly when new sites are added or network capacity is upgraded. Traditional solutions are optimized for low-cost-per-bit, fixed topologies that cannot efficiently address the operational constraints of metro and regional networks.

Metro networks face unique challenges, such as the inherent difficulty in predicting demand for services such as TDM, data, SAN, and video, or service bandwidth at 1-Gbps, 2.5-Gbps, and 10-Gbps rates. Furthermore, complexities are involved in managing metro DWDM network architectures that are typically ring topologies because of dynamic add/drop traffic patterns. Traditional solutions cannot automatically manage DWDM variables such as optical noise, dispersion, dynamics of adding and dropping wavelengths, and optical performance monitoring.

MSPP-based DWDM has been designed from the start to address these challenges. By taking advantage of the multiservice capabilities of the MSPP, it can natively transport any service-TDM, data, or wavelengths over a metro or regional network at a lower cost than traditional wavelength-only DWDM solutions. Multiservice simplifies service planning. Software intelligence simplifies operations. Management of MSPP-based DWDM is again performed with GUIs that provide intelligent optical-level wavelength monitoring and reporting. With this type of monitoring, problems can be discovered and corrected before carriers see revenue-generating services affected for carriers and enterprises experience downtime.

PART II

Designing MSPP Networks

This chapter covers the following topics:

- Traditional Service-Provider Network Architectures
- Traditional Customer Network Architectures
- MSPP Positioning in Service-Provider Network Architectures
- MSPP Positioning in Customer Network Architectures

CHAPTER 4

Multiservice Provisioning Platform Architectures

In this chapter, you will learn about various MSPP architectures. You will review traditional service-provider and customer network architectures, which help contrast the enormous benefits that today's MSPPs provide as they are positioned in service-provider network architectures and customer network architectures.

Traditional Service-Provider Network Architectures

Within Multiservice Provisioning Platform (MSPP) architectures there are traditional service-provider network architectures: the PSTN, Frame Relay/ATM, SONET IP/MPLS, and transport network types such as IOF, access, and private ring deployments. This chapter also covers heritage OSS because it plays such a large role in today's MSPP equipment providers' plans.

Public Switched Telephone Networks

Those who are familiar with packet-switched routing, which is the backbone of the Internet and uses Internet Protocol (IP), know that the Internet is the amalgamation of today's data networks (see Figure 4-1). The public switched telephone network (PSTN), shown in Figure 4-2, is analogous to the Internet, in that it is the amalgamation of the world's circuit-switched telephone networks. Although the PSTN was originally a fixed-line analog telephone system network, it has evolved into an almost entirely digital network that now includes both mobile and fixed telephones.

Just as there are many standards surrounding the Internet, the PSTN is largely governed by technical standards created by the ITU-T. It uses E.163/ E.164 addresses (known more commonly as telephone numbers) for addressing.

The PSTN is the earliest example of traffic engineering used to deliver "voice service" quality.

In the 1970s, the telecommunications industry understood that digital services would follow much the same pattern as voice services, and conceived a vision of end-to-end circuit-switched services, known as the Broadband Integrated Services Digital Network (B-ISDN). Obviously, the B-ISDN vision has been overtaken by the disruptive technology of the Internet.

Figure 4-1 *The Internet Amalgamating Numerous Data Networks*

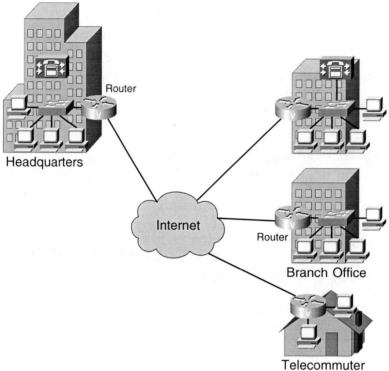

Figure 4-2 *PSTN Amalgamating Numerous PSTNs*

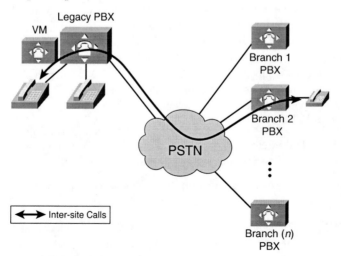

The primary section of the PSTN that still uses analog technology is the last-mile loop to the customer; however, only the very oldest parts of the rest of the telephone network still use analog technology for anything. In recent years, digital services have been increasingly rolled out to end users using services such as digital subscriber line (DSL) and ISDN.

Many pundits believe that over the long term, the PSTN will be just one application of the Internet; however, the Internet has some way to go before this transition can be made. The quality of service (QoS) guarantee is one aspect that must improve in Voice over IP (VoIP) technology. In some cases, private networks run by large companies are connected to the PSTN only through limited gateways, such as a large private automatic branch exchange/telephone (PABX) system. A number of large private telephone networks are not even linked to the PSTN and are used for military purposes.

The basic digital circuit in the PSTN is a 64-kbps channel that was originally designed by Bell Labs, called a DS0, or Digital Signal 0. To carry a typical phone call from a calling party to a called party, the audio sound is digitized at an 8 kHz sample rate using 8-bit pulse-code modulation. The call is then transmitted from the one end to the other through the use of a routing strategy.

The DS0s are the most basic level of granularity at which switching takes place in a telephone exchange. DS0s are also known as time slots because they are multiplexed together in a time-division fashion. Multiple DS0s are multiplexed together on higher-capacity circuits so that 24 DS0s make a DS1 signal. When carried on copper, this signal is the well-known T-Carrier system, T1 (the European equivalent is an E1, containing 32 64-kbps channels). In modern networks, this multiplexing is moved as close to the end user as possible, usually into roadside cabinets in residential areas, or into large business premises. Figure 4-3 shows the customer DS0 as it passes through the local loop and into the central office (CO), where it is "trunked" from one CO switch to another. From the switching network, the customer-premises DS0s can be "switched" to their destination.

Figure 4-3   *Trunking and Local Loop Architecture Relationship*

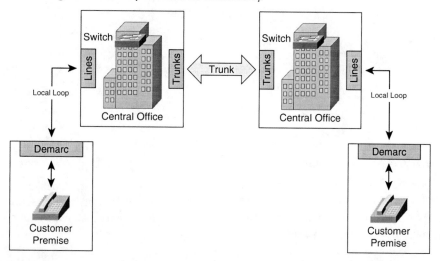

The time slots are carried from the initial multiplexer to the exchange over a set of equipment that is collectively known as the access network. The access network and interexchange transport of the PSTN use synchronous optical transmission (Synchronous Optical Network [SONET] and Synchronous Digital Hierarchy [SDH]) technology, although some parts still use the older Plesiochronous Digital Hierarchy (PDH) technology. PDH (*plesiochronous* means "nearly synchronous") was developed to carry digitized voice over twisted-pair cabling more efficiently. The local telephone companies, also known as local exchange carriers (LECs), service a given area based on geographic boundaries known as a local access and transport area (LATA). LATA is a geographic area that defines an LEC's territory. Calls that cross a LATA boundary must be carried by an interexchange carrier (IXC), as shown in Figure 4-4.

Within the access network, a number of reference points are defined. Most of these are of interest mainly to ISDN, but one—the V reference point—is of more general interest. This is the reference point between a primary multiplexer and an exchange.

Figure 4-4 *Local Telephone Company COs Connected by Long-Distance, or IXE, Carriers*

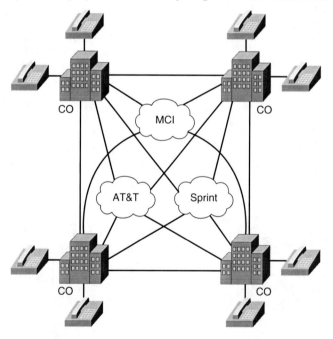

Frame Relay/ATM Networks

Frame Relay is a traditional packet-based telecommunications service that takes advantage of characteristics of today's networks by minimizing the amount of error detection and recovery performed inside the network. Streamlining the communications process results in lower delay and higher throughput.

Frame Relay offers features that make it ideal to interconnect local-area networks (LANs) using a wide-area network (WAN), as shown in Figure 4-5. Traditionally, LANs were interconnected by deploying private lines or by circuit-switching over a leased line. However, this approach has several drawbacks. The primary weakness of this legacy approach is that it becomes prohibitively expensive as the size of the network increases, in both the number of facility miles and the number of LANs. The reason for the high cost is that high-speed circuits and ports must be set up on a point-to-point basis among an increasing number of bridges. In addition, circuit-mode connectivity results in a lot of wasted bandwidth for the bursty traffic that is typical of LANs.

On the other hand, traditional X.25 packet-switched networks required significant protocol overheads and have historically been too slow at primarily supporting low-speed terminals at 19.2 kbps and lower. Frame Relay provides the statistical multiplexing interface of X.25 without its overhead. In addition, it can handle multiple data sessions on a single access line, which reduces hardware and circuit requirements. Frame Relay is also scalable, meaning that implementations are available from low bandwidths (such as 56 kbps) all the way up to T1 (1.544 Mbps) or even T3 (44.736 Mbps) speeds.

Figure 4-5   *Service Provider Frame Relay Networks Connected Through an ATM Network*

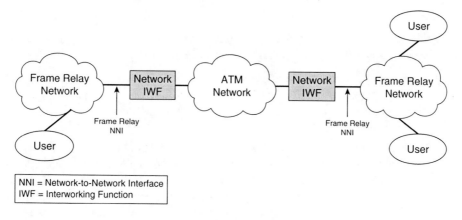

Connection to the LAN

In the past decade, significant advancements in computing and communications technology have reshaped the business milieu. With the cost of processing power falling, PCs and high-powered workstations have proliferated exponentially and are now an integral part of the end user's world. This has resulted in an explosion in the demand and use of personal computers, workstations, and LANs, and has altered the corporate information system. The major changes include the following:

- **Corporate organization**—Traditionally, information systems were arranged in a hierarchical structure with a centralized mainframe supporting a large number of users. With the emergence of today's technology, distributed computing environments based on LANs are supplementing traditional hierarchical mainframe architectures. Now information flows on a lateral level (peer to peer) both within organizations and to outside groups.

- **Network-management specifications**—The management requirements of today's networks are more complex than ever. Each network is a distinctive combination of multivendor equipment. Growth and change within a company result in constant network alterations. The network manager is pressured to find a cost-effective way to manage this complexity.

- **Rise in bandwidth demand**—LANs have grown up from the morphing of PCs and intelligent workstations, pulling along with them the workstation applications due to the user's expectation of obtaining quick response time and the capability of handling large quantities of data. The LAN pipeline, which typically runs at 10 Mbps, 100 Mbps, or 1 Gbps, must be capable of supporting these applications, so the applications typically transfer orders of magnitude more data per transaction than a typical terminal-to-mainframe transaction. Nevertheless, like their terminal-to-mainframe counterparts, the LAN applications are bursty, with long idle periods.

Benefits of Frame Relay

Frame Relay optimizes bandwidth because of its statistical multiplexing and low-protocol overhead, resulting in the following benefits of Frame Relay:

- **Reduced internetworking costs**—A carrier's Frame Relay network multiplexes traffic from many sources over its backbone. This reduces the number of circuits and corresponding cost of bandwidth in the WAN. Reducing the number of port connections required to access the network lowers the equipment costs.

- **Increased interoperability through international standards**—Frame Relay's simplified link-layer protocol can be implemented over existing technology. Access devices often require only software changes or simple hardware modifications to support the interface standard. Existing packet-switching equipment and T1/E1 multiplexers often can be upgraded to support Frame Relay over existing backbone networks.

Asynchronous Transfer Mode

Frame Relay and Asynchronous Transfer Mode (ATM) offer different services and are designed for different applications. ATM is better suited for applications such as imaging, real-time video, and collaborative computer-aided design (CAD) that are too bandwidth intensive for Frame Relay. On the other hand, at T1 speeds and lower, Frame Relay uses bandwidth much more efficiently than ATM.

ATM is a dedicated connection-switching technology that arranges digital data into 53-byte cell units and transmits them over a physical medium using digital signal technology. Individually, a cell is processed asynchronously relative to other related cells and is queued before it is multiplexed over the transmission path. The prespecified bit rates are 155.520 Mbps or 622.080 Mbps, although speeds on ATM networks can reach 10 Gbps.

Key ATM features include the following:

- Setup of end-to-end data paths using standardized signaling and load- and QoS-sensitive ATM routing
- Segmentation of packets into cells and reassembly at the destination
- Statistical multiplexing and switching of cells
- Network-wide congestion control
- Advanced traffic-management functions, including the following:
  - Negotiation of traffic policies to determine the structure that affects the end-to-end delivery of packets
  - Traffic shaping to maintain the traffic policies
  - Traffic policing to enforce the traffic policies
  - Connection admission control to ensure that traffic policies of new customers do not adversely affect existing customers

Carriers have traditionally deployed ATM for the following reasons:

- Supports voice, video, and data, allowing multimedia and mixed services over a single network. This is attractive to potential customers.
- Offers high evolution potential and compatibility with existing legacy technologies.
- Supports a wide range of bursty traffic, delay tolerance, and loss performance by implementing multiple QoS classes.
- Provides the capability to support both connection-oriented and connectionless traffic.
- Offers flexible facility options. For example, cable can be twisted pair, coaxial, or fiber optic.
- Uses statistical multiplexing for efficient bandwidth use.
- Provides scalability.
- Uses higher-aggregate bandwidth.

Service Provider SONET Networks

Not much needs to be said here because the entire focus of this book is optical networking. Essentially, SONET is a synchronous transport system that uses dual rings for redundancy. SONET, used in North America, and SDH, used in Europe, are almost identical standards for the transport of data over optical media between two fixed points. They use 810-byte frames as a container for the transport of data at speeds of up to OC–192 (9.6 Gbps).

SONET/SDH is used as the bearer layer for higher-layer protocols, such as ATM, IP, and Point-to-Point Protocol (PPP), deployed on devices that switch or route traffic to a particular endpoint, as shown in Figure 4-6. The functions of SONET/SDH in the broadband arena are roughly analogous to those of T1/E1 in the narrowband world. The SONET/SDH standards define the encapsulation of data within SONET/SDH frames, the encoding of signals on a fiber-optic cable, and the management of the SONET/SDH link.

The advantages of SONET/SDH include the following:

- Rapid point-to-point transport of data
- Standards-based multiplexing of SONET/SDH data streams
- Transport that is independent of the services and applications that it supports
- A self-healing ring structure, as shown in Figure 4-7, to reroute traffic around faults within a particular link
- A widely deployed transmission infrastructure within carrier networks
- Time-division multiplexing (TDM) grooming and aggregation from the DS0 level

Figure 4-6   *A SONET/SDH Pipe Carrying ATM*

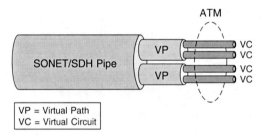

Figure 4-7   *A Service-Provider SONET Ring Using Network Elements to Add and Drop Traffic to Customers*

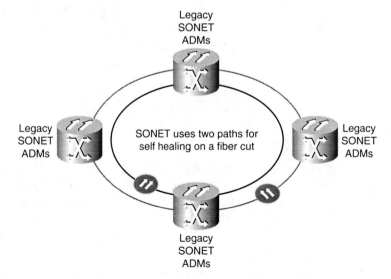

IP and MPLS Networks

Multiprotocol Label Switching (MPLS) is one of the most exciting emerging network protocols in recent years. In MPLS, a short fixed-length label is created and applied to the front end of the IP packet; that acts as a shorthand representation of an IP packet's header (see Figure 4-8). Label-switched routers make subsequent routing decisions based on the MPLS label, not the original IP address. This new technology allows core network routers to operate at higher speeds without needing to examine each packet in detail. It also allows more complex services to be deployed, which enables discrimination on a QoS basis. Thus, an MPLS network is a routed network that uses the MPLS label to add another layer of differentiation to the traffic, enabling a global class of service (CoS) so that carriers can distinguish not only between customers, but also between types of service being carried through their network.

An MPLS-based Virtual Private Network (VPN) basically uses a long IP address in which each site belongs to a VPN with an associated number. This enables you to distinguish duplicate private addresses. For example, subnet 10.2.1.0 for VPN 23 is different than subnet 10.2.1.0 for VPN 109. From the MPLS VPN provider's point of view, they are really 23:10.1.1.0 and 109:10.1.1.0, which are quite different.

Figure 4-8   *MPLS Label*

```
 0                   1                   2                   3
 0|1|2|3|4|5|6|7|8|9|0|1|2|3|4|5|6|7|8|9|0|1|2|3|4|5|6|7|8|9|0|1
┌─────────────────────────────────────┬─────┬─┬───────────────┐
│                 Label                │ EXP │S│      TTL      │
└─────────────────────────────────────┴─────┴─┴───────────────┘
```

- Label: Label Value (Unstructured), 20 bits
- Exp: Experimental Use, 3 bits; currently used as a Class of Service (CoS) field.
- S: Bottom of Stack, 1 bit
- TTL: Time to Live, 8 bits

Thus, a customer data packet has two levels of labels attached when it is forwarded across the backbone, as shown in Figure 4-9:

- The first label directs the packet to the correct provider edge router.
- The second label indicates how that provider edge router should forward the packet.

Figure 4-9   *MPLS Label Positioned Between the Data Link Layer (Layer 2) Header and Network Layer (Layer 3) Header*

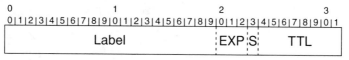

Carriers can now use the IP/MPLS customer edge (CE) and provider edge (PE), as shown in Figure 4-10. This provides them with enhanced and more tightly defined service-level agreements (SLAs), cuts their infrastructure costs, and offers new services. Carriers can also use IP/MPLS to provide VPN services to distribute traffic loads more effectively throughout their networks, and to introduce SONET-like failover times to packet forwarding without incurring SONET-like costs.

Figure 4-10 *A Carrier IP/MPLS Backbone Connecting CE Routers to PE Routers*

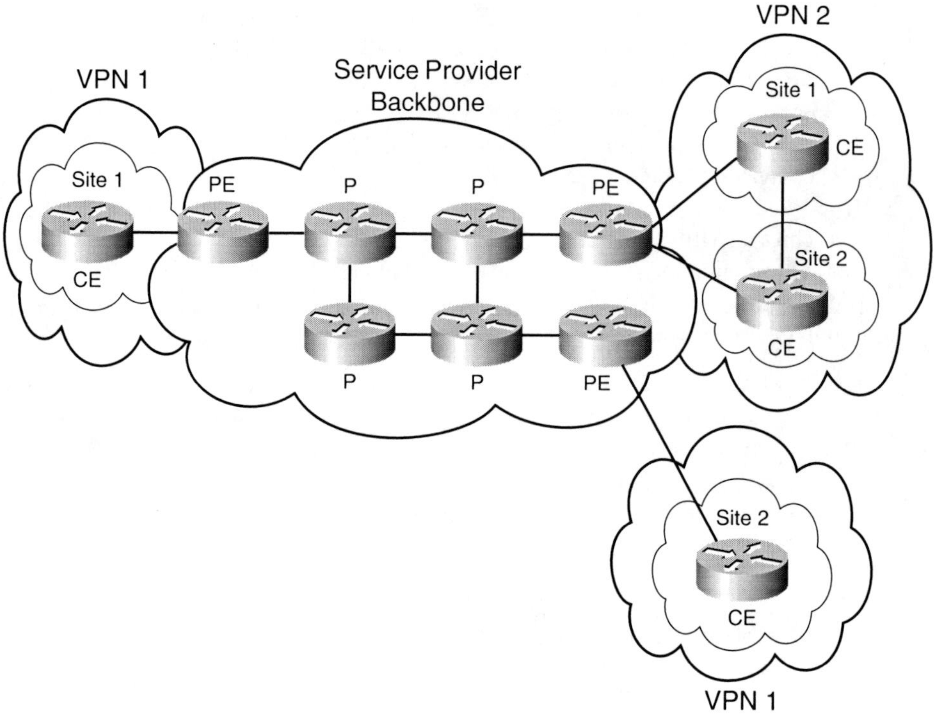

Transport Networks

Three major transport ring types exist for service providers: interoffice facilities (IOFs), access rings, and private rings. Network architectures are based on many factors, including the types of applications and protocols, distances, usage and access patterns, and legacy network topologies. In the metropolitan market, for example, point-to-point topologies might be used for connecting enterprise locations, ring topologies for connecting interoffice (IOF) facilities and for residential access, and mesh topologies for inter–point of presence (POP) connections and connections to the long-haul backbone. In effect, the optical layer must be capable of supporting many topologies. Because of unpredictable developments in this area, those topologies must be flexible.

IOF Rings

IOF rings can belong exclusively to a single carrier (as shown in Figure 4-11), as do almost all of them, or they can be a combination ring between two or more carriers, often called a midspan meet or simply a meet (shown in Figure 4-12). Some service providers consider these meets as IOF; other service providers consider the other carrier nodes on the ring as customer nodes. The main feature of IOF rings is that they are intended to carry traffic between COs, not to customers who are paying for services delivered off the ring. IOF rings are often configured as a bidirectional line switch ring (BLSR) because BLSR nodes can terminate traffic coming from either side of the ring. Therefore, BLSRs are suited for distributed node-to-node traffic applications such as IOF networks and access networks.

Figure 4-11 *IOF Ring Carrying Traffic Between COs of Same Carrier*

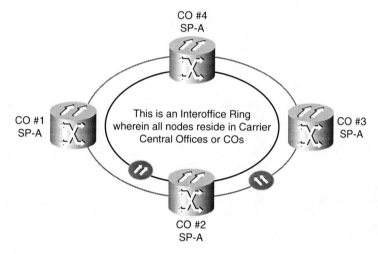

Figure 4-12 *IOF Ring Carrying Traffic Between Two Different Carriers*

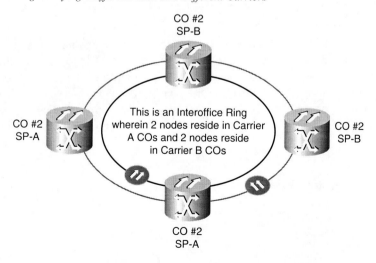

Access Rings

An access ring, shown in Figure 4-13, is a ring serviced out of a CO that provides services to the remote cabinets, controlled environmental vaults (CEVs), and huts that deliver service to the end customers. Plain old telephone service (POTS), ISDN, data services, and other special services are delivered from the access ring. Typically, the customer data is aggregated into a digital loop carrier (DLC). The DLC breaks DS1s into DS0s. The DLC DS1s are then aggregated and carried back to the CO over either T1s, asynchronous multiplexors, or an optical carrier (OC) ring.

If the traffic requirement to any specific customer is too great to be carried back over the access ring, a private ring (discussed in the next section) is deployed for that customer's specific traffic.

Thus, an access ring is differentiated from a private ring because it carries multiple customers' traffic.

Carriers today are looking for new access ring solutions to fulfill their multiservice requirements, especially as the data-networking world has exploded and emerging services such as VoIP, Internet Protocol Television (IPTV), and broadband DSL are growing in popularity. However, these alternatives must be cost-effective and must provide incremental evolutionary upgrades. Technologies such as Metro Ethernet over SONET via Route Processor Redundancy (RPR) and even course wave-division multiplexing (CWDM) are being considered alternatives for access ring technologies.

Private Rings

As mentioned in the last section, the need for a private customer ring emerges when the customer needs bandwidth that would exceed the availability of the carrier access rings.

Figure 4-14 shows a deployment of a private ring in which the customer has three locations to connect and then homed back to the carrier CO to pick up other services, such as voice dial tone and Internet access. The traditional deployments of private rings provide primarily bandwidth for connectivity; special services such as Metro Ethernet, storage, and dense wavelength-division multiplexing (DWDM) wavelength services either could not be deployed or would require a separate network altogether.

Figure 4-13 *Access Ring Carrying Traffic Between the Carrier's IOF Ring and Remote Cabinets*

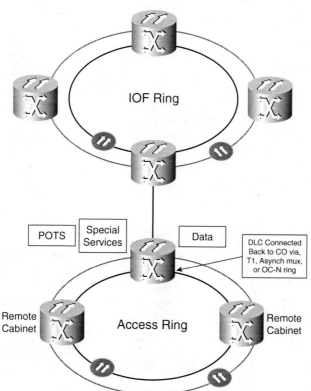

Figure 4-14 *A Private Ring Deployment Featuring Multiple Customer Sites*

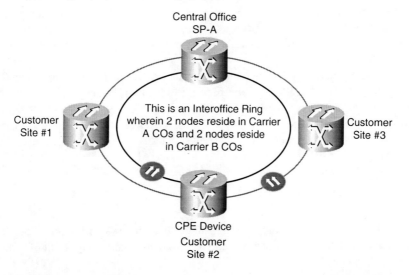

Thanks to MSPP, today numerous service types can be delivered to customers over private rings, including Ethernet, storage, and DWDM. The demand for services has become so great that service providers are replacing lost voice-revenue streams with managed service-revenue streams. In a managed service, the service provider owns all the network infrastructure and fully manages it, handing the customer a connection into the provider's network "cloud." This has been a great benefit to customers who do not have in-house expertise to manage WAN services, and it allows the customer to focus on the core competencies of their business. Service providers capitalize on the economies of scale by servicing hundreds to thousands of customers; therefore, providers can negotiate for better capital equipment prices with vendors and can draw from a huge trouble-reporting/-resolution database to resolve network problems more quickly. Thus, MSPP-based private rings offer both the service provider and the customer benefits that neither could obtain without this next-generation technology.

Heritage Operational Support System

Before leaving traditional carrier network architectures, it is only appropriate that you consider some of the legacy OSS components. Not only have these components affected the large incumbent local exchange carrier (ILEC) networks in the past, but they are still greatly influencing their architectural decisions today as they try to evolve and transform their networks while being "attached" to these legacy OSS components. These systems have hindered transformation because they are so ubiquitously integrated into the network that changing any device in the network, or deploying a new technology, must accommodate their rules and conventions.

This has greatly affected telecom equipment providers. If the providers want to sell their products to a Regional Bell Operating Company (RBOC), they must successfully complete the Operations Systems Modifications for the Integration of Network Elements (OSMINE) certification process. Many years ago, this process was only an internal process operated by the RBOCs. The RBOCs exclusively commissioned Telcordia Technologies, formerly Bellcore, to perform interoperability testing and device integration. Today Telcordia Technologies is the certification body that guarantees that a Network Equipment Provider's device is interoperable with a heritage Telcordia OSS, such as Trunk Integrated Record Keeping System (TIRKS), Transport Element Activation Manager (TEMS), and Network Monitoring and Analysis (NMA). Thus, Telcordia is the door through which equipment providers must pass to get a license to sell in the RBOC space. Telcordia also ensures service providers that the equipment providers' device interface conforms to the Telcordia-defined TL1 standards. TIRKS, TEMS, and NMA are covered in the next few sections.

TIRKS

TIRKS is an inventory record-keeping and provisioning system for interoffice trunk facilities. TIRKS has traditionally made record-keeping and assignment of central office equipment and interoffice facilities an efficient and easy process by accomplishing the following:

- Significantly minimizing manual intervention and analysis time related to circuit provisioning

- Facilitating efficient planning and use of interoffice facilities

- Allowing more cost-effective ordering and inventorying of equipment in multivendor environments

- Interfacing with other systems involved in backbone network provisioning

TEMS

TEMS is Telcordia's Transport Network Element (NE) Activation Manager. TEMS (or Transport) is an element-management system used to provision and examine transport network elements. Additionally, TEMS provides memory-management functions and interfaces with upstream Telcordia OSSes, such as TIRKS.

NMA

NMA is a fault-management OSS in service-provider networks. NMA is a device-monitoring and surveillance OSS that collects alarm and performance data from network elements. It continually receives alarm information from the equipment and correlates these alarms. NMA performs root-cause analysis by correlating multiple related alarm messages and then outputting a single trouble ticket. NMA tracks trouble tickets and uses these to identify the equipment and other facilities that require service restoration and maintenance by network technicians.

To correlate alarms, NMA must understand the equipment architecture and the relationships between the equipment facilities. Therefore, for NMA certification, the chief output from OSMINE includes configuration files, known as NMA templates, which NMA uses to manage the equipment. Telcordia produces diagrams of the equipment's containment and support hierarchy. These diagrams are profiles of the multiple configurations of the equipment's racks, shelves, and circuit packs. Telcordia delivers a "Methods & Procedures" document to the equipment provider's customers that includes these diagrams, and instructs service providers on how to build and modify the hierarchies for the NE.

Traditional Customer Network Architectures

In customer network architectures, the customer owns and operates the equipment on the ring. In some cases, the customer also owns and operates the fiber or copper facilities that tie together the network elements.

ATM/Frame Relay Networks

You have already learned about the service-provider ATM and Frame Relay networks. The difference between a customer ATM/Frame Relay network and a carrier network is based on who owns the equipment. If the customer decides to lease its own fiber and DS1s or DS3s, the customer could buy the equipment itself and deploy its own private ATM or Frame Relay network. For a large customer with many sites, this might prove more cost-effective than acquiring the ATM or Frame Relay service from a carrier.

The customer has a couple of added benefits when privately deploying ATM/Frame Relay:

- **Remote access**—For remote-access devices, access line charges can be lowered by reducing the number of physical circuits needed to reach the networks.

- **Increased performance with reduced network complexity**—By reducing the amount of processing (as compared to X.25) and efficiently using high-speed digital transmission lines, Frame Relay can enhance the performance and response times of customer applications.

- **Protocol independence**—Frame Relay can easily be configured to combine traffic from different networking protocols, such as IP, Internetwork Packet eXchange (IPX), and Systems Network Architecture (SNA). Cost reduction is achieved by implementing Frame Relay as a common backbone for the different kinds of traffic, thus unifying the hardware and reducing network management.

Customer Synchronous Optical Networks

Many enterprise businesses are deploying their own private SONET networks by leasing dark fiber and "lighting" it themselves with their own capital SONET equipment. They deploy a SONET ring just as a service provider would; however, they are responsible for maintaining the network if it goes down. A disadvantage of the private SONET ring deployment is the troubleshooting process, which can take longer because the customer does not have the expansive trouble history database that a carrier with thousands of rings has. The Technical Assistance Center (TAC) personnel of the carriers see many more issues in a day, and TAC records can identify trends because of the large data pool of trouble tickets.

The big advantage, of course, is money. The business case is strong for private ownership of SONET rings, especially because many geographies possess a fiber "glut" and lease fiber relatively inexpensively. Payback periods using legacy SONET equipment are nowhere near that of MSPPs but are still reasonable and justifiable, depending on the time frame. Legacy platforms were more rigid and costly, so there were not many private deployments; as a result, not much expertise was available among customer personnel to deploy legacy private SONET customer rings. Return on investment (ROI) for the SONET capital equipment stemmed from saving monthly SONET service charges from a carrier's service.

End users must evaluate whether they have the talent on staff to monitor and maintain the SONET rings.

IP and MPLS Networks

As mentioned in the section on service-provider IP/MPLS, IP MPLS networks enable a global CoS data service across multiple IP VPN backbones to extend the reach and service capabilities for enterprises without purchasing additional customer premises equipment (CPE).

The requirement is that each router in the backbone must be MPLS enabled; MPLS adds a label, which is a tag, at the edge of the network to each packet so that they are switched, not routed. This label contains a VPN identifier and a CoS identifier, and allows all packets to follow the same path and receive the same treatment.

In a private deployment of IP/MPLS, such as in a carrier deployment, differentiating among customers is not a requirement because only the customer's traffic rides the backbone. However, a private IP/MPLS deployment creates a secure VPN for the customer with the following benefits:

- **Connectionless service**—MPLS VPNs are connectionless. They also are significantly less complex because they do not require tunnels or encryption to ensure network privacy.

- **Centralized service**—VPNs in Layer 3 privately connect users to intranet services and allow flexible delivery of customized services to the user group represented by a VPN. VPNs deliver IP services, such as multicast, QoS, and telephony support within a VPN, as well as centralized services, such as content and Web hosting.

- **Scalability**—MPLS-based VPNs use a Layer 3 connectionless architecture and are highly scalable.

- **Security**—MPLS VPNs provide the same security level as connection-based VPNs. Packets from one VPN cannot accidentally go to another VPN. At the edge of a network, incoming packets go to the correct VPN. On the backbone, VPN traffic remains separate.

- **Easy creation**—Because MPLS VPNs are connectionless, it is easy to add sites to intranets and extranets, and to form closed user groups. A given site can have multiple memberships.

- **Flexible addressing**—MPLS VPNs provide a public and private view of addresses, allowing customers to use their own unregistered or private addresses. Customers can freely communicate across a public IP network without network address translation (NAT).

Figure 4-15 shows an example of a customer-deployed IP/MPLS network. Here the customer owns the routers and configures the VPNs. The links between the routers are most often leased from providers; however, some companies, utilities, or public-sector entities own their own fiber or copper facilities, so even the links are privately owned.

Figure 4-15 *Customer-Deployed IP/MPLS Network Where Links Between Routers Are Leased from Service Providers*

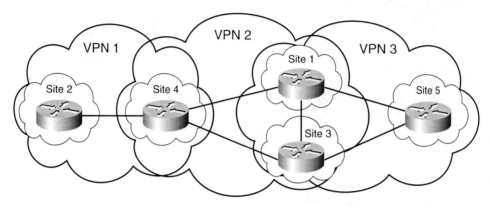

MSPP Positioning in Service-Provider Network Architectures

MSPP brings with it the capability to deploy traditional unidirectional path-switched ring (UPSR) and BLSR ring architectures. It also has introduced the capability to use the MSPP in what is called a path-protected meshed network (PPMN), shown in Figure 4-16.

Figure 4-16 *PPMN Network That Uses a Meshed Protection Scheme to Protect the Network*

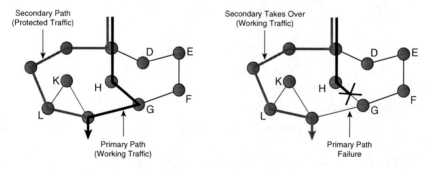

NOTE   Although this chapter focuses on the traditional UPSR and BLSR ring deployments of MSPP in the forthcoming sections, a brief introduction to PPMN is relevant because PPMN offers a valuable alternative to the traditional ring types.

PPMN allows network designers to design a mesh that uses both protected and unprotected spans at various line rates. If one route fails, connection is reestablished through another path in the mesh in less than 50 ms. This offers network designers the flexibility they can use for mesh networks today.

A meshed network refers to any number of sites randomly linked together with at least one loop. In other words, each network element has two or more links connected to it.

How MSPP Fits into Existing Networks

The magnificence of MSPPs is that they offer physical Layer 1 services, media access Layer 2 services, and even network Layer 3 features such as QoS. This allows MSPPs to fulfill a number of service needs while integrating various service types into one platform that can aggregate and backhaul these services through either a carrier's network or even a customer's private network. The next sections look at how MSPPs are integrated into the networks and the value they add.

MSPP IOF Rings

Enough cannot be said of the value MSPP has brought to carrier IOF rings. To review the differences between legacy optical platforms and today's highly enhanced MSPPs, take a look at Chapter 1, "Market Drivers for Multiservice Provisioning Platforms." Next, you will look at two major IOF deployments: DWDM and SONET. You will see that MSPP has revolutionized each by decreasing costs and provisioning time, while increasing flexibility and scalability. Figure 4-17 shows an IOF ring using MSPPs.

Figure 4-17 *MSPP Used as the Backbone for IOF*

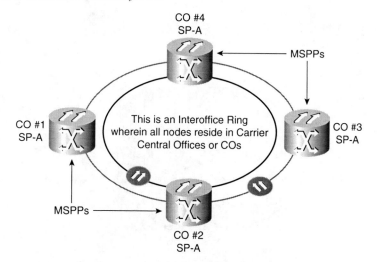

DWDM Transport

As shown in Chapters 1, "Market Drivers for Multiservice Provisioning Platforms," and 3, "Advanced Technologies over Multiservice Provisioning Platforms," DWDM transport is now integrated into today's MSPPs, allowing for wavelengths to be launched from the MSPP. IOF DWDM transport types are deployed as either passive or active DWDM from the MSPP. Because DWDM uses specific wavelengths or lambdas for transporting data, the lambdas provisioned must be the same on both ends of any given connection. The International Telecommunications Union (ITU) has standardized on a grid with spacings of 100 GHz; however, vendors use wider spacing, sometimes at 200 GHz or narrower. In addition, different vendors that do use the same grid might not use the same lambda-numbering scheme. That is, wavelength A on one vendor's equipment could be assigned a different wavelength number from wavelength B on another vendor's equipment. Hence, it is important to be aware of these potential interoperability problems that stem from different grid alignments.

IOF passive DWDM transport uses a combination of ITU wavelength optics in the MSPP and "off-the-shelf" filters in what is typically called a passive MSPP DWDM deployment because the filters do not reside on the MSPP. IOF active DWDM transport is a deployment in which the filters reside on the MSPP and can be integrated into a complete DWDM ring using MSPPs, as shown in Figure 4-18.

Figure 4-18 *MSPP-Based Active DWDM Example*

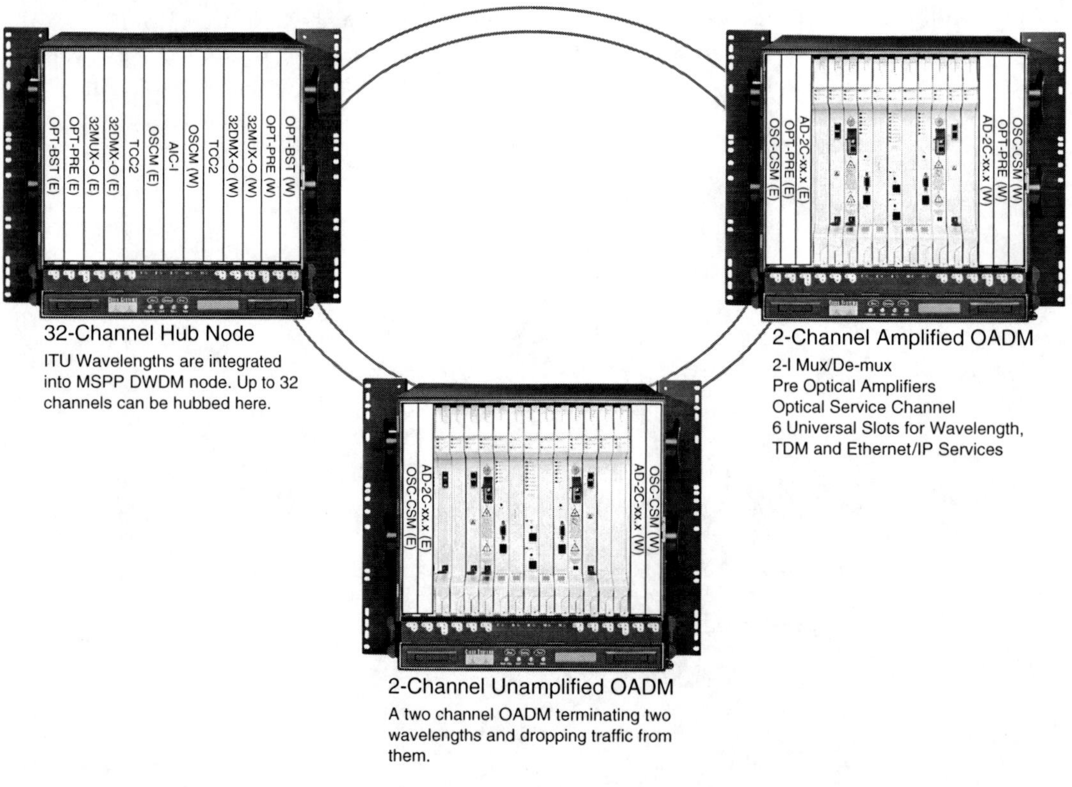
OPT-BST (E)
OPT-PRE (E)
32MUX-O (E)
32DMX-O (E)
TCC2
OSCM (E)
AIC-I
OSCM (W)
TCC2
32DMX-O (W)
32MUX-O (W)
OPT-PRE (W)
OPT-BST (W)
32-Channel Hub Node
ITU Wavelengths are integrated into MSPP DWDM node. Up to 32 channels can be hubbed here.
OSC-CSM (E)
OPT-PRE (E)
AD-2C-xx.x (E)
AD-2C-xx.x (W)
OPT-PRE (W)
OSC-CSM (W)
2-Channel Amplified OADM
2-l Mux/De-mux
Pre Optical Amplifiers
Optical Service Channel
6 Universal Slots for Wavelength, TDM and Ethernet/IP Services
OSC-CSM (E)
AD-2C-xx.x (E)
AD-2C-xx.x (W)
OSC-CSM (W)
2-Channel Unamplified OADM
A two channel OADM terminating two wavelengths and dropping traffic from them.

Before the emergence of MSPPs, two separate networks had to be built: a DWDM overlay network and a SONET ring to groom the traffic from the DWDM wavelengths. This is no longer necessary because the MSPP optical add/drop multiplexers (OADMs) can both terminate the wavelength and carve out the SONET traffic to be delivered from the MSPP. Thus, in an IOF MSPP DWDM transport application, one integrated DWDM/SONET platform can take the place of what used to be two or more devices. If an MSPP does not have DWDM capabilities, it can be upgraded, thus making it scalable and protecting the carrier's investment.

SONET Transport

Today's SONET IOF rings benefit greatly from the flexibility and density, footprint and power requirements, scalability, and ease of provisioning and management of MSPPs. Descriptions of each of these benefits as they pertain to the SONET IOF are as follows:

- **Flexibility and density**—For IOF needs, MSPPs integrate numerous optical and electrical cards within the same shelf. The port density of these cards is much greater than that of legacy platforms, so they consume much less rack space than legacy systems.

- **Footprint and power requirements**—Because of the decrease in shelf size, carriers can place up to four MSPPs in a single bay. This is a vast improvement over legacy platforms. Also, because of the evolution in optics cards, power requirements are greatly reduced. This translates into dollar savings for carriers with thousands of network elements.

- **Scalability**—The need to scale the IOF is as great as ever. With new higher-bandwidth services such as Ethernet emerging as access technology, an enormous need exists to rapidly add IOF trunking to carry these services. The capability to upgrade in service from one optical speed to another, such as OC-12 to OC-48, provides the carrier with rapid scalability.

- **Provisioning and management**—Today's MSPPs take advantage of all the development of modern-day computing and software. Graphical user interfaces (GUIs) enable the technicians and operators to provision and manage the network much more efficiently. Procedures that took hours in the past, such as circuit provisioning and even changing ring types such as UPSR to BLSR, can take seconds to minutes now. Remote access to these MSPPs is much more feasible because many MSPPs are IP based, capitalizing on the ubiquitous nature of IP.

MSPP Private Architectures

Just as MSPP has been deployed for the service-provider IOF rings, it is being deployed today with great success and growing demand as the infrastructure of choice for private rings, which are used to deliver services such as Ethernet over SONET, storage, TDM services, and wavelength services. The architectures for each ring type are covered in the next few sections.

Ethernet

Looking at Figure 4-19, you can see once again that MSPP nodes act to terminate the fiber from the service provider that owns the fiber and the MSPPs in the CO, and even at the customer premises.

Figure 4-19 *MSPP Used to Deliver Ethernet Services*

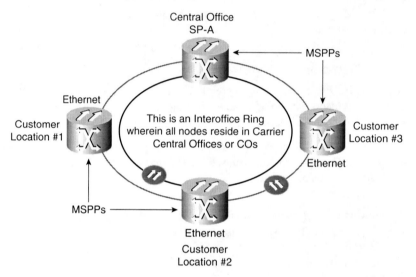

Regardless of the type of service deployed from the MSPP—for example, Ethernet Private Line (EPL), Ethernet Wire Service (EWS), Ethernet Relay Service (ERS), Ethernet Multipoint Service (EMS), Ethernet Relay Multipoint Service (ERMS), or Resilient Packet Ring (RPR)—the ring architecture is the same. The only difference among services being offered lies in the card type and the provisioning of the software. The ports off the MSPP connect to the customer-owned premises equipment. In a traditional SONET mapping application in which the Ethernet data is mapped into the SONET payload, multiport cards in the MSPP are used to terminate the Ethernet connection to the CPE, as shown. Figure 4-20 shows MSPP and CPE in the customer premises, with MSPP owned by the carrier.

In an Ethernet RPR configuration, the ring nodes act as packet switches implementing a Layer 2 protocol for media access and bandwidth sharing. RPR uses both primary and back-up rings for data. Although this capability to use both primary and back-up rings is a capacity advantage for RPR, it comes at a price. Fault recovery, in the case of RPR, is always accompanied by loss of service (LoS). While the faulty ring is being repaired, all data transmission on that ring is disrupted, and service to the customer is stopped until the fault is isolated and fixed. This might be acceptable for casual network use, but revenue-generating services cannot meet their service-level requirements without a prioritization scheme. Thus, in the case of a fiber cut or a node failure, RPR networks must continue to provision bandwidth for high-priority traffic, such as voice, while applying fairness and buffering for low-priority data traffic, such as e-mail or file transfers.

Figure 4-20 *MSPP and CPE in the Customer Premises*

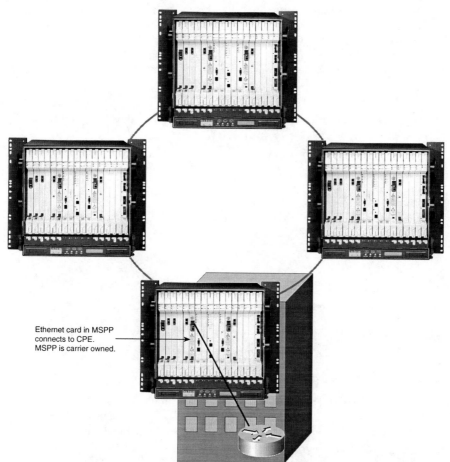

Ethernet card in MSPP
connects to CPE.
MSPP is carrier owned.

In a ring architecture, when restoration occurs after a single fiber cut, unidirectional rings and 4-Fiber bidirectional rings continue sustaining the same amount of bandwidth after restoration as before. This is made possible because the same amount of bandwidth required to restore service after the cut was being held in reserve before the cut.

For the 2-Fiber BLSR, the network throughput remains the same, except when the number of nodes is even and the demand is not split. In this case, curiously, the demand supported before the cut is less than the demand that can be supported after the cut.

The RPR ring is a bidirectional dual counter-rotating ring and, thus, is similar to a 2-Fiber BLSR. But because the entire bandwidth is used for working traffic (that is, no bandwidth is set aside for protection unlike in a 2-Fiber BLSR), the network throughput is equal to that of a 4-Fiber BLSR. After a fiber cut/restoration, the wrapped ring is exactly the same as a wrapped 2-Fiber BLSR. Therefore, after restoration, network throughput of an RPR ring drops by half.

Storage-Area Networking

Storage-area networking private rings offered by service providers are relatively new, given that the technology required to deliver this service has not been available until recently. Figure 4-21 shows a typical SAN private ring application.

Figure 4-21 *SAN Private Ring Offered by a Service Provider*

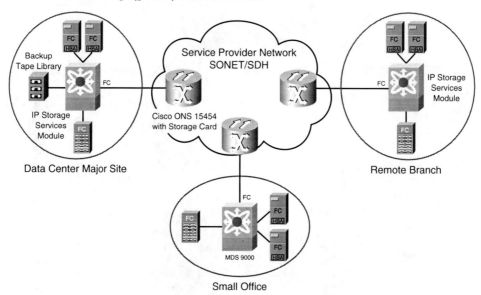

Looking at Figure 4-21, as in the case with private Ethernet rings, the carrier owns and operates the SONET MSPPs. The storage cards with interfaces reside within the MSPP chassis, and they offer the termination points for the Fibre Channel (FC) connections. The ring can be configured in any one of a number of ring topologies:

- UPSR/Call Control application programming interface (CCAT)
- 2F- and 4F-BLSR/virtual concatenation (VCAT) and contiguous concatenation (CCAT)
- Automatic protection switching <$I~>(APS)/(1+1 uni- or bidirectional)
- PPMN
- Unprotected (0+1)

In a 4-Fiber BLSR, two fibers are reserved as a protection ring, and the other two other fibers are used as the working ring. In a 2-Fiber BLSR, data between two neighboring ring nodes is sent in both directions because it is bidirectional. But because no additional pair of fibers exists to serve as a protection ring, half the time slots on each fiber are reserved as protection bandwidth. Regardless of the private ring service being offered—Ethernet, storage, TDM, or optical services—the carrier has the option of using a 2- or 4- Fiber deployment for the ring architecture.

TDM

TDM traffic over SONET was the primary driver behind traditional SONET platform growth long before MSPPs arrived on the scene. Today MSPPs continue to carry a plethora of TDM circuits throughout the SONET networks.

The SONET equipment is most often deployed in rings for TDM traffic. The ring allows services to be routed either way around the ring so that if a fiber is cut or an SONET OADM fails, the service can be rerouted via the other half of the ring.

Point-to-point connections are also allowed and can be protected using two pairs of fibers. Finally, a linear topology is another option, with a sequence of MSPPs interconnected in a chain. Leased lines can be provided as protected or unprotected services. Different levels of protection exist, depending on the type of protection mechanism that is provided in the network devices.

The ring architectures that a carrier can use to deliver TDM traffic to customers in a private ring deployment are similar to these familiar ones:

- UPSR
- 2F- and 4F-BLSR
- Automatic Protection Switching/Subnetwork Connection Protection (APS/SNC) (1+1 uni- or bidirectional)
- PPMN
- Unprotected (0+1)

For TDM private rings, UPSR rings dominate the landscape.

UPSRs are popular topologies in lower-speed local exchange and access networks, particularly where traffic is primarily point to multipoint—that is, from each node to a hub node, and vice versa. The UPSR is an attractive option for TDM applications because of its simplicity and lower cost. No specified limit exists on the number of nodes in a UPSR or on the ring length. In practice, the ring length is limited by the fact that the clockwise and counterclockwise paths that a signal takes have different delays associated with them; this affects the restoration time in case of failure. UPSRs are fairly easy to implement because their protection scheme is simple: It requires action only at the receiver, without any complicated signaling protocols. As in the case of Ethernet private rings and SAN private rings, the carrier owns the MSPPs and connects to the CPE equipment, which is titled to the customer. A TDM ring can be exclusively DS1s, DS3s, or a combination of the two. With MSPP, the underlying ring architecture is not a factor for the TDM service type.

Wavelength Services

The ring architecture for DWDM has the same options as the other private ring deployments. UPSR, 2-Fiber BLSR, and 4-Fiber BLSR are the most common deployments. In a wavelength service, the carrier owns the MSPPs and hands off a wavelength from the MSPP to the customer premises equipment, as Figure 4-22 shows.

Figure 4-22 *MSPPs Owned by Service Provider and Handing Off Wavelength to Customer*

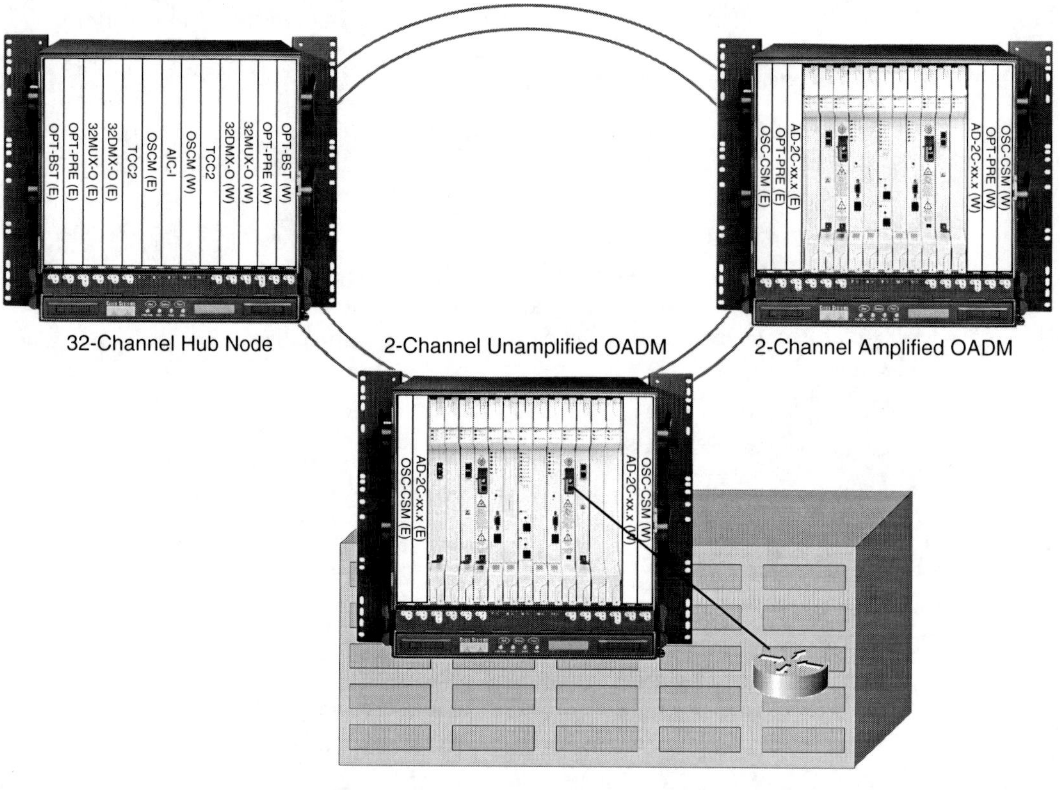

32-Channel Hub Node   2-Channel Unamplified OADM   2-Channel Amplified OADM

MSPP Access Rings

Many service providers are gaining incredible benefit by deploying MSPPs in their access network. With many customers and service types being aggregated into this network, MSPPs offer incredible flexibility, scalability, and the capability to rapidly launch new services from the access rings to the end users. Next you will take a look at how MSPPs have enhanced the ring architecture for carrier access rings.

MSPPs in the access network can be used to deliver services to customers using either SONET, CWDM, or DWDM infrastructures, or a combination of these technologies off the same shelf.

SONET Access Ring Architecture

The proliferation of MSPP SONET access rings is growing rapidly. The same features and cost benefits found in an MSPP IOF ring for a carrier are available in the access part of the network. The only difference is the size of the MSPP. Equipment vendors typically offer their flagship MSPP, which provides the highest number of slots and is the largest footprint; then they offer a smaller version, or "MSPP Jr.," which is scaled down for smaller applications. This gives the service provider a lower-cost price point to justify the system architecture from a business standpoint, allowing them to recoup their initial capital expenditure investment in a shorter period of time. An example of this is the Cisco ONS 15454, which is the flagship MSPP, and the ONS 15327 (shown in Figure 4-23), the smaller version of it. Both platforms use the same management system, the Cisco Transport Controller (CTC); network nodes for each platform can be accessed and provisioned simply by changing the IP address.

Figure 4-23 *Cisco ONS 15327*

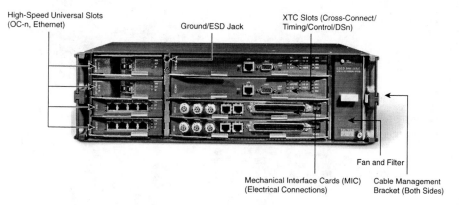

As far as the ring architecture go, MSPPs can use point-to-point, UPSR, BLSR, or mesh topologies to deliver SONET access services.

CWDM Access Ring Architecture

Broadband CWDM access technologies enable service providers to quickly deploy metro services to customers at unprecedented price points. CWDMs are one of the most cost-effective access solutions, and they can be deployed either as point-to-point feeders or as access rings with hubbed traffic patterns. Service providers add bandwidth and wavelengths as needed. In access ring applications, advanced features, such as single-wavelength add/drop and low-cost path protection, are available. This enables the service provider to offer metro services with maximum flexibility and high reliability.

In a CWDM application, fiber to the buildings must be available to terminate the fiber in the customer premises. If it is not available to the building, an intermediate terminal site must exist to drop the wavelength and then carve the services out of the MSPP to be delivered to the customer premises over a copper facility. In this case, the CWDM access ring acts as a core aggregator of services coming from the customer premises; then these services are backhauled into the CWDM core, with the MSPP embedding the STSs into the CDWM wavelength.

A CWDM ring built on MSPPs, shown in Figure 4-24, enables service providers to do the following:

- Rapidly add new services over existing fiber

- Easily implement hubbed collector rings to feed the metro core

- Implement advanced optical features, such as single-wavelength add/drop and path protection, at very low price points

- Flexibly support all data, storage, voice, and video applications

Figure 4-24 *Carrier CWDM Ring*

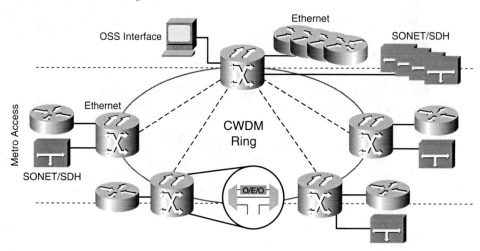

DWDM Access Ring Architecture

A DWDM access ring architecture is just like that of a CWDM access ring: The only difference is the channel spacing of the wavelengths on the ring. The cost of a DWDM access ring has been traditionally prohibitive; however, with the emergence of MSPP, it is becoming more viable to implement in a cost-effective manner. Nonetheless, a business case for a DWDM access ring would still require a highly dense population of high-bandwidth users within a given geographic area, such as a major metropolitan area with many high-rises in which the DWDM wavelengths could be terminated in the buildings and the MSPP can distribute services to multiple tenants. At the same time, bandwidth demand is growing among end users, with applications such as voice, video, and storage requiring larger pipes to carry the data. Figure 4-25 shows a carrier DWDM access ring using MSPP, as already shown. In the next section, you take a look at a three MSPP-based DWDM service provider access architectures, including point-to-point, ring, and mesh architectures.

Figure 4-25 *Carrier DWDM Ring*

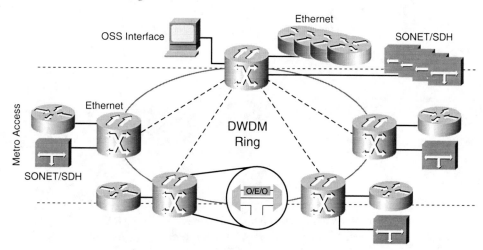

Point-to-Point MSPP-Based DWDM Service Provider Access Architecture

Point-to-point DWDM topologies can be implemented with or without OADM. These networks are characterized by ultra-high-channel speeds of 10 Gbps to 40 Gbps, high-signal integrity and reliability, and fast-path restoration. In long-haul networks, the distance between transmitter and receiver can be several hundred kilometers, and the number of amplifiers required between endpoints is typically less than 10. In metropolitan-area networks (MANs), amplifiers are often not needed.

Protection in point-to-point topologies can be provided in a couple ways. In legacy equipment, redundancy is at the system level. Parallel links connect redundant systems at either end. Switchover in case of failure is the responsibility of the client equipment (a switch or router, for example); the DWDM systems themselves just provide capacity.

In next-generation MSPPs, redundancy is at the card level. Parallel links connect single systems at either end that contain redundant transponders, multiplexers, and central processing units (CPUs). Here protection has migrated to the DWDM equipment, with switching decisions under local control. One type of implementation, for example, uses a 1+1 protection scheme based on SONET APS, as shown in Figure 4-26. This can be used for access from a high-rise to a service provider's central office.

MSPP-Based DWDM Service-Provider Access Ring Architecture

DWDM access rings are the most common DWDM architecture found in metropolitan areas; they span a few tens of kilometers. The fiber ring might contain as few as four wavelength channels and typically fewer nodes than channels. Bit rate is in the range of 622 Mbps to 10 Gbps per channel.

Figure 4-26 *Point-to-Point Architecture*

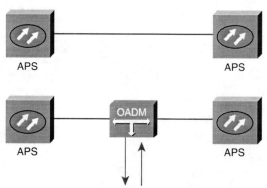

Ring configurations can be deployed with one or more DWDM systems, supporting any-to-any traffic, or they can have a hub node and one or more OADM nodes, as shown in Figure 4-27. As the hub node traffic originates, it is terminated and managed, and interfaces with other networks are established. At the OADM nodes, selected lambdas are dropped and added, while the others pass through transparently. In this design, ring architectures allow nodes on the ring to provide access to customer premises equipment, such as routers, switches, or servers, by adding or dropping wavelength channels in the optical domain. With an ever-increasing number of OADMs, however, comes an ever-increasing signal loss, and amplification can be required.

Figure 4-27 *DWDM Hub and OADM Ring Architecture*

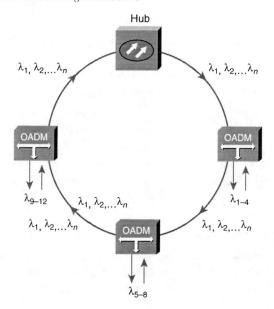

Candidate networks for DWDM deployments in the metropolitan area are often already based on SONET ring designs with 1+1 fiber protection. Thus, schemes such as UPSR and BLSR can be reused for DWDM implementations. Figure 4-28 shows a UPSR scheme with two fibers. Here, a central office hub and nodes send data on two counter-rotating rings, but all equipment normally uses the same fiber to receive the signal—hence, the name unidirectional. If the working ring fails, the receiving equipment switches to the other pair of fibers. Although this provides full redundancy to the path, no bandwidth reuse is possible because the redundant fiber must always be ready to carry the working traffic. This scheme is most commonly used in service-provider MSPP-based DWDM access networks.

Figure 4-28 *UPSR Protection on a DWDM Ring*

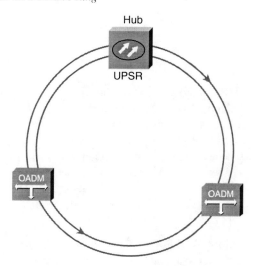

Other schemes, such as BLSR, allow traffic to travel from the sending node to the receiving node by the most direct route. As you have seen in our IOF discussion, BLSR is considered preferable for core SONET rings, especially when implemented with four fibers, which offers complete redundancy.

Mesh Topologies

Mesh architectures are the future of optical networks. As MSPPs have emerged, rings and point-to-point architectures still have a place, but mesh promises to be the most robust topology. From a design standpoint, there is an elegant evolutionary path from point-to-point to mesh topologies. By beginning with point-to-point links equipped with MSPP nodes at the outset for flexibility, and subsequently interconnecting them, the network can evolve into a mesh without a complete redesign. Additionally, mesh and ring topologies can be joined by point-to-point links, as shown in Figure 4-29.

Figure 4-29 *Mesh, Point-to-Point, and Ring Architectures*

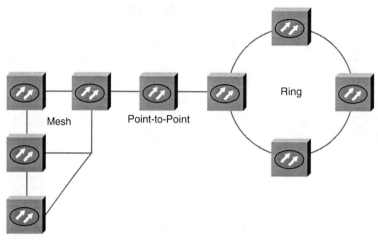

Service-provider MSPP-based DWDM meshed access network architectures, which consist of interconnected all-DWDM nodes, require the next generation of protection. Where previous protection schemes relied upon redundancy at the system, card, or fiber levels, redundancy now must transition to the wavelength itself. This means, among other things, that a data channel might change wavelengths as it makes its way through the network, due either to routing or to a switch in wavelength because of a fault. The situation parallels that of a virtual circuit through an ATM cloud, whose virtual path identifier (VPI)/ virtual channel identifier (VCI) values can be altered at switching points. In optical/DWDM networks, this concept is occasionally called a light path.

Service-provider MSPP DWDM meshed access networks therefore require a high degree of intelligence to perform the operations of bandwidth management and protection, including fiber and lambda switching. This flexibility and efficiency are profitable. For example, fiber use, which can be low in ring solutions because of the requirement for protection fibers on each ring, can be enhanced in a mesh design. Protection and restoration can be based on common paths, thereby requiring fewer fiber pairs for the same amount of traffic and not wasting unused wavelengths.

Finally, service-provider MSPP DWDM meshed access networks are highly dependent upon software for management. MPLS can now support routed paths through an all-optical network.

Next-Generation Operational Support Systems

MSPP next-generation operational support systems (NGOSS) have been designed to deliver multilevel operations, administration, maintenance, and provisioning (OAM&P) capabilities to support distributed and centralized operation models. With the objective of providing rapid response to service demands while achieving low initial cost and ongoing operational costs, the MSPP NGOSS requirements need to complement legacy and other NGOSS architectures by providing an array of options to support the diverse customer platform needs.

OAM&P Functionality

MSPP NGOSS has expanded the industry's management functionality by combining add/drop multiplexor (ADM), digital cross-connect, DWDM, and data capabilities. This management approach provides a number of options, which complement the legacy management architectures while providing more functionality at each level. The most significant OAM&P options that NGOSS supports include these:

- Craft Management System
- Simple Network Management Protocol (SNMP) and TL1 interface
- OSMINE certification
- Element Management System

Craft Management System

MSPP NGOSS expands the level of functionality of traditional craft systems by including extensive element- and network-level management capabilities. These include fault management, configuration management, accounting management, performance management, and security management (FCAPS), as well as circuit management across multiple data communication channels (DCC)–connected and data communication network (DCN)–connected nodes. A significant feature called "A to Z Provisioning" allows provisioning across a domain of MSPPs, including optional auto-routing of circuits.

Most of today's MSPP OSSes provides instant-on, anywhere management with a user-friendly, point-and-click GUI. MSPP OSSes also connect to network elements (NEs) over a Transmission Control Protocol/Internet Protocol (TCP/IP) DCN and over the SONET DCC.

For example, CTC, the Cisco ONS 15454 MSPP OSS, can autodiscover DCN and DCC-connected ONS 15454 and 15327 nodes. It provides graphical representations of network topology, conditions, and shelf configurations. CTC is launched from any platform running a JDK 1.3–compliant Java Web browser, including both Microsoft Windows PCs and UNIX workstations.

Operations staff and craft personnel in the Network Operations Center (NOC), the CO, or the field primarily use the craft interface functionality of MSPP OSS as a task-oriented tool. Its main functions are the installation and turn-up of NEs, provisioning of NE and subnetwork resources (such as connections/circuits within the subnetwork), maintenance of NE and subnetwork resources, and troubleshooting or repair of NE faults.

MSPP NGOSS Craft Interface Functionality   The craft interface of MSPP OSS provides extensive coverage for FCAPS as follows:

- **Network surveillance**—NGOSS craft interfaces give the user access to real-time alarm and event reporting, as shown in Figure 4-30. The reports include flexible sorting based upon date, severity, node, slot, port, service effect, condition, and description. Status notification is embedded within the shelf and topology views.

Figure 4-30   *Network Surveillance*

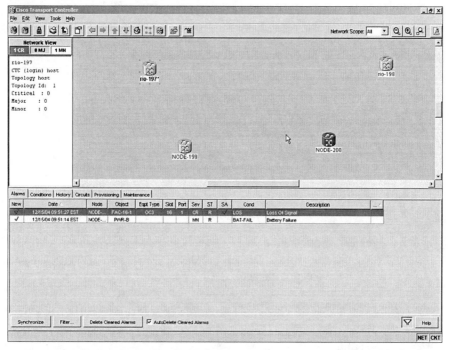

- **Equipment configuration**—Many NGOSS craft interfaces provide single-screen equipment configuration with easy-to-use point-and-click and drop-down menu selection of service parameters, as shown in Figure 4-31.

- **Circuit management**—Many MSPP NGOSSes allow authorized users to create, delete, edit, and review circuits on a selected subnetwork, as shown in Figures 4-32 and 4-33. The path of an individual circuit can be viewed on the route display. The A–Z circuit-provisioning wizard of CTC for example, provides automatic or manual circuit path selection and supports linear, hub-and-spoke, UPSR, BLSR, and interconnected ring configurations. CTC provides the capability to auto-provision multiple circuits. This simplifies and speeds the creation of bulk circuits. Many MSPP NGOSSes provide point-and-click activation of terminal and facility loopbacks on certain line cards (OCn, DS1, DS3), allowing for network testing, troubleshooting, and maintenance.

Figure 4-31 *Circuit Management Node View*

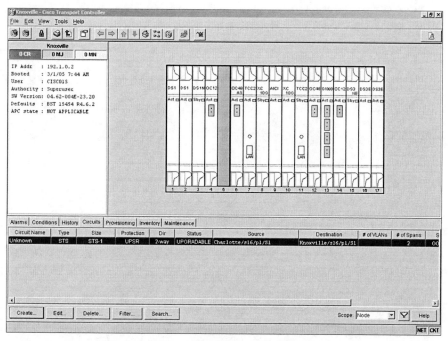

Figure 4-32 *Circuit Management Ring View*

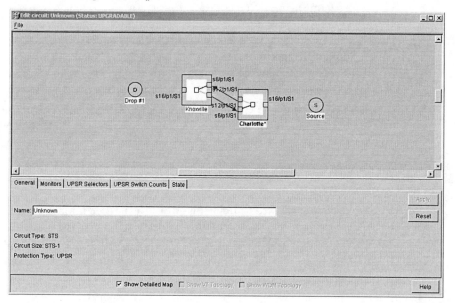

Figure 4-33 *Statistics (DS3 and Gigabit Ethernet Cards)*

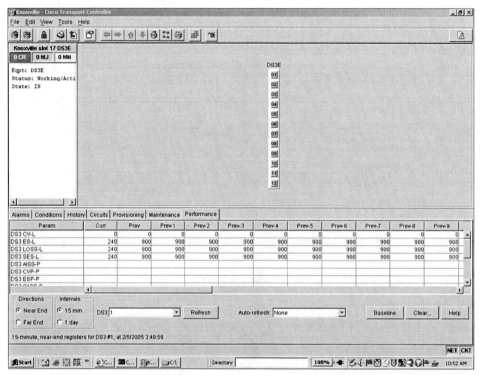

- **Performance monitoring**—NGOSS can retrieve near-end and far-end performance statistics on a configured interval, in addition to reporting threshold-crossing alerts (TCAs). This allows proactive, continuous performance monitoring of SONET facilities. Additionally, performance statistics are supported on the data interfaces (10-/100-/1000-Mbps Ethernet) to provide information on packet/frame counts as well as other performance metrics, as shown in Figure 4-33.

- **Security**—NGOSS supports various levels of user privileges, allowing the system administrator to control personnel access and the level of system interaction or access.

SNMP and TL1 Interfaces

Many NGOSSes include standard SNMP interfaces that can be used for fault management, including autonomous alarm and event reporting, of the ONS 15454. These interfaces support gets and traps using SNMPv1 and SNMPv2c. The SNMP interface is based on Internet Engineering Task Force (IETF) standard Management Interface Blocks (MIBs).

NGOSSes also include a standard TL1 interface that can be used for command-line configuration and provisioning, as well as autonomous alarm and event reporting.

OSMINE Certification

Telcordia (formerly Bellcore) uses the OSMINE process to certify network element compatibility with their OSSes. NGOSSes that are placed in operation within RBOCs that require vendor equipment to be "OSMINEd" have completed the OSMINE process. Common Language Equipment Identification (CLEI) and function codes exist for all equipment. TIRKS, NMA, and TEMS have been integrated into the NGOSSes.

EMS

Despite the powerful functionality of MSPP OSS craft interfaces, they are not EMSes; they are complementary to but are not a replacement to an EMS. The following are some of the traditional EMS functions that MSPP NGOSSes provide:

- Support for multiple concurrent users
- Network-wide coverage
- Continuous surveillance
- Continual storage in a standard database
- Consolidated northbound interface to an NGOSS
- Scalability for large networks

Additionally, MSPP NGOSS EMSes provide GUIs for easy point-and-click management of the MSPP network. Comprehensive FCAPS support includes inventory management, in which the inventory data is stored in the MSPP NGOSS database and is available for display, sorting, searching, and exporting through the GUIs.

Finally, MSPP NGOSSes provide seamless integration with existing and next-generation network-management system OSSes by using open standard interfaces, which offer the ultimate management flexibility for all the service-provider telecommunications network-management layers.

Multiservice Switching Platforms

An evolving technology platform that has emerged from the proliferation of MSPPs is the Multiservice Switching Platforms (MSSPs).

The undeniable success of MSPPs has created another revolution in bandwidth and traffic patterns, creating the need for a new switching platform optimized for the metropolitan area to aggregate and switch that higher-bandwidth traffic. As already shown, higher-bandwidth services are now starting to dominate the metro, and the management of bandwidth has transitioned to STS levels, in contrast to DS0s and T1s, as shown in Figure 4-34. The introduction of the MSSP has taken this design approach one step further. The MSSP provides far more efficient scaling in large metropolitan areas. The MSSP also enables more bandwidth in the metro core for an even greater density and diversity in higher-bandwidth services. Finally, the MSSP unleashes the additional service potential found within the MSPP multiservice traffic originating at the edge of the metro network, and can now be aggregated through a single, scalable, multiservice network element at metro hub sites, as shown in Figure 4-35.

Figure 4-34 *Growth in Metropolitan Traffic*

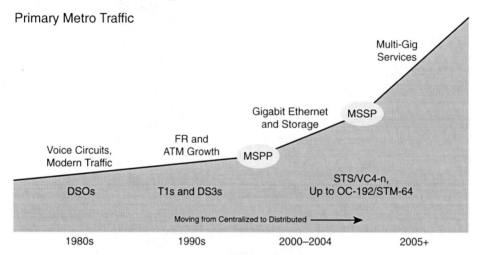

Figure 4-35 *MSSP at Core of Optical Network*

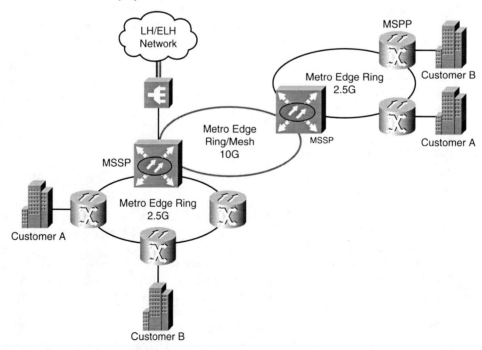

The MSSP is a true multiservice platform. MSSPs have interfaces, such as OC-48/STM-16 and OC-192/STM-64 for the high-bandwidth metro aggregation, as well as interfaces for Ethernet and integrated DWDM. This multiservice capability allows carriers to use their existing SONET infrastructure while supporting current TDM services, and carriers can move the benefits of next-generation services, such as Ethernet, into the central office. The multiservice functionality also gives the MSSP and the MSPP tighter integration, allowing the service provider to carry the strengths and benefits of an MSPP from one end of the optical network through the metro core and out to the other end of the network, as shown in Figure 4-36.

Figure 4-36 *Illustration of the Optical Network and the Placement MSSPs*

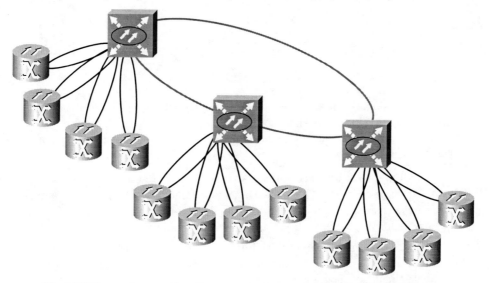

The MSSP must be capable of leveraging integrated DWDM functionality in addition to its data-switching capabilities. Integrated DWDM allows carriers to accomplish more in a single switching platform by minimizing the need to purchase another appended transponder to place traffic onto the DWDM infrastructure. By offering integrated DWDM, Ethernet, and STS switching capabilities in a single switching platform, a service provider can place the MSSP in the central office and use it not only for today's STS switching and interoffice transport demands, but also for generating additional high-margin services as they are requested.

MSSPs need to support variable topologies and network architectures, just as MSPPs do. Thus, MSSPs support 1+1 APS, UPSR, BLSR, and PPMN, as shown in Figure 4-37.

To aggregate the numerous high-speed metropolitan rings, the MSSP needs a high port density—in particular, OC-48 and OC-192, the predominant metro core interfaces today. MSSPs also remove intershelf matrix connections to achieve a footprint that is greatly reduced compared to that of legacy broadband cross-connect systems. The small footprint of the MSSP not only provides savings for the service provider, but it also authenticates that the MSSP is a huge technological advancement.

Figure 4-37 *Topologies Supported by MSSP*

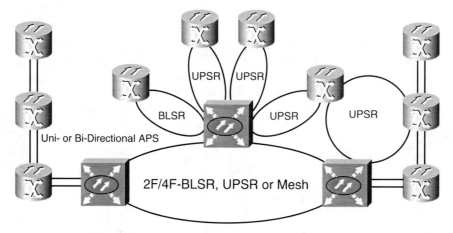

MSSP provisioning times are comparable to those on today's MSPP network. Integration with existing MSPP provisioning software is key to reducing provisioning times and providing a common management look and feel for network operators. Some of the key functions are as follows:

- **GUI and CLI interface options**—MSSPs have a GUI similar to that of an MSPP. This type of interface has gained acceptance among service providers, allowing technicians to perform OAM&P functions intuitively and with less training. Additionally, MSSPs are equipped with TL1 command-line interfaces (CLIs).

- **Cross-network circuit provisioning**—MSSP provisioning software provides the capability to provision circuits across network elements without having to provision the circuit on a node-by-node basis.

- **Procedure wizards**—MSSPs management and provisioning software uses wizards. Wizards provide step-by-step procedures for complex operations. Wizards dramatically reduce the complexity of many of the OAM&P tasks, such as span upgrade, software installations, and circuit provisioning.

MSPP Positioning in Customer Network Architectures

With regard to the architecture of MSPPs in a customer's private network deployment, there is no great difference in architecture between a customer's private deployment and a service provider's private ring deployment for the customer, typically offered to the customer as a "managed service." The difference is not architectural; it is in who owns the equipment— that is, in whose name the equipment is titled. The customer also has an advantage in management flexibility: Customers do not have to tie the network-management system of the MSPP to legacy service-provider systems, such as TIRKS, NMA, or Transport. What is of more interest, and what we cover in this section, is how a network manager might position an MSPP in the network to justify it in an ROI business case.

Several issues are at the top of corporate executives' minds today:

- "We still need to lower costs."
- "We need to do more with less."
- "We need speed of provisioning."
- "We need to be secure and prepared, with a goal of 100 percent uptime."

First and foremost for many organizations is business continuance, which is a real concern. Executives are asking themselves, "Is my company prepared to survive a disaster?" In a variety of organizations, many who think they are prepared aren't. This "unpreparedness" has prompted the government to step in and develop standards mandating safeguards to protect consumer information, such as banking records, trading records, and other critical financial data.

Second, profitability is back. The days of a "sexy" 15-page business plan with no substantial demonstration of the capability to turn a profit just won't get you $10 million in start-up financing anymore. Everyone is focused on profitability. Thus, established organizations are looking to do this on a number of fronts, including, but not limited to, lowering costs while simultaneously raising productivity. Companies are still spending money, but for the most part, they are doing it to save money or increase employee productivity.

Third, organizations are looking at new approaches to doing business. Whether they are looking to lower costs or increase employee productivity, many are using technology to achieve their goals. Technologies and applications such as IP telephony, wireless communications, video on demand, network collaboration applications such as "webinars," and more are driving these changes.

So business continuance, profitability, and new ways of doing business are making companies rely even more on their networks and networking infrastructures. An unreliable, unstable, or inefficient network foundation is just not adequate when applications are producing gigabytes of data. This data has to be readily available to employees, customers, and business partners if something "disrupts" the fabric of that infrastructure.

For example, if a major power outage take place in Seattle, retail outlets throughout the United States and the world can't just stop selling hot beverages and scones. They must seamlessly switch operations from their main Seattle systems to redundant systems located outside the Seattle area. This is where MSPPs play a key role, as shown in Figure 4-38.

Figure 4-38 *Issues in Executives' Minds, with MSPP as a Core Component of Their Solutions*

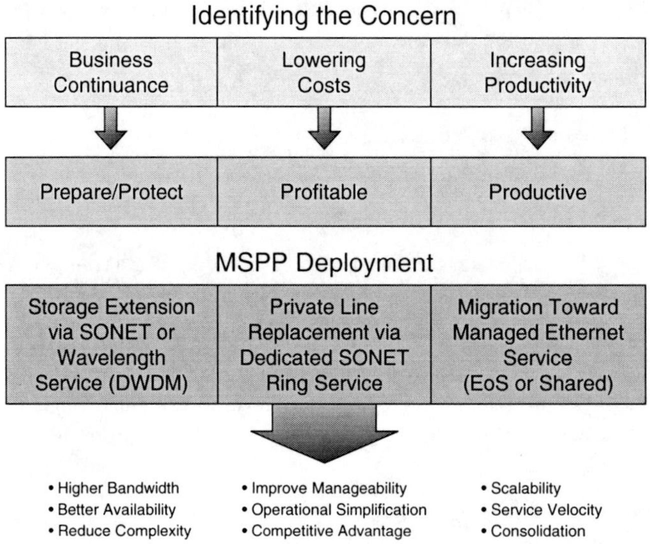

Identifying the Concern

Business Continuance	Lowering Costs	Increasing Productivity

Prepare/Protect	Profitable	Productive

MSPP Deployment

Storage Extension via SONET or Wavelength Service (DWDM)	Private Line Replacement via Dedicated SONET Ring Service	Migration Toward Managed Ethernet Service (EoS or Shared)

- • Higher Bandwidth
- • Better Availability
- • Reduce Complexity

- • Improve Manageability
- • Operational Simplification
- • Competitive Advantage

- • Scalability
- • Service Velocity
- • Consolidation

Figure 4-39 shows the key application drivers of organizations and their effect on creating demand for varied and higher-bandwidth services, such as Ethernet, shown in Figure 4-40. This again sets the stage for an MSPP architecture private deployment.

Figure 4-39 *Application Drivers That Are Creating Demand for Higher-Bandwidth Service of Ethernet in the MAN*

Application Drivers of Ethernet Adoption	Key Driver Today	Key Driver in 12 Months	Key Driver in 24 Months	Not a Key Driver
LAN-to-LAN	86%	14%	0%	0%
Content Delivery	79%	7%	0%	14%
Internet Access	79%	7%	0%	14%
Large File Transfer/Sharing (e.g., Medical Imaging, Seismic Data)	64%	7%	7%	21%
IP VPN-Related (Extranet, Intranet)	57%	21%	7%	21%
Storage	50%	29%	7%	14%
IP Telephony	36%	50%	7%	7%
IP Video	14%	29%	29%	29%
Others	0%	0%	0%	0%

n = 14 carriers answering this question

Figure 4-40 *Various Services Within an MSPP-Based Architecture*

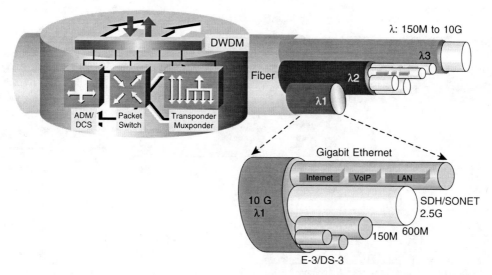

A customer who has the capability to acquire dark fiber between desired locations can develop a business case for using MSPPs as a means to carry traffic from site to site. The business cases are typically strong, with payback periods on the capital expense required to purchase the MSPPs from 6 months to 18 months, as shown in Figure 4-41. These costs need to be considered beyond the dark fiber lease:

- A maintenance contract on fiber, or associated costs of internal personnel to maintain fiber.
- Right-of-way fees for fiber.
- Additional last-mile fiber built out to sites, whether aerial or underground. Using aerial fiber is typically considered less expensive than trenching and laying new fiber underground.
- MSPP equipment costs.
- Monthly OAM&P.

Figure 4-41 *Corporate Business Case Analysis for Private MSPP Deployment of MSPP Example*

Dark Fiber Case Study—10 Mile Route (IRU cost amortized over 20 yrs)

Fiber Build Miles	One Time Fiber Trenching Cost	Annual R of W MRC	Annual Main Cost MRC	Fiber IRU MRC (10 Miles)	MSPP Equip	Monthly OAM&P	Leased OC48 Lit Service Per Month	Payback Period (months)	Total Lifetime Savings (20yrs) ($M)	MRC Savings
0	$0 $0k/M	$250	$416	$80,000@ 8%AIR $669/mth	$60k	$0.2k	$20k	<4	$4.3	$18,465 + Interest Write-off
1	$60,000 @8%AIR $500/mth	$250	$416	$80,000 @8%AIR $669/mth	$60k	$0.2k	$20k	<4	$4.25	$17,965 + Interest Write-off
3	$180,000 @8%AIR $1500/mth	$250	$416	$80,000 @8%AIR $669/mth	$60k	$0.2k	$20k	<4	$4	$16,965 + Interest Write-off
5	$300,000 @8%AIR $2500/mth	$250	$416	$80,000 @8%AIR $669/mth	$60k	$0.2k	$20k	4	$3.77	$15,965 + Interest Write-off

The business case is strong for private deployments of MSPP architecture to satisfy corporate business objectives. The MSPP architecture solution provides the following:

- MSPP technology: SONET + TDM + Ethernet + DWDM + storage
- Gigabit Ethernet (GigE) at speeds of:
    - 10 Mb
    - 100 Mb
    - 1000 Mb
    - Variable-rate Ethernet
- Ethernet with 100 percent Diverse Fiber Access
- Elimination of Ethernet to TDM latency
- Simplification of customer equipment
- Savings in capital expenses
- Readiness for 10 GE
- Readiness for SAN

Summary

A variety of traditional service-provider and customer network architectures exist, including the PSTN, Frame Relay/ATM, and SONET IP/MPLS. Transport network types include IOF, access, and private ring deployments. The emergence and insertion of MSPPs into these service-provider and customer networks has transformed them tremendously. They provide physical Layer 1 services, Media Access Layer 2 services, and even network Layer 3 features such as QoS. This enables MSPPs to fulfill a number of service needs. MSPPs also integrate various service types into one platform that can be aggregated and backhauled through either a carrier's network or even a customer's private network. They offer capital expense reduction, flexibility in service types offered, a simplified architecture requiring less space and power, vastly improved scalability, and a much easier and more cost-effective OAM&P option. Additionally, MSPPs offer next-generation services, such as DWDM and storage-area networking, which allow customers to capitalize on the advantages of SONET self-healing rings in the MAN while deploying these services.

Another enormous advantage of MSPPs over traditional network architectures is the use of NGOSS craft interfaces, which provide the following benefits:

- Instant-on anywhere management
- Reduced cost
- Simplified upgrades
- Familiar GUI
- Telcordia OSMINE compliance

This chapter covers the following topics:

- MSPP Network Design Methodology
- Protection Design
- Network Timing Design
- Network Management Considerations
- MSPP Network Technologies
- Linear Networks
- UPSR Networks
- BLSR Networks
- Subtending Rings
- Subtending Shelves
- Ring-to-Ring Interconnect Diversity
- Mesh Networks

CHAPTER 5

Multiservice Provisioning Platform Network Design

You must consider many factors when you plan to design Multiservice Provisioning Platform (MSPP) networks. Some are obvious and are a function of the traffic requirements and growth forecast for the network: fiber cable sizing, span bandwidth(s), circuit allocation, and so on. However, you must consider other factors that are foundational to the design and that help determine the extent of the network's flexibility and resiliency, as well as its operational characteristics. This chapter examines many of these design components, including protection options, synchronization (timing) design, and network management. In addition, this chapter presents a survey of the various MSPP network topologies, such as linear, ring, and mesh configurations.

MSPP Network Design Methodology

Network designers must address three important questions during the transport network planning process:

- What level of reliability must the transport system provide to meet end-user availability requirements?

- How will the network be synchronized to ensure that all transmissions from each of its elements are precisely timed?

- How will the network be managed to facilitate fault management, provisioning, and performance evaluation of the system?

This section covers these design parameters and the factors to consider when making decisions related to these.

Protection Design

A key feature of MSPPs is the capability to provide a very high level of system reliability because of the multiple forms of redundancy or protection provisionable in the platform. These protection mechanisms are available to the MSPP network designer:

- Redundant power feeds
- "Common control" redundancy

- "Tributary" interface protection
- Synchronization source redundancy
- Cable route diversity
- Multiple shelves, or chassis
- Protected network topologies (rings)

Redundant Power Feeds

MSPPs require a power source with a -48 V potential (within a certain tolerance) to operate. Two separate connections for power supply cabling are generally provided on the rear of an MSPP chassis. These are typically referred to as Battery A and Battery B. The internal power distribution is designed so that the system can continue to function normally if one of the supply connections is removed or fails. If only one of these supply connections is cabled to the power source, which is typically an alternating current (AC)–to direct current (DC) rectifier plant, an alarm is raised in the system's software (such as "Battery Failure B").

For maximum protection from service interruption from power issues, the MSPP design should include connections for both the A and B terminals. Power feed cable sizing and the associated fusing should follow the local electrical codes. If possible, separate power supplies should provide these redundant feeds, to avoid losing both feeds in a single-supply failure. Finally, it is strongly recommended that you use power supplies that include battery back-up systems, to maintain service if a commercial power outage occurs.

Common Control Redundancy

Most fully featured MSPP systems employ a modular architecture. This means that various cards (also called blades or plug-ins) can be inserted into the chassis to provide various functionality or service interfaces. Typically, a subset of the cards installed in the chassis is known as common control cards. These cards provide functions that all the installed interface cards need, and also control and monitor the operation of the system. Consider some examples of the functions that the common control cards in an MSPP provide:

- System initialization
- Configuration control
- Alarm reporting and maintenance
- System and network communications
- Synchronization
- Diagnostic testing
- Power monitoring
- Circuit cross-connect setup and maintenance

Because many of these functions are critical to the proper operation of the system, the cards that perform these functions typically must be installed in pairs. For example, the Cisco ONS 15454 MSPP reserves five chassis slots for common control cards; four of these are used to house two redundant pairs. The Timing, Communications, and Control (TCC) card and the Cross-connect (XC) card are essential to the system and are always placed in an active/standby pair. If the active card fails, the standby card takes over. Failure to install the secondary common control card(s) normally results in an alarm condition in the MSPP, such as a "Protection Unit Not Available" alarm.

Tributary Interface Protection

Interface cards that connect an MSPP to external equipment, such as routers, Ethernet switches, and private branch exchange (PBX) switches, are sometimes referred to as tributary interfaces. MSPP cards can be protected to ensure service continuity in case of card failure. MSPPs provide protection in many ways, which include one or more of the following:

- **1:1 protection**—A single protect, or standby, interface card protects a single working, or active, service interface card. This type of protection can be provided for electrical time-division multiplexing (TDM) interfaces, such as DS1, DS3, or EC-1. If the working card fails or is removed, traffic switches from the working card to the protection card. The standby card slot is linked through the MSPP chassis backplane to the working card slot so that it can reuse the cabling (such as coax cables for a DS3 interface) from the working slot if a protection switch occurs.

- **1:N protection**—In this protection scheme, a single standby card is used to protect multiple working cards (such as a single DS3 interface card used to protect up to five working DS3 cards). This is an advantage economically because fewer total cards are required to protect the working service interfaces. In addition, non-revenue-generating protection cards consume less valuable chassis real estate. As in 1:1 protection, some MSPP systems provide this type of redundancy for electrical TDM interfaces.

- **1+1 protection**—Optical (OC-N) interfaces can use this protection type in a failover scenario. 1+1 protection implies that a standby card or port protects a single working optical card (or port on a multiport card). Each of the optical cards must be cabled to the external equipment to provide this functionality.

- **Higher-layer protection**—Data interfaces, such as native Ethernet cards, are typically unprotected. However, if required, you can protect these interfaces by installing separate cards, with each having a 100-Mbps or Gigabit Ethernet (GigE) connection to a switch or router. This external networking equipment can then be configured to provide protection at Layer 2 or Layer 3. As an example, separate GigE links with load balancing can be configured between the MSPP Ethernet interface and the external switch or router.

- **0:1 protection**—This is also known as unprotected operation. This can be an option with certain designs, as in the case of Ethernet or Storage-Area Network (SAN) extension services, and when Layer 3 is used to protect data traffic.

Figure 5-1 gives an example of an MSPP (a Cisco ONS 15454) with multiple service interfaces configured with the various tributary protection methods. For critical service requirements that cannot be subjected to outage caused by the failure of a single interface card, it is strongly recommended that you provide tributary interface protection. In addition to avoiding traffic loss, these protection mechanisms can be used to defer time-consuming repair visits (truck rolls) for card replacements until regularly scheduled maintenance visits.

Figure 5-1   *MSPP with Multiple Tributary Protection Methods*

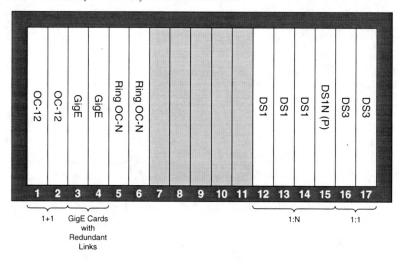

Synchronization Source Redundancy

Timing is an essential element of Synchronous Optical Network (SONET)/Synchronous Digital Hierarchy (SDH) networks. MSPP systems enable network operators to provision multiple synchronization sources to provide redundancy. Primary timing can be provided for a node either externally, through wired connections to a digital clocking source, or from a connected OC-N interface. In either case, MSPP systems also must be equipped with internal clocking sources (usually contained within the common control cards) for failover in case the primary sources fail.

For MSPP locations that have a clock source installed, such as a Cesium clock or Global Positioning System (GPS) clock receiver, the MSPP typically provides redundant clock source connection terminals. These are normally referred to as Building Integrated Timing Supply (BITS) connections. These should be wired to diverse timing interfaces on the BITS clock system for maximum redundancy. In this type of arrangement, the network operator can configure three timing sources for the MSPP node and rank them in order of preference to use. For example, the first-choice timing reference source would be the BITS-1 connection. The second choice would be the BITS-2 connection (in case the BITS-1 connection fails), followed by the internal system clock as a last resort.

Similarly, MSPP locations that do not have an external clocking source can be configured with three timing source options. The first two options would be the East and West OC-N ring interfaces, which can be traced back to another node's clocking source. The third option is the internal clock.

Cable Route Diversity

Physical cable route diversity should be considered to ensure service continuity in case of a cable break or inadvertent disconnection. Cable route diversity can be provided on both a system-level and a network-level basis.

At the system or node level, diverse pathing and ducting for direct system connections can be provided during shelf installation, with one cable lead routed away from the shelf on the left side of the network bay frame or cabinet, and the other cable lead on the right. Connections that can provide diversity in this manner include power feeders, timing leads, optical fibers, and Category 5 Ethernet cables.

At the network level, optical cable diversity can be provided for MSPP nodes configured in ring topologies. This protection can be provided both inside the building in which the MSPP is located and in the exterior (or outside-plant) cabling that is used to connect to other node locations. For example, a carrier operating an MSPP network might choose to use diverse routing for each pair of fibers connecting MSPP nodes in a two-fiber ring, such as a unidirectional path switched ring (UPSR). This could include separate underground conduit runs leaving a building and diverse geographical cable routes. If only one route exists between two locations, such as to a "spur" site on a ring, diversity can be provided by using two fibers in a buried fiber cable and two fibers in a cable lashed to a utility pole line. This would be the case when a particular site is limited to a single physical-access route (such as one roadway) from the remaining sites on the network.

Multiple Shelves

In an MSPP network that is used to transport critical services, such as medical or E911 applications, consideration should be given to providing chassis-level protection for service-termination nodes. For example, a hospital using a resilient SONET ring to carry real-time video for remote robotic-assisted surgery might elect to install MSPP nodes for interface diversity in separate data rooms or closets in a building or campus, to protect against a power outage, fire, or flooding hazard.

Protected Network Topologies (Rings)

Protected network topologies, or rings, can be used to protect a network if a single node or fiber span fails. Rings allow protected traffic to be rerouted automatically over an alternate path if the active path becomes unavailable. Ring topologies are covered in detail in the sections "UPSR Networks" and "BLSR Networks."

Network Timing Design

SONET MSPPs rely on highly accurate clocking sources to maintain proper network synchronization. Each element in the network should be timed from a Stratum Level 3 (or better) source traceable to a primary reference source (PRS). The PRS is typically a Stratum Level 1 clock.

Timing Sources

For timing design, the first consideration should be the type of timing source available at each location. As discussed previously, the possible timing sources for MSPPs include the following:

- External sources, such as a Cesium or GPS clocks
- Line sources, which include the SONET optical ports
- Internal clock, which is normally a Stratum 3 clock built into a system common control card

Based on these different timing source types, a network designer has several options to choose from when deciding how the network will be synchronized. The following are the most common recommended designs:

- **External/line-timed configuration**—In this configuration, one MSPP in the network is timed from an external clocking source, such as a BITS clock; the other MSPP nodes are line-timed from their optical interfaces. The clocking source for the externally timed node is traceable to the PRS. This is a typical timing design for a carrier access ring, in which one (or more than one) node is located in a telco central office and the other nodes are located at remote locations. Figure 5-2 shows this configuration.

- **Externally timed configuration**—If external clocking sources are available at all MSPP locations, each node can be cabled and configured as externally timed. Figure 5-3 shows an example of this type of design. Because this requires clocks at each site, this type of configuration is normally seen in interoffice transport applications, in which each MSPP is located in a telco central office. If all clock sources are traceable to a single primary reference source, the network is said to be synchronous. However, if the various local clocks are traceable to two or more primary reference sources with nearly the same timing (such as an MSPP network that spans more than one carrier), the network is referred to as plesiochronous.

Figure 5-2   *External/Line-Timed Network Synchronization Configuration*

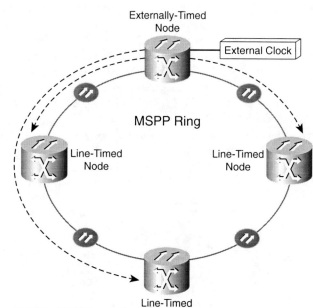

Figure 5-3   *Externally Timed Network Synchronization Configuration*

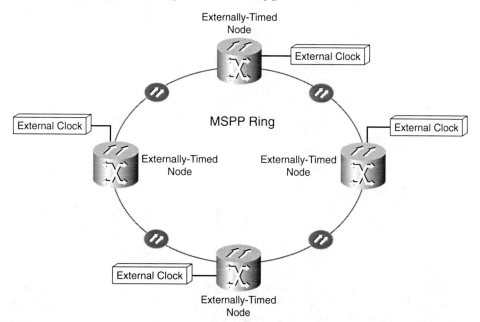

- **Internal/line-timed configuration**—For networks in which no external clocking source is available, such as a private corporate network, a single MSPP node can be configured for internal timing from its embedded Stratum 3 clock, with the remaining nodes deriving their timing from their connected optical interfaces. Some MSPP vendors refer to configured Internal clocking as "free-running" mode. This type of design is unacceptable for rings in which SONET connections (such as OC-3 or STS-1) are required to other networks. Also, additional jitter is introduced on low-speed add/drop interfaces, such as DS1s. This can cause problems if these interfaces are connected to external equipment that is sensitive to such jitter. One example is a voice switch that contains a Stratum 3 clock of its own. Figure 5-4 depicts a ring configured for the internal/line-timed configuration.

Figure 5-4 *Internal/Line-Timed Network Synchronization Configuration*

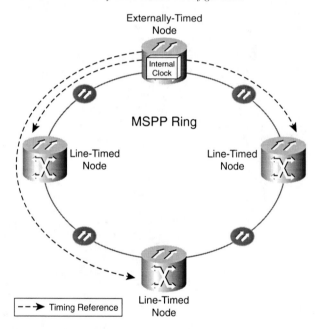

Timing Reference Selection

After selecting the timing configuration, the network designer must assign the prioritized list of timing references to be provisioned for each MSPP node. The MSPP system software uses this prioritized list to determine its primary reference source for timing and the order of failover switching in case the primary source becomes unusable or unavailable.

Typically, this list can be set up in the MSPP node to be revertive or nonrevertive. Revertive means that, if the primary reference fails and a switch to the secondary reference occurs, the node switches back to the primary source after the problem that forced the switch is corrected. A countdown timer is implemented for this reversion switch so that the network operator can set an amount of time (such as 5 minutes) to allow the primary reference to stabilize before the node switches back. This prevents multiple clock source switches from a "flapping" primary timing source.

The order of prioritization for timing sources varies based on whether the node is externally timed, line-timed, or internally (free-running) timed. For externally timed nodes, in which the MSPP is synchronized to DS1 timing leads cabled to the chassis from an external clocking source, the primary and secondary options for timing should be the redundant pair of connections, usually referred to as BITS-1 and BITS-2. Generally, it is recommended that you use the internal Stratum 3 clock as the third option so that if both BITS inputs fail, the MSPP fails over to internal timing. However, some MSPP systems allow for another option, which is to provision an optical port for the third selection. Cisco calls this mixed timing mode. This is an option for a network that is configured in the external/line-timed configuration, in which more than one node uses external timing. Using mixed timing can be tricky, however, as you will see in the examples later in this chapter. Although the Cisco ONS 15454 allows for mixed timing configuration, Cisco does not recommend its use and urges caution if you are implementing it.

For line-timed nodes, a typical recommended prioritization list includes the "main ring" optical ports, East and West, for the first and second options, with the internal clock as the third option. Internally timed nodes should be provisioned to use the internal clock for all three references.

Synchronization Status Messaging

To maintain adequate timing in an MSPP network, SONET provides a protocol known as Synchronization Status Messaging (SSM) to allow for communication related to the quality of timing between nodes. This information is carried over 4 bits in the S1 byte of the line overhead, with each node transmitting its current status to the adjacent line-terminating equipment (LTE). Recall that an LTE is a piece of SONET equipment that can read and modify the line overhead bytes. These messages enable each MSPP in the network to select alternate timing sources, as defined in the prioritized reference list, when events in the network make a change necessary. Two sets of SSM codes are in use today: The older and more widely used version is Generation 1; Generation 2 is a newer version that defines additional quality levels. Table 5-1 defines the message set for Generation 2.

Table 5-1   *Synchronization Status Messaging Generation 2 Message Set*

Message Description	Message Acronym	S1 Bits 5–8	Quality Level
Stratum 1 Traceable	PRS	0001	1
Synchronized–Traceability unknown	STU	0000	2
Stratum 2 Traceable	ST2	0111	3

continues

Table 5-1   *Synchronization Status Messaging Generation 2 Message Set (Continued)*

Message Description	Message Acronym	S1 Bits 5–8	Quality Level
Transit Node Clock	TNC	0100	4
Stratum 3E Traceable	ST3E	1101	5
Stratum 3 Traceable	ST3	1010	6
SONET Minimum Clock Traceable	SMC	1100	7
Stratum 4 Traceable	ST4	—	8
Do Not Use for Synchronization	DUS	1111	9
Provisionable by Network Operator	PNO	1110	User assignable

To understand how these SSM messages are used in a SONET MSPP network, consider the example network shown in Figure 5-5. This network is configured in the externally timed/ line-timed arrangement, with Node 1 being externally timed from a collocated BITS clock and Nodes 2–6 being line-timed traceable back to the timing source at Node 1. Because Node 1 is being timed by a Stratum 1 clock, the last 4 bits of the S1 byte transmitted in both the East and West directions are set to the value 0001, indicating the primary reference source.

NOTE   Node 1 has the two BITS inputs as its primary and secondary timing sources, with the internal clock as the last resort.

Proceeding clockwise in the network, Nodes 2, 3, 4, 5, and 6 have been provisioned to select the timing from the OC-N card installed in their West slot (in Figure 5-5, this is Slot 6) as their first choice for timing.

Note that each of these nodes receives the SSM value of PRS on their Slot 6, indicating that the timing being received is of Stratum 1 quality. Each node passes that along in the daisy chain to the next node in the ring.

In the reverse direction (counterclockwise), each MSPP node transmits the SSM value of DUS (Don't Use for Synchronization) back in the direction from which primary timing is received. This is required to prevent a timing loop from occurring somewhere in the network if the primary timing source fails. Note that at the Node 6–to–Node 1 connection, the PRS value is being both transmitted and received by each of the two network elements. This is a normal scenario. Node 1 always ignores the PRS being received on its Slot 6 OC-N card because that optical line is not among its possible timing references. Meanwhile, Node 6 makes use of the SSM information received on its Slot 12 only if its primary reference source fails.

Figure 5-5   *SSM Operation*

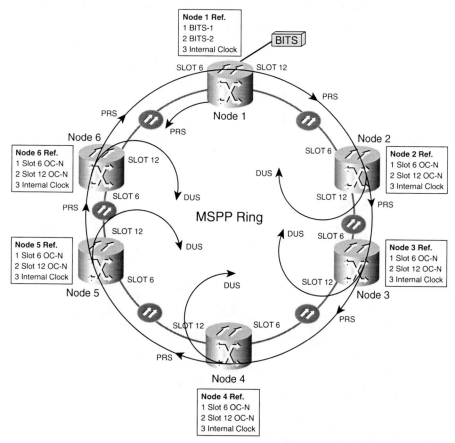

The example shown in Figure 5-5, in which each of the line-timed nodes has the same Reference 1, is one typical method of timing configuration in externally timed/line-timed designs. Another method for this type of network is to provision the ring OC-N interface that is closest to the primary reference source as Reference 1. For example, in Figure 5-5, Nodes 2 and 3 would have the OC-N interface in Slot 6 as their Reference 1, whereas Nodes 5 and 6 would have Slot 12 for their first selection. Node 4 could be provisioned either way, since it is the same number of "hops" in either direction back to the BITS-connected Node 1. Cisco recommends the latter selection criteria for their MSPP, the ONS 15454.

Network Management Considerations

Network management is an important consideration in MSPP network design. The designer has various options, depending upon the type of network environment in which the equipment is deployed. Fault detection, configuration and provisioning, performance monitoring, and security management are all functions of the network-management system.

In an enterprise campus or privately owned metro MSPP network, each MSPP node can have the capability to connect to the local LAN for remote management. Figure 5-6 shows this type of scenario. From the standpoint of fault recovery, this is an advantageous arrangement because no single network connection failure will isolate the network from the network-management system. Each MSPP node has its own connectivity for management purposes.

Figure 5-6   *An MSPP Network with LAN Connections to Each Node for Network Management*

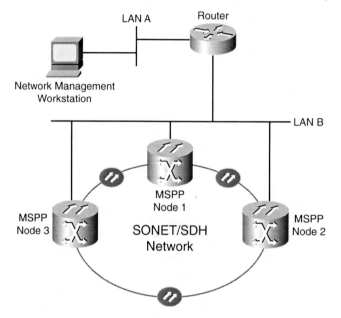

Figure 5-7 shows a more typical service provider scenario. Service provider MSPP networks typically have one or more network elements located in secure, company-owned locations, such as a telco central office. The other nodes are normally located on the premises of the service provider's customers or in common locations serving multiple customers. The node(s) located in the service provider's central office (sometimes referred to as a point of presence, or POP) can be linked back to the network operations center (NOC). This node is known as the gateway network element (GNE). The other nodes, called external network elements (ENEs), use the SONET in-band management channel, known as the data communications channel (DCC), to communicate to the NOC via the GNE. In this example, a single MSPP node (Node 1, the GNE) is attached to the management network; the other nodes, ENEs, use the DCC for management connectivity. If additional MSPP nodes in the network are located in facilities that the service provider owns and can also be used as GNEs, the single point of failure for network-management connectivity shown in this example could be eliminated.

Figure 5-7 *An MSPP Network with a LAN Connection to a Single GNE Node for Network Management*

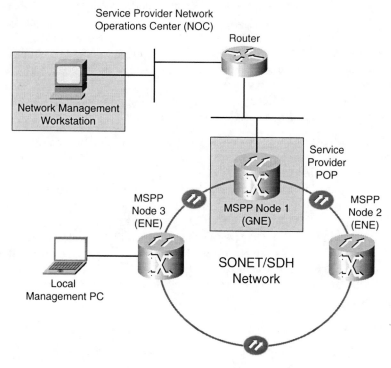

MSPP Network Topologies

Network-protection characteristics and total system payload capacity are determined largely by the type of network topology deployed. SONET MSPPs provide support for multiple available topologies, with varying levels of resiliency and traffic-carrying capabilities. These topologies include linear, ring, and mesh networks. This section covers these topologies and describes their operation and applications.

Linear Networks

One topology option for MSPP network design is the linear network. Linear networks are typically used when all the traffic to be serviced by the network is routed in a point-to-point manner, or point-to-point with intermediate add/drop locations. Linear networks that transport all traffic between two nodes are commonly referred to as point-to-point or terminal mode networks. If some traffic needs to be inserted into the network at intermediate sites, a linear add/drop network, sometimes called a linear add/drop chain, can be used.

Figure 5-8 shows an example of a terminal mode MSPP network. All traffic is routed from Node A to Node B. The connection between the nodes can be protected against facility failures, or it can be designed as unprotected if protection is not a requirement or is provided through some other network connection. If the terminal mode network will be unprotected, a single optical interface card, such as an OC-12 or OC-48, should be installed at each node (or terminal) and a single pair of optical fibers should be used to connect between them. In this case, a failure on the fiber cable span or of one of the optical interface cards causes traffic to be interrupted, and information will be lost unless it can be routed between the sites through some alternate connection.

In the protected scenario, two optical interfaces are needed in each terminal. These can be provided either on separate cards or on different ports of the same multiport card. A type of protection known as 1+1 is provided for this type of network. This means that one of the optical ports on each end (and the associated fiber connection) is set up as "working" or "active"; the secondary ports are the "protect" or "standby" ports. If a port failure or fiber facility failure occurs on the working link, the network switches the traffic to the protection link. To decrease the chances of a failure-related service interruption, terminal mode networks can be designed so that the working and protect ports are on different interface cards in each terminal, and the working and protect fiber pairs are physically routed in separate cable sheaths following diverse paths between the two endpoints.

Figure 5-8 *A Terminal Mode MSPP Network*

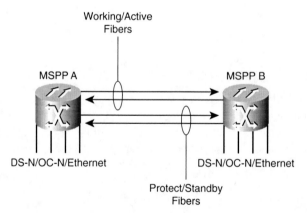

In a linear add/drop chain MSPP network, the intermediate nodes each have a connection to their neighboring MSPP nodes, one facing East and the other facing West. Again, these connections can be either 1+1 protected or unprotected. In the protected scenario, the intermediate add/drop nodes require two pairs of optical interfaces, with a pair facing East and a pair facing West. Figure 5-9 shows an example of a linear add/drop chain network.

Figure 5-9 *A Linear Add-Drop Chain MSPP Network*

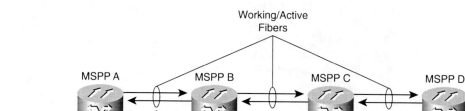

In modern MSPP networks, linear systems are rarely deployed. This is because of the availability of ring-based networks, which are more fault tolerant and often more economical to deploy. In a ring network, a single-node failure causes a service disruption only for traffic that terminates in that particular node. However, in a linear network, a single node failure can disrupt traffic servicing additional network locations. For example, in Figure 5-9, if MSPP C experienced a failure affecting the entire node (such as a power supply failure), traffic routed from MSPP A to MSPP D would be lost because this traffic transited MSPP C. In a ring network, this transit traffic would be rerouted via an alternate path. This network also can be provided with facility protection much more economically in a ring-based architecture. MSPP C would require a total of four costly optical interfaces if it were installed in a protected linear add/drop configuration, with two interfaces for East and two for West. However, a ring network would require only a single interface in each direction. Ring networks are covered in detail in the following sections.

UPSR Networks

A UPSR is a SONET protection topology used to maintain service across the optical network in case of a failure condition, such as a fiber span break or optical interface card failure. In a UPSR, traffic between MSPP nodes is safeguarded by carrying duplicate copies of the signals around diverse East and West paths from origin node to destination node, ensuring a swift failover to the opposite path. Because of this method of operation, a UPSR is sometimes referred to as a dual counter-rotating, self-healing ring.

UPSR Operation

To understand the operation of a UPSR, consider Figure 5-10, which shows a typical four-node OC-12 UPSR network. In this example, a client signal is being transported across the network from MSPP A to MSPP C. In the case of a bidirectional signal, such as a time-division multiplexing (TDM) circuit (such as a DS3), the signal consists of a transmit side and a receive side at each termination point; however, for clarity, Figure 5-10 shows only one direction. In the transmit direction from A to C, the signal enters MSPP A through a traffic interface card and is internally routed to the node's cross-connect matrix. Within the cross-connect matrix, the signal is bridged, or duplicated, and connected to both the East-facing and West-facing UPSR optical interface cards for transmission across the optical network. The circuit is assigned to travel on the UPSR on STS-1 path number 1, on both the East and West interfaces. At the intermediate MSPP nodes B and D, the signals are passed through. At MSPP C, a selector switch locks to the highest-quality signal from the two available STS-1 paths.

Figure 5-10 *UPSR Operation*

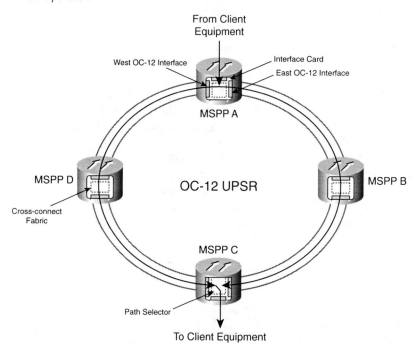

If a problem occurs with the selected signal, such as a high bit-error rate, the selector mechanism switches to the back-up path. Figure 5-11 shows the resulting path switch from a facility failure.

Because each signal transported on the UPSR consumes the same amount of bandwidth on both sides of the ring between the source and destination, the total amount of traffic carried (in Synchronous Transport Signal [STS] equivalents) is limited to the ring's transmission bandwidth, regardless of the origin and destination of the circuits. Therefore, an OC-12 UPSR is limited to transporting 12 total STS-1s, whereas an OC-48 UPSR can handle a total of 48 STS-1s.

Figure 5-11 *UPSR Path Switch Example*

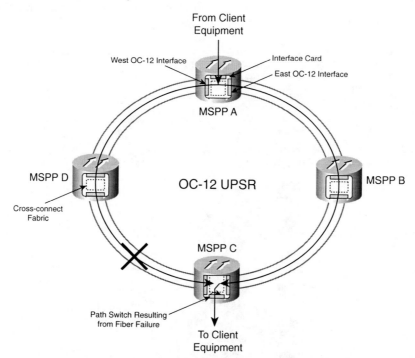

For example, in the OC-12 network in Figure 5-10, if an STS-3c signal was being carried between MSPP A and MSPP C, and was assigned to use the first three STS-1s (STS-1-1, STS-1-2, and STS-1-3), it would traverse the East path through MSPP B, as well as the West path through MSPP D on these same three STS-1s. This is a major difference between UPSR and bidirectional line switch ring (BLSR) topologies. Depending on the number of nodes in the network and the traffic pattern, a BLSR can accommodate more traffic than the transmission bandwidth.

Within the cross-connect matrix of the MSPP, a UPSR-protected circuit consumes three ports on the source and destination nodes, and two ports on the intermediate, "pass-through" nodes. Figure 5-10 shows this. At MSPP A, which is a source/destination node, the DS3 signal is mapped into an STS-1 and enters the cross-connect matrix on a single input/output (I/O) port. Because copies of this signal are to be transported around both the East and West sides of the ring, the signal is sent to two different I/O ports leaving the matrix. One of these will connect to the East optical line interface, the other to the West.

Meanwhile, at MSPP B, which is a pass-through node for this circuit, the signal enters one optical interface, gets passed through the matrix by entering one I/O port and exiting another, and then exits through the opposite optical interface. Figure 5-12 shows the details of how this works at each of the circuit-termination nodes. At each of these sites, three STS-1 I/O ports on the cross-connect matrix are consumed for this drop/add circuit.

For example, if this were a 60-Gbps cross-connect fabric, with a total of 1,152 STS-1 ports (1,152 × 51.84 Mbps = 59.71968 Gbps), this connection would use 3 of the 1152 ports, or approximately 0.26 percent of its STS-1 capacity.

Figure 5-12 *MSPP Cross-Connect Matrix Use at a UPSR-Protected Circuit Source or Destination Node*

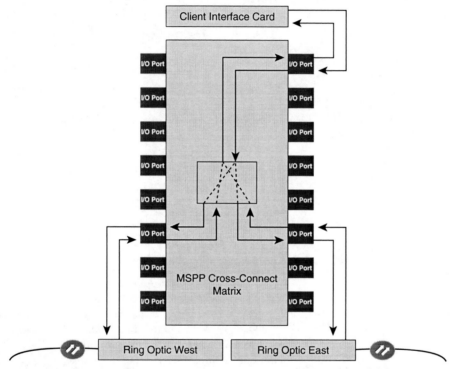

UPSRs can be built using any SONET line rate, including OC-3, OC-12, OC-48, OC-192, and OC-768. In an MSPP configured for UPSR operation, you need two optical interface cards, one designated East and the other designated West. A node design using a multiport optical interface card with separate ports used for East and West would be ill advised because a card-level failure (such as a card power supply failure) would result in losing both ring interfaces. That would cause service disruption on any traffic terminating at that node.

SONET MSPPs are typically configured so that an East-designated interface on one node is connected via the fiber-optic network links to the West-designated interface on the next adjacent node, and vice versa. For example, in an ONS 15454 MSPP OC-48 UPSR, a network designer might choose to designate an OC-48 interface card in chassis Slot 5 as the West interface, and an OC-48 interface card in chassis Slot 6 as the East interface. The card in Slot 5 would connect to a similar card in Slot 6 of the adjacent node to the west of this node, and the card in Slot 6 would connect to a similar card in Slot 5 of the adjacent node to the east. Figure 5-13 shows such a design. For simplicity, it is usually recommended that you keep this scheme consistent throughout the network.

Figure 5-13 *Typical UPSR Fiber Cable Interconnection*

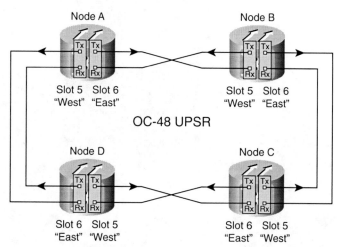

UPSR Applications

UPSRs are best suited for networks in which the traffic requirements call for most circuits to have a single common endpoint, or hub location. Some examples of this type of network requirement include the following:

- A local exchange carrier access ring in which voice and data traffic from residential or commercial areas is required to be carried to the local carrier central office for connection to the public switched telephone network (PSTN) or interoffice facilities

- A private corporate network in which Ethernet connectivity is needed from various branch offices to a main location or server farm

- A storage-area network extended via a SONET backbone for linking physically diverse locations to a common data/storage center

Figure 5-14 shows one such example. A carrier has deployed an OC-12 ring to service a neighborhood called East Meadow. This MSPP ring needs to service three locations: a residential area, a wireless provider cellular tower, and a business park.

The residential area, served by the Fairview Avenue Remote Terminal Cabinet, needs DS1 circuits for connecting a digital loop carrier (DLC) system and a DS3 for the trunk feeder for a remote digital subscriber line access multiplexer (DSLAM). The DLC is used to provide plain old telephone service (POTS) for the neighborhood. The remote DSLAM provides broadband DSL service for high-speed Internet connections. The POTS DS1s is transported back to the Oak Street central office for connection to the PSTN via a Class 5 switch. The DSLAM DS3 must connect to the Asynchronous Transfer Mode (ATM) network equipment at Oak Street.

Figure 5-14 *East Meadow OC-12 UPSR Network*

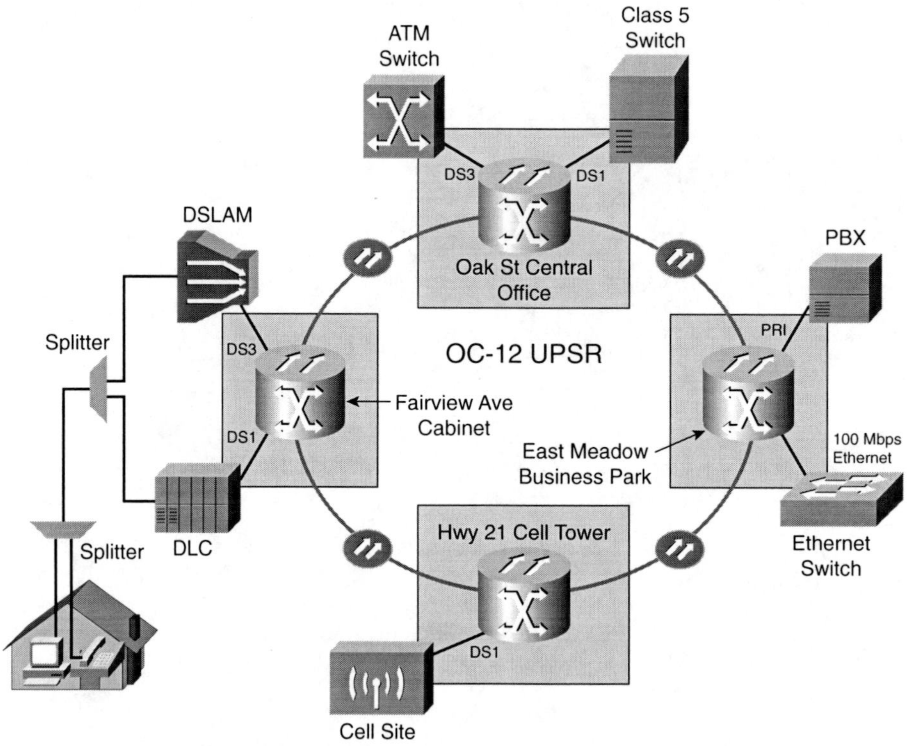

The Highway 21 cellular tower site has a requirement for multiple DS1 circuits. The wireless provider, which purchases this service from the carrier, needs to connect its cell site equipment to its mobile telephone switching office (MTSO). The carrier will connect the DS1s at the Oak Street office to interoffice facilities that route to the MTSO. Customers served by the East Meadow Business Park Site require a variety of services, including primary rate ISDN (PRI) circuits for PBX trunks and 100-Mbps Ethernet links for Internet service provider (ISP) connectivity. These circuits will also need to be routed back to the Oak Street central office to interface with the PSTN or the Internet.

Because all traffic for this network needs to be tied back into the Oak Street central office, a UPSR is an excellent choice for the MSPP network topology.

BLSR Networks

As in a UPSR, a BLSR network is designed to allow traffic continuity in case of a fiber span or MSPP node failure. The key differentiator versus UPSR, however, is that the BLSR is not capacity-limited to the transmission bandwidth of the ring. For example, depending on the number of nodes in the network and the traffic distribution, a two-fiber OC-192 BLSR can be used to transport more than 1000 STS-1 equivalents. BLSRs accomplish this via a method called spatial reuse. The same bandwidth, or traffic channels, can be used to transport different signals on different parts of the network. This is shown for both 2-Fiber and 4-Fiber BLSRs in the next sections.

2-Fiber BLSR Operation

In a 2-Fiber BLSR, a pair of optical fibers is used to connect the MSPP nodes on the ring, just as is the case with a UPSR. However, unlike in a UPSR, active traffic is connected through the ring in only one direction from source to destination: either the East route or the West route. In the opposite direction, spare capacity is held in reserve as "protection" bandwidth, in case a failure occurs in the assigned route. This is the case on every fiber link in a 2-Fiber BLSR: Exactly half the SONET bandwidth is used for assigned, working traffic; the other half is held in reserve for protection.

Consider the example four-node 2-Fiber OC-192 BLSR shown in Figure 5-15. Each fiber link in the ring, such as the one between Node 1 and Node 2, consists of two fibers transmitting in opposite directions. The bandwidth carried on each fiber is divided equally, with the first sequential half transporting production traffic and the second half reserved for protection. For example, in the link from Node 1 to Node 2 OC-192, the clockwise fiber path can use the first 96 STS-1s to map SONET circuits; STS-1s 97-192 are designated as the protection path for use during a fault-initiated switch. The counterclockwise fiber works in the same manner while transmitting the OC-192 signal in the reverse direction.

Figure 5-15 *A 2-Fiber BLSR Example*

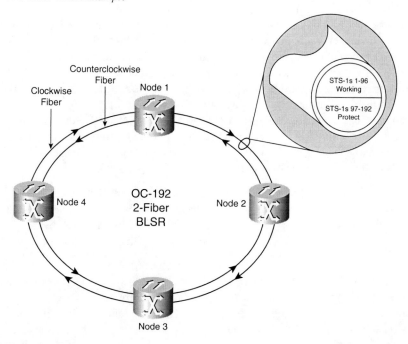

To illustrate the fault-recovery process for a BLSR, reference the scenario shown in Figure 5-16; this is the same network from Figure 5-15, with an STS-3c circuit mapped from Node 1 to Node 3 through the route passing through Node 2 using STS-1s 1 to 3.

Figure 5-16 *A 2-Fiber BLSR with an STS-3c Circuit-Routed from Node 1 to Node 3*

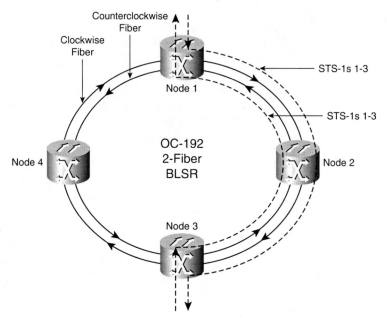

Suppose that a fiber failure occurs on the span between Node 1 and Node 2, with both clockwise and counterclockwise fibers affected as shown. In this case, the traffic would switch as shown in Figure 5-17. The original route for the transmit side (from Node 1 to Node 3) for the signal was rerouted from the clockwise fiber (STS-1s 1 to 3) to the counterclockwise fiber (STS-1s 97 to 99), reaching Node 3 through Node 4. The opposite side of the signal, which is the Node 3–to–Node 1 transmit direction, traverses the counterclockwise fiber (STS-1s 1 to 3) from Node 3 to Node 2. However, because of the fiber facility failure that has occurred between Node 1 and Node 2, this side of the signal is switched to STS-1s 97 to 99 on the clockwise fiber at Node 2, and reaches Node 1 through Node 3 and Node 4, as shown in the figure.

To facilitate the protection switching in a BLSR, the network elements in the ring use the K1 and K2 bytes in the line overhead to communicate protection-switch requests and status. These bytes are known collectively as the Automatic Protection Switch (APS) bytes. The switching is accomplished by the nodes that are adjacent to the failure (optical line or network node).

BLSR operation and protection-switch performance requirements are defined in a Telcordia (formerly BELLCORE) document called GR-1230-CORE. In this document, the total completion time for protection switching and traffic restoration is defined to be less than or equal to 50 ms, not including detection time. This requirement is for networks consisting of a maximum of 16 nodes with up to 1200 km of fiber.

Figure 5-17 *A 2-Fiber BLSR Line-Switch Example*

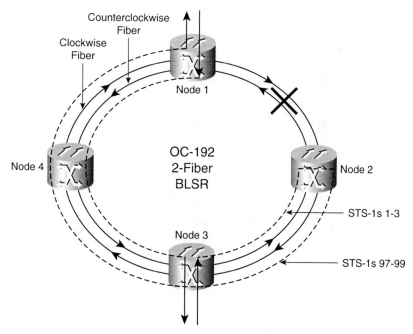

2-Fiber BLSR System Capacity

Because of the protection scheme employed in the 2-Fiber BLSR architecture, the total capacity of a fiber span between two network nodes in this configuration is equal to half the transmission bandwidth, or $n/2$, where n is the ring's STS-1 capacity (for example: $n = 48$ in an OC-48 ring, so the capacity of a single OC-48 span is equal to 48/2, or 24). In addition, the protection channel(s) held in reserve on the side of the ring opposite to the assigned working link is defined as $c + (n/2)$, where c is the assigned working STS-1 channel (or first STS-1 in a concatenated signal, such as a STS-12c) and n is the ring STS-1 capacity. For example, in an OC-48 ($n = 48$) 2-Fiber BLSR, the protection channel for a signal using STS-1 5 ($c = 5$) would be 5 + (48/2), or 29.

Because the total number of STS-1 equivalents that a 2-Fiber BLSR can carry depends on the number of spans, or nodes, in the ring, you need to define an equation for calculating this capacity. The variables in the equation are the number of nodes in the network, the bandwidth, and the traffic pattern for the ring, which is defined in the equation as the number of nonterminated, "pass-through" STS-1s transiting any node in the network. If you use X as the number of nodes, N as the network bandwidth, and A as the number of pass-through STS-1s in the ring, the capacity (or number of circuits that can be carried) for the ring is defined as $(N/2) \times X - A$. For each possible 2-Fiber BLSR configuration, Table 5-2 shows the capacity equations.

Table 5-2 *2-Fiber BLSR STS-1 Capacity Equations*

2F Ring Bandwidth	N	Working STS-1 Channels	Protect STS-1 Channels	Ring Capacity
OC-12	12	1–6	7–12	$(12/2) \times X - A$
OC-48	48	1–24	25–48	$(48/2) \times X - A$
OC-192	192	1–96	97–192	$(192/2) \times X - A$
OC-768	768	1–384	385–768	$(768/2) \times X - A$

To illustrate the application of these equations, consider the example in Figure 5-18. This is a representation of a four-node, 2-Fiber, OC-48 BLSR, with the traffic pattern as shown in the accompanying chart. The double-ended arrowed lines in the chart indicate the termination points for the various circuits shown. For example, the first line in the chart indicates that there are 10 STS-1 circuits routed on the ring from Node 1 to Node 2 (directly, with no pass-through nodes); the fifth line in the chart indicates that eight STS-1 circuits are riding the ring from Node 1 to Node 3, passing through Node 2.

Figure 5-18 *Traffic Routing in a 2-Fiber BLSR Example*

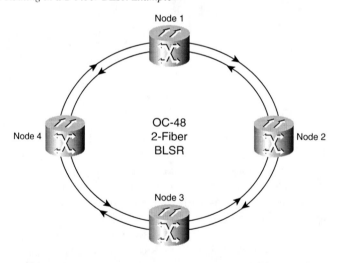

Circuits	Node 1	Node 2	Node 3	Node 4
10 STS-1	|←——————→|		
12 STS-1		|←——————→|	
6 STS-1			|←——————→|
18 STS-1	——————→|		|←——
8 STS-1		←···········PT············→		
4 STS-1			←·······PT········→	

PT = Pass Through

To calculate the circuit capacity of this ring, you can apply the formula $(N/2) \times X - A$, where $N = 48$ (because the ring is an OC-48), $X = 4$ (number of nodes), and $A = 12$. You arrive at the value of A by adding the number of pass-through circuits for each pair of source-destination defined circuit sets. Proceeding through the traffic chart from top to bottom, then, yields $0 + 0 + 0 + 0 + 8 + 4 = 12$ total STS-1 pass-through circuits. Substituting these values into the equation, you get a total possible ring capacity, given the current traffic mapping, of $(48/2) \times 4 - 12 = 96 - 12 = 84$ STS-1 circuits. Because the ring is currently carrying 58 STS-1s (the summation of all the current services), a network planner can determine that there is space for up to 26 additional STS-1 circuits on this network, depending on their future routing.

Protection Channel Access

In their normal mode of operation, BLSRs inherently have some unused capacity in the form of the traffic channels held in reserve for protection switching. For instance, in an OC-48 2-Fiber BLSR, channels 25 to 48 on all ring spans are normally unused. Most protection-switch events (such as fiber cable cut and human provisioning error) are unplanned and unpredictable; however, it is theoretically possible that a network will operate in a normal, switch-free state for many months, or even years. Because of this, many network operators choose to use the BLSR protection channels to carry live production traffic. In the case of telecom carrier networks, this capability, referred to as protection channel access (PCA), represents an opportunity to generate additional revenue.

The caveat to using PCA circuits is that they are typically unprotected, or preemptible. This is because, if a line switch occurs, these PCA circuits will be dropped in favor of the working (protected) traffic. Carriers (which use this arrangement to provide service) typically sell the bandwidth to their customers at a lower rate and offer a more lenient service-level agreement (SLA) to accommodate the possible service interruptions from line switches. This might be a perfectly acceptable arrangement to an end user, who is willing to forgo the reliability of SONET-protected traffic in favor of the economic benefits, or who may not require 99.999 percent network availability.

For example, a company that requires a 1-Gbps connection for 4 hours per day for regularly scheduled data backup might be a good candidate for a PCA-based service. The carrier could use a STS-24c or STS-1-21v PCA circuit on an OC-192 BLSR with Fibre Channel interfaces to provide this service to a customer, and still be capable of carrying a normal "full load" of protected traffic. Figure 5-19 shows this application. The STS-24c connection for a Fibre Channel application between Nodes 3 and 4 uses a PCA circuit on STS-1 paths 97 to 120, which is the protection bandwidth for the GigE link between Nodes 1 and 2. Those nodes use STS-1 paths 1 through 24. Of course, STS-1s 1 through 24 can also be used on the other links for carrying traffic because this is a BLSR network. If a line switch occurs on the GigE link, the Fibre Channel circuit is preempted to allow protection for the GigE.

Figure 5-19 *PCA Use in an OC-192 2-Fiber BLSR for a SAN Extension Application Example*

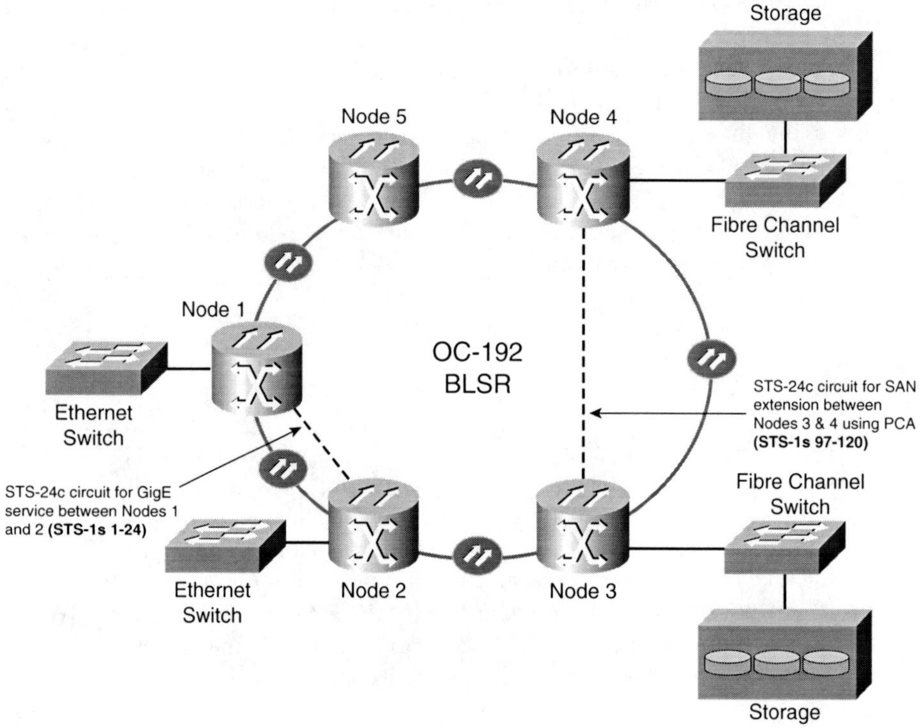

4-Fiber BLSR Operation

4-Fiber BLSRs provide twice the available bandwidth of a bandwidth-equivalent 2-Fiber BLSR, but they require additional costs and fiber facility resources to implement. As the name implies, four fiber strands (or two fiber pairs) are required between adjacent network elements in a 4-Fiber BLSR; two fibers are designated as working and two fibers are designated as protect. The entire bandwidth of the working pair can be assigned, whereas all the channels of the protection pair are held in reserve. This is shown in the 4-Fiber OC-192 BLSR in Figure 5-20, in which 192 STS-1 protected channels are available for assignment on the working fiber pair.

Figure 5-20 *A 4-Fiber BLSR Network Example*

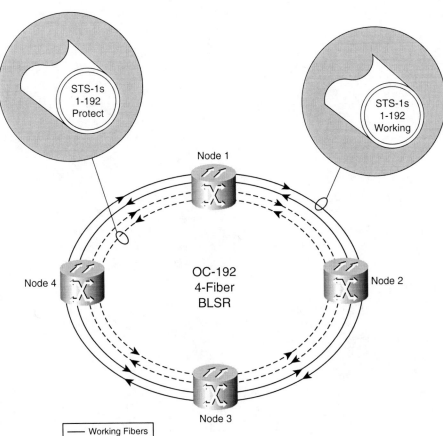

The 4-Fiber BLSR adds a layer of protection over the 2-Fiber version. In a 4-Fiber BLSR, a facility failure, which affects the working fibers, can be restored using a span switch. Figure 5-21 shows this. If the failure affects the working and protect fibers in the same segment, such as with a fiber cable cut, which severs all four strands, a ring switch occurs, as shown in Figure 5-22.

Figure 5-21 *A 4-Fiber BLSR Span Switch Example*

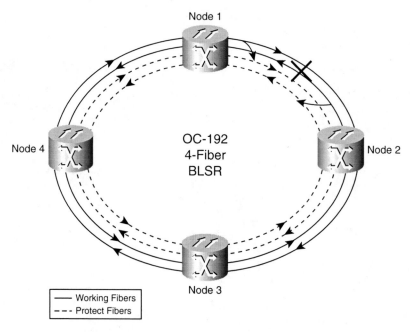

Figure 5-22 *A 4-Fiber BLSR Ring Switch Example*

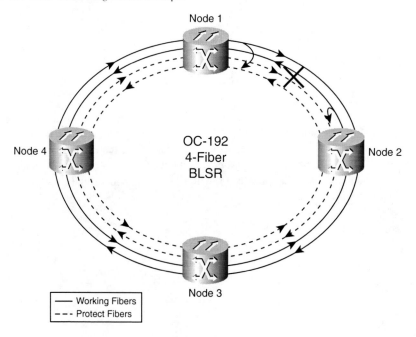

Although a 4-Fiber BLSR yields twice the bandwidth and offers enhanced protection from facility failures, it requires twice the number of optical fibers and also increases the number of optical interfaces required by a factor of 2. This topology is generally used in high-capacity long-distance carrier networks, such as those owned by U.S. interexchange carriers (IXCs). Typical MSPP implementations of 4-Fiber BLSRs are at the OC-48 and OC-192 levels.

4-Fiber BLSR System Capacities

The equation for determining the maximum channel-carrying capability of a 4-Fiber BLSR is similar to the 2-Fiber version, except that the ring bandwidth (N) is not halved in the formula. Table 5-3 shows the equations for performing these calculations, where X is the number of nodes in the 4-Fiber BLSR ring and A is the number of pass-through STS-1 circuits:

Table 5-3   *4-Fiber BLSR STS-1 Capacity Equations*

4F Ring Bandwidth	N	Working STS-1 Channels (Fiber Pair 1)	Protect STS-1 Channels (Fiber Pair 2)	Ring Capacity
OC-48	48	1–48	1–48	$(48 \times X) - A$
OC-192	192	1–192	1–192	$(192 \times X) - A$
OC-768	768	1–768	1–768	$(768 \times X) - A$

BLSR Applications

BLSR networks are well suited for distributed, nonhubbed traffic patterns because they can handle traffic requirements that exceed their transmission bandwidth. This characteristic of BLSRs makes it the topology of choice for metro or regional carrier networks in which TDM circuits for PSTN switch trunk links and data circuits for Ethernet or ATM networks are needed among multiple central offices.

Subtending Rings

In legacy SONET environments, traffic routing between two or more rings is typically accomplished using low-speed drop connections at digital cross-connect (DSX) panels, or through a separate device called a Digital Crossconnect System (DCS). This is because the older equipment was very limited in terms of ring terminations and cross-connect fabric sizes. With the current MSPP systems in the marketplace, these restrictions no longer apply. In many cases, a single MSPP add-drop multiplexer (ADM) can replace racks of cumbersome DSX panels or expensive DCS gear. This is done by terminating (or "closing") multiple SONET rings on a single MSPP chassis, also known as creating subtending rings.

In a large MSPP network, a carrier can use a single MSPP node to aggregate multiple access rings into a single backbone interoffice ring, as shown in Figure 5-23. In this figure, Node A is used to terminate and interconnect several subtended OC-12/48 UPSR access rings into a core OC-192 BLSR for metro or regional transport. These access rings can consist of the same type of MSPP chassis as the core ring, or they can include compatible miniature MSPP systems, which many manufacturers include in their portfolio for the access and customer premise market. For example, the Cisco ONS 153xx series includes several models for both the SONET and SDH markets, which are geared toward OC-3 or OC-12 (and even OC-48) access rings for aggregation of TDM and Ethernet connections. The 153xx series, which includes the 15302, 15305, 15327, and 15310, is compatible with the ONS 15454, which is the Cisco flagship SONET/SDH MSPP.

Figure 5-23 *An OC-192 Metro Core Ring with Multiple Subtended Rings*

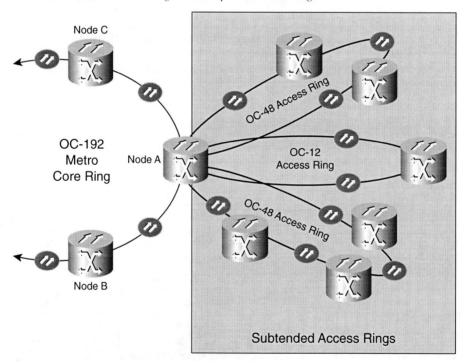

This MSPP capability can also be leveraged (in some systems) to bridge large backbone rings without costly interconnection solutions, as in the case of two OC-192 BLSRs terminated in a single MSPP chassis (see Figure 5-24).

Figure 5-24 *Two OC-192 Metro Core Rings Interconnected at a Single MSPP Node*

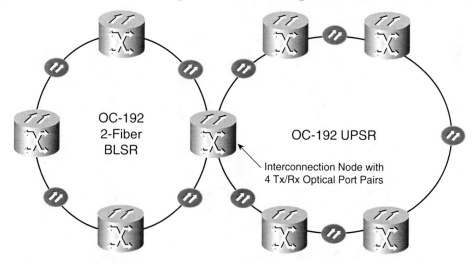

Subtending Shelves

Because of the relatively high cost of high-bandwidth and long-reach optical interfaces (for example, OC-192 long-reach 1550-nm cards are typically some of the most expensive MSPP blades), it is sometimes advantageous to stack or subtend additional shelves from a primary shelf at an MSPP site. For example, if a particular site has a heavy concentration of DS1 interfaces, along with other traffic requirements, it might be more economical to place the DS1 cards in a second shelf and use lower-rate optics, such as OC-3 or OC-12, to interconnect to the primary shelf for routing the DS1s onto the network. Figure 5-25 shows such an arrangement, with a primary node participating in an OC-192 ring with a subtended shelf, interconnected to the primary OC-192 node through OC-12 optical interfaces, used for a high concentration of DS1 drops. This allows the interface slots in the primary shelf to be used for other services, such as DS3, OC-3/12/48, and GigE.

Figure 5-25   *An OC-192 MSPP Ring Node with a Subtended OC-12 Chassis for DS1 Handoffs*

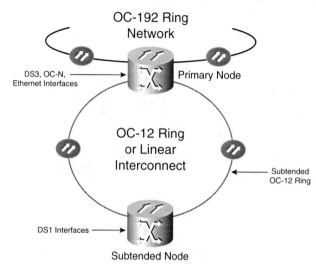

Ring-to-Ring Interconnect Diversity

One potential issue with connecting two or more rings at a single site (refer to Figure 5-24) is the creation of a nonredundant point of failure—the interconnection node. If a serious problem, such as a loss of power or natural disaster, occurs at this location, traffic passing between the ring networks through this node will be lost. A method of connecting rings to avoid this issue is known as Dual Ring Interconnect (DRI). In a DRI configuration, rings can be connected at two nodes to eliminate the single point of failure. If these two interconnection points are geographically diverse, such as in different carrier central offices, an even greater level of fault protection is provided. DRI can be achieved with UPSRs, BLSRs, or a combination of the two. DRI can be accomplished in two methods (see Figure 5-26):

- **Traditional DRI**—Uses four nodes to perform the interconnection, with two nodes on each of the separate rings optically linked to form the protection paths.

- **Integrated DRI**—Uses only two nodes, with the two nodes at the junction points participating simultaneously in each of the two rings. This requires a pair of optical links between the two interconnect nodes.

  Figure 5-27 shows an example of how this is installed with optical interfaces and fibers. Because the integrated DRI solution uses fewer nodes and optical interfaces, it can be installed with less capital outlay and with fewer network elements to manage.

Figure 5-26 *Traditional and Integrated DRI Examples*

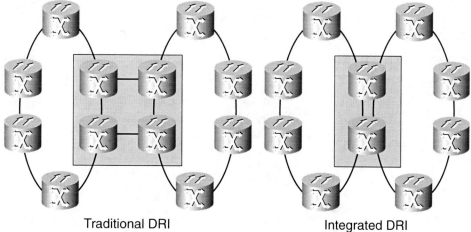

Traditional DRI   Integrated DRI

Figure 5-27 *Shelf Fiber Interconnections for Integrated DRI Example*

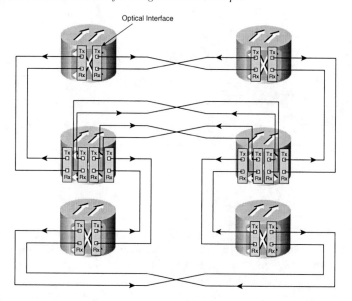

Figure 5-28 shows the operation of DRI in a UPSR, in the integrated DRI configuration. Nodes 4 and 5 are the interconnection nodes for the two rings. As shown in the figure, a signal entering Node 1 and bound for Node 8 is bridged in the cross-connect fabric at Node 1. At Node 5, the path selector chooses the primary signal. Using a circuit-creation method known as drop and continue, this signal is also bridged to Node 4 for protection. Similarly, the second copy of the signal bridged at Node 1 is used as the secondary or backup and is dropped to the path selectors at Nodes 4 and 5. This ultimately allows for two copies, a working signal and a protect signal, to be present at the path-selector switch on the receive side at Node 8. This duplication of signals at the interconnect nodes avoids service interruption from the failure of either of these network elements.

Figure 5-28 *Integrated DRI Configuration*

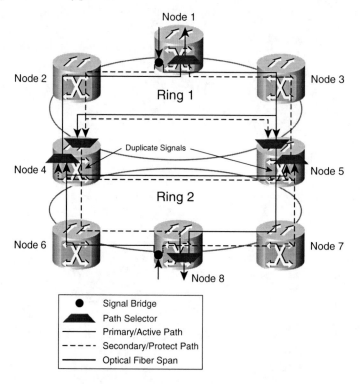

Mesh Networks

A mesh network is any set of interconnected locations (for example, MSPP nodes) with at least one loop pathway. Each site in the mesh can be reached from any other site by at least two completely separate routes. Any network other than a point-to-point or purely linear network can be considered a mesh, even a UPSR or BLSR network. However, some of the distinctive features and advantages of a mesh network include these:

- Mesh networks can contain spans with varying SONET optical rates, or even dense wavelength-division multiplexing (DWDM) spans for high-capacity routes.

- Reduced dedicated protection bandwidth requirements because of distributed backup capacity.

- Automated, algorithmic route optimization, requiring the network operator to specify only the ingress and egress points of traffic connections to establish service.

Figure 5-29 shows an example of a mesh network. The interconnecting fiber spans can contain a variety of SONET (OC-N) line rates and DWDM links.

Figure 5-29 *A Mesh Network*

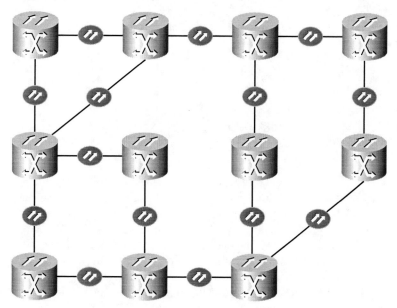

Cisco provides a supported mesh networking topology on the ONS 15454 MSPP known as a path-protected mesh network (PPMN). Coupled with the automatic circuit-routing feature of the 15454, PPMN provides an effective and efficient means of rapidly provisioning protected traffic across complex meshed networks. Each working and protection path pair effectively forms a virtual UPSR for the circuit. Figure 5-30 shows an example of this.

Figure 5-30   *Protected Circuit Routing in a Mesh Network Example*

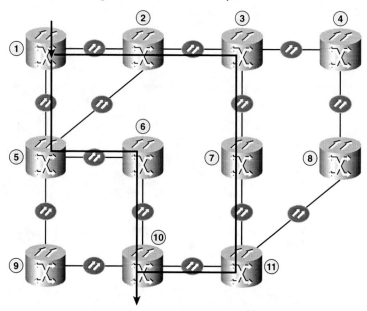

In this case, a protected circuit is specified to connect Nodes 1 and 10. The network software determines the shortest path between those nodes and routes the circuit accordingly, automatically creating the pass-through cross-connections at Nodes 5 and 6. Traffic between the terminating nodes initially traverses this link. The optimal protection path is calculated to be through Nodes 2, 3, 7, and 11, and pass-through cross-connections are created at those locations as well. If the primary link into Node 10 fails, the MSPP performs a protection switch and selects traffic on the protection link from Node 11 instead of the primary link from Node 6. Note that the loop formed by Nodes 1, 2, 3, 7, 11, 10, 6, and 5 is a UPSR for this particular circuit.

Summary

In this chapter, you learned some considerations in the design of MSPP networks, and you explored the operation and application of various MSPP network topologies.

With regard to network resiliency and protection, multiple mechanisms and design features are available with MSPPs to ensure high availability for critical services. These include the use of redundant power feeds, common control cards, tributary interfaces, synchronization sources, and MSPP chasses. In addition, system reliability can be substantially improved through diverse cable routing and ring network topologies.

MSPP network topologies include linear networks, UPSRs, BLSRs (both 2-Fiber and 4-Fiber), mesh networks, subtended rings and nodes, and diverse ring interconnections. Whereas linear networks are not commonly used in current MSPP networks, UPSR and BLSR networks are standard configurations in service-provider and enterprise networks. UPSRs are advantageous because of their design simplicity and fast protection switch times, but they are inefficient in terms of bandwidth use for meshed traffic patterns between MSPP network sites. UPSRs are typically used in access networks, where all traffic hubs to a central site (such as a telco central office). BLSRs, which have slower protection-switching times because of the required internode switching protocol, can accommodate a much higher amount of traffic in a meshed traffic pattern, such as a service provider core network.

This chapter covers the following topics:

- ONS 15454 Shelf Assembly
- Electrical Interface Assemblies
- Timing, Communications, and Control Cards
- Cross-Connect Cards
- Alarm Interface Controller Card
- SONET/SDH Optical Interface Cards
- Ethernet Interface Cards
- Electrical Interface Cards
- Storage Networking Cards
- MSPP Network Design Case Study

CHAPTER 6

MSPP Network Design Example: Cisco ONS 15454

The Cisco ONS 15454 is a highly flexible and highly scalable multiservice Synchronous Optical Network (SONET)/Synchronous Digital Hierarchy (SDH)/dense wavelength-division multiplexing (DWDM) platform. Service providers and enterprise customers use the ONS 15454 to build highly available transport networks for time-division multiplexing (TDM), Ethernet, storage extension, and wavelength services. In this chapter, you will learn the major components of the ONS 15454 system, including these:

- Shelf assembly
- Common control cards
- Electrical interface cards
- Optical interface cards
- Ethernet interface cards
- Storage networking cards

In addition, an example network demand specification is used throughout this chapter to demonstrate Multiservice Provisioning Platform (MSPP) network design using the ONS 15454.

ONS 15454 Shelf Assembly

The ONS 15454 Shelf Assembly is a 17-slot chassis with an integrated fan tray, rear electrical terminations, and front optical, Ethernet, and management connections. Slots 1–6 and 12–17 are used for traffic interface cards; Slots 7–11 are reserved for common control cards. Slots 1–6, on the left side as you face the front of the shelf, are considered Side A; Slots 12–17 are considered Side B. This distinction is important for planning backplane interface types (for electrical card terminations), as well as protection group planning. These issues are covered later in this chapter. The bandwidth capacity of each of the 12 traffic slots varies from 622 Mbps to 10 Gbps, depending upon the type of cross-connect card used. See the section titled "Cross-Connect Cards" later in this chapter for a discussion of the various types available. Table 6-1 summarizes the card slot functions and bandwidth capacities for the ONS 15454 shelf assembly.

Table 6-1 *ONS 15454 Shelf Assembly Slot Functions and Bandwidth Capacities*

Slot Number	Shelf Side	Slot Use	Slot Bandwidth (XCVT System)	Slot Bandwidth (XC10G or XC-VXC-10G System)
1	A	Multispeed high-density slot	622 Mbps/STS-12	2.5 Gbps/STS-48
2	A	Multispeed high-density slot	622 Mbps/STS-12	2.5 Gbps/STS-48
3	A	Multispeed high-density slot; *N*-protection slot for 1:*N* protection groups	622 Mbps/STS-12	2.5 Gbps/STS-48
4	A	Multispeed slot	622 Mbps/STS-12	2.5 Gbps/STS-48
5	A	High-speed slot	2.5 Gbps/STS-48	10 Gbps/STS-192
6	A	High-speed slot	2.5 Gbps/STS-48	10 Gbps/STS-192
7	—	TCC Slot	—	—
8	—	XC Slot	—	—
9	—	AIC Slot	—	—
10	—	XC Slot	—	—
11	—	TCC Slot	—	—
12	B	High-speed slot	2.5 Gbps/STS-48	10 Gbps/STS-192
13	B	High-speed slot	2.5 Gbps/STS-48	10 Gbps/STS-192
14	B	Multispeed slot	622 Mbps/STS-12	2.5 Gbps/STS-48
15	B	Multispeed high-density slot; *N*-protection slot for 1:*N* protection groups	622 Mbps/STS-12	2.5 Gbps/STS-48
16	B	Multispeed high-density slot	622 Mbps/STS-12	2.5 Gbps/STS-48
17	B	Multispeed high-density slot	622 Mbps/STS-12	2.5 Gbps/STS-48

ONS 15454 Shelf Assembly Backplane Interfaces

Backplane interfaces of the ONS 15454 chassis can be divided into four general areas:

- Power terminal area
- Alarm and chassis ground area

- Side A Electrical Interface Assembly

- Side B Electrical Interface Assembly

Figure 6-1 shows a diagram that identifies the location of each of these areas as you face the rear of the shelf assembly. Electrical Interface Assemblies (EIAs) are required for terminating electrical traffic signals, such as DS1s and DS3s, on the shelf. EIAs are covered in the next section.

Figure 6-1   *ONS 15454 Backplane Interfaces*

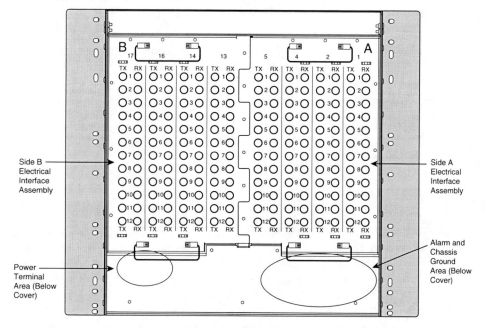

The power terminal area consists of four power terminal screws on the lower-left side. The RET1/BAT1 terminals are for the A power connection; the RET2/BAT2 terminals are for the B connection. These connections are redundant; either can power the entire shelf. The A and B designations do not refer to the A and B sides of the shelf.

The alarm and chassis ground area is located in the rear of the chassis on the lower right side. It has the following terminations:

- **Frame ground terminals**—Two terminals with kepnuts are provided for ground-wire lug connection. This connection ensures that the shelf assembly is at the same electrical potential with the office ground.

- **BITS**—(Building Integrated Timing Supply) This consists of four wire-wrap pin pairs for connection to a BITS or for wiring out to external equipment when the BITS-Out feature is used to supply timing from the ONS 15454.

- **LAN**—(Local-area network) A LAN connects the MSPP node to a management workstation or network. It consists of four wire-wrap pin pairs; typically, two pairs are used.

- **Environmental alarms**—Sixteen wire-wrap pin pairs are provided for external alarms and controls. The Alarm Interface Controller card, covered later in this chapter, is required to use these connections.

- **ACO**—Alarm cutoff, a wire-wrap pin pair that is used to deactivate the audible alarms caused by the contact closures on the shelf backplane. This operation is described in the "Timing, Communications, and Control Cards" section, later in this chapter.

- **Modem**—Four pairs of wire-wrap pins are provided for connecting the ONS 15454 to a modem for remote management.

- **Craft**—Two wire-wrap pin pairs are provided for a TL1 craft-management connection. VT100 emulation software is used to communicate with the system by way of this connection.

- **Local alarms**—Eight wire-wrap pin pairs are used for Critical, Major, Minor, and Remote audible and visual alarms.

EIAs

EIAs are backplane connector panels that must be equipped on the ONS 15454 chassis if you will provide DS1, DS3, or EC-1 services from the node. EIAs are made to fit on either Side A or Side B of the upper section of the backplane. Because Side A consists of the slots that are on the left as you face the front of the shelf assembly, the Side A EIA is installed on the right side as you face the rear. Likewise, the Side B EIA is installed on the left side facing the rear of the shelf assembly.

Various EIAs are available from Cisco, and each side of the shelf can be independently equipped with any type. If no electrical terminations are required for a shelf side, it can be equipped with a blank backplane cover. If there is initially no requirement for electrical connections, and the requirement appears later, the blank backplane cover can be removed and replaced with an EIA while the shelf is powered and in service. You select an EIA based on the type and quantity or density of connections required. These EIA types are available:

- **BNC (Bayonet Neill-Concelman) EIA**—Used for DS3 (clear channel), DS3 Transmux (channelized DS3), and EC-1 services. This EIA type has been made obsolete by newer versions, and Cisco no longer sells it.

- **High-Density (HD) BNC EIA**—Offers the same services as the BNC EIA, with twice the number of available connections.

- **HD mini-BNC EIA**—Offers the same services as the BNC and HD BNC EIAs, with twice as many connections as the HD BNC EIA and four times as many as the BNC EIA. Either the HD mini-BNC EIA or one of the Universal Backplane Interface Connector (UBIC)-type EIAs are required for cabling the HD DS3/EC1 cards.

- **AMP Champ EIA**—Used for DS1 services only.

- **SMB (Sub-Miniature B) EIA**—Can be used for any type of electrical termination, including DS1, DS3, DS3 Transmux, and EC-1.

- **UBIC-V and UBIC-H EIAs**—Can be used for any type and any density electrical interface card termination. These are the most flexible EIA type. The difference between the two is in the orientation of the cable connectors, either vertical (UBIC-V) or horizontal (UBIC-H). Either the UBIC-V or the UBIC-H is required for cabling the high-density DS1 cards.

Table 6-2 provides a summary of the available EIAs, with their associated connector types, supported working shelf slots, and supported electrical interface cards.

Table 6-2   *ONS 15454 Electrical Interface Assemblies*

EIA Type	Cards Supported	A-Side Connectors	A-Side Slots	B-Side Connectors	B-Side Slots
BNC	DS3-12E DS3N-12E DS3XM-6 DS3XM-12 EC1-12	24 pairs of BNC connectors (2 slots; 12 pairs/slot)	Slot 2 Slot 4	24 pairs of BNC connectors (2 slots; 12 pairs/slot)	Slot 14 Slot 16
High-Density BNC	DS3-12E DS3N-12E DS3XM-6 DS3XM-12 EC1-12	48 pairs of BNC connectors (4 slots; 12 pairs/slot)	Slot 1 Slot 2 Slot 4 Slot 5	48 pairs of BNC connectors (4 slots; 12 pairs/slot)	Slot 13 Slot 14 Slot 16 Slot 17
High-Density mini-BNC	DS3-12E DS3N-12E DS3XM-6 DS3XM-12 DS3/EC1-48 EC1-12	96 pairs of mini-BNC connectors	Slot 1 Slot 2 Slot 4 Slot 5 Slot 6	96 pairs of mini-BNC connectors	Slot 12 Slot 13 Slot 14 Slot 16 Slot 17

continues

Table 6-2 *ONS 15454 Electrical Interface Assemblies (Continued)*

EIA Type	Cards Supported	A-Side Connectors	A-Side Slots	B-Side Connectors	B-Side Slots
AMP Champ	DS1-14 DS1N-14	6 AMP Champ connectors (1 connector per slot)	Slot 1 Slot 2 Slot 3 Slot 4 Slot 5 Slot 6	6 AMP Champ connectors (1 connector per slot)	Slot 12 Slot 13 Slot 14 Slot 15 Slot 16 Slot 17
SMB	DS1-14 DS1N-14 DS3-12E DS3N-12E DS3XM-6 DS3XM-12 EC1-12	84 pairs of SMB connectors (6 slots; 14 pairs/slot)	Slot 1 Slot 2 Slot 3 Slot 4 Slot 5 Slot 6	84 pairs of SMB connectors (6 slots; 14 pairs/slot)	Slot 12 Slot 13 Slot 14 Slot 15 Slot 16 Slot 17
UBIC-V	DS1-14 DS1N-14 DS1-56 DS3-12E DS3N-12E DS3XM-6 DS3XM-12 DS3/EC1-48 EC1-12	8 pairs of SCSI connectors—vertical orientation	Slot 1 Slot 2 Slot 3 Slot 4 Slot 5 Slot 6	8 pairs of SCSI connectors—vertical orientation	Slot 12 Slot 13 Slot 14 Slot 15 Slot 16 Slot 17
UBIC-H	DS1-14 DS1N-14 DS1-56 DS3-12E DS3N-12E DS3XM-6 DS3XM-12 DS3/EC1-48 EC1-12	8 pairs of SCSI connectors—horizontal orientation	Slot 1 Slot 2 Slot 3 Slot 4 Slot 5 Slot 6	8 pairs of SCSI connectors—horizontal orientation	Slot 12 Slot 13 Slot 14 Slot 15 Slot 16 Slot 17

Figure 6-2 shows two examples of EIAs, the HD BNC and the AMP Champ.

Figure 6-2   *EIA Examples: HD BNC and AMP Champ*

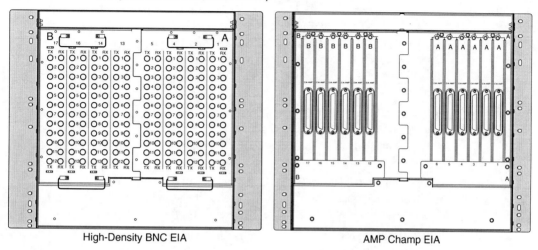

High-Density BNC EIA   AMP Champ EIA

Timing, Communications, and Control Cards

The Timing, Communications, and Control (TCC) cards are required for operation of the ONS 15454 MSPP system and are installed in a redundant pair in shelf Slots 7 and 11. Two current versions are available from Cisco: the Advanced Timing, Communications, and Control card (TCC2), and the Enhanced Advanced Timing, Communications, and Control card (TCC2P). Both perform the same basic functions, but the TCC2P is an updated version of the TCC2 and includes some security enhancements and additional synchronization options that are not available in the TCC2. Both cards have a purple square symbol on their faceplates, which corresponds to matching symbols on the front of the ONS 15454 shelf assembly. This serves as an aid in easily identifying the correct location to install the card. TCC cards are the only card type allowed in Slots 7 and 11, and both slots should always be equipped. Cisco does not support the operation of the ONS 15454 MSPP system with only a single TCC card installed. Although the system technically can function with only a single card, the second card is necessary for redundancy and to allow for continuity of system traffic in case of a failure or reset of the primary card. The system raises the "Protection Unit Not Available" (PROTNA) alarm if the secondary TCC card is not installed.

Two earlier versions of the TCC2 and TCC2P cards exist, called simply the Timing, Communications, and Control (TCC) card and the TCC Plus (TCC+). These older-version TCC cards provide similar functionality to the current cards, but they are much more limited in processing power. Although they may be installed in some existing systems, Cisco no longer produces the TCC or TCC+ versions.

The TCC2 and TCC2P cards perform a variety of critical system functions, which are as follows:

- **System initialization**—The TCC2s/TCC2Ps are the first cards initially installed in the system and are required to initialize system operation.

- **Data communications channels (DCCs) termination and processing**—The DCCs, which allow for communications and remote management between different MSPP network elements, are processed by the TCC2/TCC2P card. The TCC card automatically detects DCC-connected nodes.

- **Software, database, and Internet Protocol (IP) address storage**—The node database, system software, and assigned system IP address (or addresses) are stored in nonvolatile memory on the TCC2/TCC2P card, which allows for quick restoration of service in case of a complete power outage.

- **Alarm reporting**—The TCC2/TCC2P monitors all system elements for alarm conditions and reports their status using the faceplate and fan tray light-emitting diodes (LEDs). It also reports to the management software systems.

- **System timing**—The TCC card monitors timing from all sources (both optical and BITS inputs) for accuracy. The TCC selects the timing source, which is recovered clocking from an optical port, a BITS source, or the internal Stratum level 3 clock.

- **Cell bus origination/termination**—The TCC cards originate and terminate a cell bus, which allows for communication between any two cards in the node and facilitates peer-to-peer communication. These links are important to ensure fast protection switching from a failed card to a redundant-protection card.

- **Diagnostics**—System performance testing is enabled by the TCC cards. This includes the system LED test, which can be run from the faceplate test button on the active TCC or using Cisco Transport Controller (CTC).

- **Power supply voltage monitoring**—An alarm is generated if one or both of the power-supply connections is operating at a voltage outside the specified range. Allowable power supply voltage thresholds are provisionable in CTC.

Figure 6-3 shows a diagram of the faceplate of the TCC2/TCC2P card.

The two card types are identical in appearance, with the exception of the card name labeling. The cards have 10 LEDs on the faceplate, including the following:

- **FAIL**—This red LED that is illuminated during the initialization process. This LED flashes as the card boots up. If the LED does not extinguish, a card failure is indicated.

- **ACT/STBY**—Because the TCC cards are always installed as a redundant pair, one card is always active while the other is in standby state. The active TCC2/TCC2P has a green illuminated ACT/STBY LED; the standby card is amber/yellow.

- **PWR A and B**—These indicate the current state of the A- and B-side power-supply connections. A voltage that is out of range causes the corresponding LED to illuminate red; an acceptable level is indicated with green.

Figure 6-3   *TCC2/TCC2P Faceplate Diagram*

- **CRIT, MAJ, and MIN**—These indicate the presence of a critical (red), major (red), or minor (amber) alarm (respectively) in the local ONS 15454 node.

- **REM**—This LED turns red if an alarm is present in one or more remote DCC-connected systems.

- **SYNC**—This green SYNC lamp indicates that the node is synchronized to an external reference.

- **ACO**—The Alarm Cutoff lamp is illuminated in green if the ACO button on the faceplate is depressed. The ACO button deactivates the audible alarm closure on the shelf backplane. ACO stops if a new alarm occurs. If the alarm that originated the ACO is cleared, the ACO LED and audible alarm control are reset.

The faceplate also has two push-button controls. The LAMP TEST button initiates a brief system LED test, which lights every LED on each installed card and the fan tray LEDs (with the exception of the FAN FAIL LED, which does not participate in the test).

The RS-232 and Transmission Control Protocol/Internet Protocol (TCP/IP) connectors allow for management connection to the front of the ONS 15454 shelf. TCP/IP is an RJ-45 that allows for a 10Base-T connection to a PC or workstation that uses the CTC management system. A redundant local-area network (LAN) connection is provided via the backplane LAN pins on the rear of the shelf. The RS-232 is an EIA/TIA-232 DB9-type connector used for TL1 management access to the system. The CRAFT wire-wrap pins on the shelf backplane duplicate the functionality of this port.

Cross-Connect Cards

The cross-connect (XC) cards are required for operation of the ONS 15454 MSPP system and are installed in a redundant pair in shelf Slots 8 and 10. Three versions currently are available from Cisco: XCVT, XC10G, and XC-VXC-10G. Each version has both a high-order (STS-N) cross-connect fabric and a low-order virtual tributary (VT1.5) fabric. All three perform the same basic functions, but they feature varying cross-connect capacities. Table 6-3 summarizes the high-order and low-order cross-connect capacities of the XCVT, XC10G, and XC-VXC-10G cards. The meanings of these capacities are covered later in this section.

Table 6-3   *ONS 15454 XC Card Capacities*

Cross-Connect Card	High-Order (STS-1) Capacity	Low-Order (VT1.5) Capacity
XCVT	288	672
XC10G	1152	672
XC-VXC-10G	1152	2688

The XCVT, XC10G, and XC-VXC-10G have green cross symbols on their faceplates, which correspond to matching symbols on the front of the ONS 15454 shelf assembly. This serves as an aid in easily identifying the correct location to install the cards. The XC card types are the only cards allowed in Slots 8 and 10 for the MSPP, and both slots should always be equipped. Cisco does not support the operation of the ONS 15454 MSPP system with only a single XCVT, XC10G, or XC-VXC-10G installed. Although the system will technically function with only a single card, the second card is necessary for redundancy and to allow for continuity of system traffic.

An earlier version of the cross-connect card is called simply the XC card. This older version provides a 288 STS-1 fabric, which is the same size as the high-order fabric in the Cross-Connect Virtual Tributary (XCVT). However, the XC card does not support low-order (VT1.5) grooming, and systems that are equipped with the XC card cannot drop DS1 circuits. Although it can be installed in some existing systems, Cisco no longer produces the XC card.

The XCVT, XC10G, and XC-VXC-10G cards have only two faceplate LEDs. The red FAIL LED illuminates during a reset and flashes during the boot process to indicate that the card's processor is not ready for operation. If the FAIL LED does not extinguish, this is an indication that the card has failed and needs to be replaced. The ACT/STBY LED indicates whether the card is the active (green) or standby (amber) card in the redundant pair.

Cross-Connect Card Bandwidth

Each of the three cross-connect card types has a high-order (STS-1) and low-order (VT1.5) capacity, as shown in Table 6-3. For example, the XC10G card has an STS-1 capacity of 1152 STS terminations. Each STS-1 circuit requires at least two terminations, one for entering (ingress) and one for exiting (egress) the cross-connect matrix. Therefore, a single Bidirectional Line Switch Ring (BLSR) circuit, a pass-through circuit, or an unprotected circuit consumes two terminations of the available capacity. In a Unidirectional Path-Switched Ring (UPSR) circuit-termination node, an STS-1 circuit consumes three matrix terminations because of the signal bridging that occurs to enable UPSR protection. As an example, a DS3 circuit in a UPSR termination node would use three STS-1 terminations (of the available 1152 for the XC10G or XC-VXC-10G, or the available 288 for the XCVT).

VT1.5-Level cross-connections are made via logical STS ports in the VT matrix of the various cross-connect cards. The XCVT and XC10G VT matrices have 24 logical STS ports (24 STS ports × 28 VT1.5/port = 672 VT capacity); the XC-VXC-10G has 96 logical STS ports (96 STS ports × 28 VT1.5/port = 2688 VT capacity). To fully use the VT matrix capacity, each STS port must carry 28 VT1.5 circuits. Because of this, stranded capacity can occur when using, for example, a DS1-14 card as a circuit source/destination. Because the 14 DS1s from the DS1-14 card's 14 ports are carried to the cross-connect matrix on an STS-1, the remaining 14 VT1.5 capacity within the STS-1 is unused on the VT cross-connect matrix.

To further aid in understanding the way the cross-connect matrixes operate on the ONS 15454, see Figures 6-4 and 6-5. Figure 6-4 shows a VT1.5 circuit from a DS1-14 card in a BLSR termination node; Figure 6-5 shows the same circuit in a UPSR termination node. Note the matrix use information shown for each of the figures.

NOTE   The transition connections between the STS (high-order) matrix and the VT (low-order) matrix are not counted when calculating ports used on the STS (high-order) matrix.

Figure 6-4   *VT Matrix Use for a DS1 Circuit in a BLSR Termination Node*

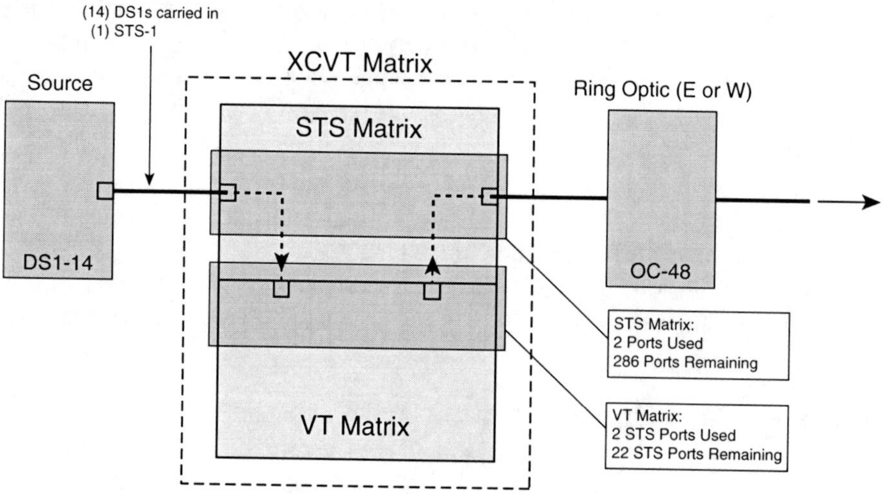

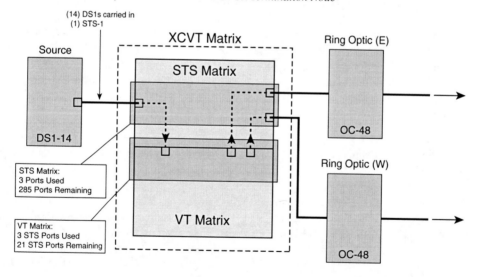

Figure 6-5   *VT Matrix Use for a DS1 Circuit in a UPSR Termination Node*

Alarm Interface Controller Card

The Alarm Interface Controller (AIC-I) card is an optional circuit pack that is installed in shelf Slot 9. The faceplate of the card is marked with a red diamond, corresponding to the symbol marked on the front of the ONS 15454 shelf assembly. This serves as an aid in easily identifying the correct location to install the card. For MSPP sites where the AIC-I is not required, a BLANK/FILLER is required to maintain proper airflow through the system while operating without the front door, and also to allow the system to meet Network Equipment Building Standards (NEBS), electromagnetic interference (EMI) standards, and electrostatic discharge (ESD) standards.

When is the AIC-I card required? The card provides four main capabilities to the network operator:

- Environmental alarm connection and monitoring
- Embedded voice-communication channels, known as orderwires
- A-Side and B-Side power supply input voltage monitoring
- Access to embedded user data channels

You examine each of these major functions, as well as the associated card faceplate LEDs and cabling connectors, in this section. Figure 6-6 shows the faceplate layout of the AIC-I.

An earlier version of the Alarm Interface Controller is called the AIC (no -*I*). This older version provides a more limited environmental alarm-monitoring capacity and does not provide user data channel access or input voltage monitoring. Although they may be installed in some existing systems, Cisco no longer produces the AIC version.

Similarly to all ONS 15454 common control cards, the AIC-I has a FAIL LED and ACT LED on the upper part of the card faceplate, just below the top latch. The FAIL LED is red and indicates that the card's processor is not ready for operation. This LED is normally illuminated during a card reset, and it flashes during the card boot-up process. If the FAIL LED continues to be illuminated, this is an indication that the card hardware has experienced a failure and should be replaced. The ACT (Active) LED is green and illuminates to indicate that the card is in an operational state. Unlike the XC cards and TCC cards, the ACT LED does not have a standby (STBY) state because there is no secondary or back-up card to protect the active AIC-I card. If the card fails, the system can continue to operate normally, with the exception of the functionality provided by the AIC-I.

Environmental Alarms

Environmental alarms are associated with events that affect the operation of the system and are specific to the surrounding environment and external support systems at an MSPP node location. These alarms are usually provisioned and monitored at locations other than those staffed and maintained by a carrier (for example, a central office). This can include an end-user customer's telecom equipment room or an outside plant location, such as a controlled-environment vault (CEV) or concrete hut. Some examples of these alarms include power system performance degradation or failure, hazardous condition alarms (for example, smoke, heat, rising water, and so on), and intrusion alarms (for example, unauthorized entry into a secured area). The ONS 15454 can use the alarm-monitoring capability of the AIC-I to report alarms via the SONET overhead back to the network operations center for trouble resolution or dispatch of maintenance personnel. Figure 6-7 shows an example of this application. A pair of LEDs, labeled as INPUT and OUTPUT, are included on the card faceplate, and illuminate when any input alarm or output control are active.

Figure 6-6 *AIC-I Card Faceplate Diagram*

Figure 6-7   *Environmental Alarms Reported Using AIC-I Card Interfaces*

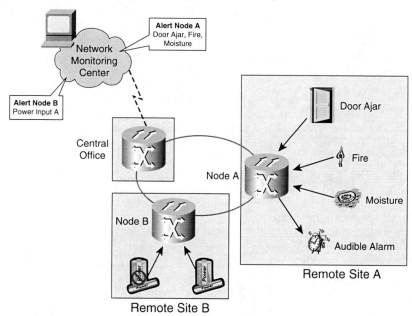

CTC enables the user to provision several parameters related to the operation of the environmental alarms, including an assigned severity (Critical, Major, Minor, or Not Reported), an alphanumeric alarm description, and the capability to set the alarm to be raised upon detection of an "open" or "closed" condition across the alarm contacts. The AIC-I card provides 12 alarm input connections and 4 additional connections that are provisionable as either inputs or outputs. An output is used to control operation of an external device, such as an alarm-indication lamp or a water pump. The backplane of the ONS 15454 chassis has 16 wire-wrap pin pairs for connection to the external equipment to be monitored or controlled. Figure 6-8 shows these connections.

Figure 6-8   *Backplane Environmental Alarm Connections*

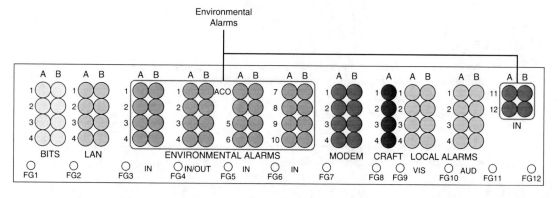

By using an additional piece of hardware, called the Alarm Expansion Panel (AEP), the AIC-I can actually be used to provide up to 32 alarm inputs and 16 outputs, for a total of 48 connections. The AEP is a connector panel that is wired to a subset of the environmental alarm wire-wrap pins and attached to the backplane. Cables can then be installed from the AEP to an external terminal strip for connecting alarm contacts to the system.

One interesting application that involves the use of both an environmental alarm and a control is referred to as a "virtual wire." A virtual wire enables the user to consider the activation of an incoming environmental alarm as triggering the activation of a control. Figure 6-9 encourages this: One such scenario is shown in Figure 6-9, where the activation of an alarm at the remote location of Node A causes a control to activate an audible alarm at the staffed location of Node B. A virtual wire is used to associate the alarm with the control activation.

Figure 6-9 *Virtual Wire Operation*

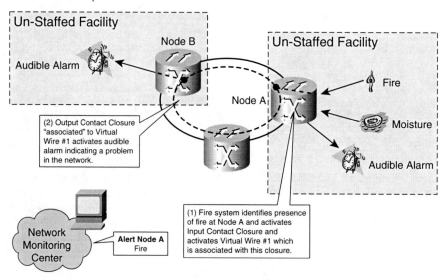

Orderwires

Orderwires allow technicians to attach a phone to the faceplate of the AIC-I card and communicate with personnel at other ONS 15454 MSPP sites. The AIC-I provides two separate orderwires, known as local and express. These can be used simultaneously, if desired. The local orderwire uses the E1 byte in the Section overhead to provide a 64-kbps voice channel between section-terminating equipment, while the express orderwire uses the E2 byte in the Line overhead to provide a channel between line-terminating equipment. Both orderwires operate as broadcast channels, which means that they essentially behave as party lines. Anyone who connects to an orderwire channel can communicate with everyone else on the channel.

Phone sets are connected to the AIC-I using the two standard RJ-11 jacks marked LOW (Local Orderwire) and EOW (Express Orderwire). A green LED labeled RING is provided for each jack. The LED lights and a buzzer/ringer sounds when the orderwire channel detects an incoming call.

Power Supply Voltage Monitoring

The AIC-I monitors the A and B power supply connections to the ONS 15454 for the presence of voltage, under-voltage, and over-voltage. Two bicolor LEDs are provided on the AIC-I faceplate for visual indication of either normal (green) or out-of-range (red) power levels. These LEDs are marked as PWR A and PWR B, and are located on the upper portion of the faceplate between the FAIL and ACT LEDs. The TCC2 and TCC2P controller cards also monitor the A and B power supplies for the chassis, and will override this feature of the AIC-I if installed in the same shelf. The TCC2/TCC2P force the power monitor LEDs on the AIC-I faceplate to match the state of their power-monitor LEDs. Because the older TCC+ controller cards do not include the power-monitoring feature, this feature of the AIC-I is more useful when installed with them.

User Data Channels

Four point-to-point data communications channels are provided for possible network operator use by the AIC-I, with two user data channels (UDC-A and UDC-B) and two data communications channels (DCC-A and DCC-B). These channels enable networking between MSPP locations over embedded overhead channels that are otherwise typically unused. The two UDCs are accessed using a pair of RJ-11 faceplate connectors; the two DCCs use a pair of RJ-45 connectors.

The UDC-A and UDC-B channels use the F1 Section overhead byte to form a pair of 64-kbps data links, each of which can be routed to an individual optical interface for connection to another node site. The DCC-A and DCC-B use the D4-D12 line-overhead bytes to form a pair of 576-kbps data links, which are also individually routed to an optical interface.

SONET/SDH Optical Interface Cards

All current industry-standard SONET/SDH interface types are available for the ONS 15454 platform, including OC-3/STM-1, OC-12/STM-4, OC-48/STM-16, and OC-192/STM-64. These interface cards are typically distinguished by bandwidth, wavelength, and number of ports; however, with the newer interfaces based on Small Form Factor Pluggable (SFP/XFP) technology, these parameters can vary from port to port on the same interface card.

Therefore, we briefly discuss the available card types in terms of two categories: fixed optics interfaces and modular optics interfaces.

Fixed optics interfaces are those for which the bandwidth (for example, OC-12/STM-4), the wavelength (for example, 1310 nm), and the number of equipped ports (for example, a four-port OC-12/STM-4 interface card) are predetermined parameters that cannot be field-modified. Table 6-4 gives a listing of these interfaces, including card name, SONET/SDH bandwidth for each port, transmitter wavelength, the number of ports included on the interface card, and the quantity and type of optical fiber connectors on the card's faceplate.

NOTE   Each port has two associated connectors, one for the transmitter and one for the receiver.

Table 6-4   *ONS 15454 Fixed Optics Interfaces*

Card Name	Per-Port Bandwidth (Mbps)	Wavelength (nm)	Number of Ports	Connectors
OC3 IR 4/STM1 SH 1310	155.52	1310	4	8 SC
OC3 IR/STM1 SH 1310-8	155.52	1310	8	16 LC
OC12 IR/STM4 SH 1310	622.08	1310	1	2 SC
OC12 LR/STM4 LH 1310	622.08	1310	1	2 SC
OC12 LR/STM4 LH 1550	622.08	1550	1	2 SC
OC12 IR/STM4 SH 1310-4	622.08	1310	4	8 SC
OC48 IR/STM16 SH AS 1310	2488.32	1310	1	2 SC
OC48 LR/STM16 LH AS 1550	2488.32	1550	1	2 SC
OC48 ELR/STM16 EH 100 GHz	2488.32	Various*	1	2 SC
OC192 SR/STM64 IO 1310	9953.28	1310	1	2 SC
OC192 IR/STM64 SH 1550	9953.28	1550	1	2 SC
OC192 LR/STM64 LH 1550	9953.28	1550	1	2 SC
OC192 LR/STM64 LH ITU 15xx.xx	9953.28	Various*	1	2 SC

*Multiple different cards are included in this "family" of cards, with wavelengths corresponding to the ITU DWDM frequency grid.

A card name that includes the IR (and SH) designations indicates a card designed for intermediate-reach (short-haul) applications. A card with the LR/LH designations indicates a long-reach/long-haul card. Similarly, ELR/EH (extended long reach/extended haul) and SR (short reach) are indicative of the card's transmission distance specifications.

Modular optics interfaces are those for which the bandwidth, the wavelength, and possibly even the number of equipped ports are parameters that are made flexible through the use of SFPs or XFPs. SFPs and XFPs are essentially electrical-to-optical signal converters that provide a modular interface between a port on an interface card and the external fiber-optic cabling. SFPs and XFPs are similar in design; the primary difference is the module size (XFPs are larger). Modular optics are available for various wavelengths, optical reaches, and technologies, such as SONET/SDH and Gigabit Ethernet (GigE). An SFP or XFP is inserted into the required card port faceplate receptacle and provides a transmit/receive pair of LC fiber connectors. Figure 6-10 shows a group of SFPs, as well as an enlarged view of an SFP connector end.

Figure 6-10 *SFP Interface Modules*

Currently, two SONET/SDH interfaces take advantage of SFP/XFP technology to offer user flexibility and maintenance spare savings:

- **OC-192/STM-64 Any Reach**—This is a 10G interface card with a single XFP port capable of housing an SR, IR, or LR XFP module. Having a single card to stock (while maintaining a maintenance inventory of 10G XFPs) allows a carrier customer to realize efficiencies in maintenance sparing.

- **MRC-12**—The 12-port multirate card (MRC) contains 12 modular SFP receptacles, with the various ports capable of being equipped for OC-3/STM-1, OC-12/STM-4, or OC-48/STM-16 operation. In addition to savings related to maintenance sparing, this card enables the network operator to equip ports for SONET/SDH services on demand. This flexibility also allows for a much more efficient use of chassis slots. For example, a single MRC-12 card can be equipped with one or more OC-48/STM-16 ports, one or more OC-12/STM-4 ports, and one or more OC-3/STM-1 ports, and to use only a single card slot in the chassis instead of a minimum of three slots if fixed interface card types were used. Combinations of each port type can be used, up to the maximum available slot bandwidth, which varies based on the equipped cross-connect card type.

Ethernet Interface Cards

Ethernet interface cards are used in the ONS 15454 platform to enable a service provider or network operator to integrate Ethernet into the SONET/SDH bandwidth. This allows data traffic to share the same transport platform as time-division multiplexing (TDM) links. Ethernet interface cards can be used for 10-Mbps, 100-Mbps, and 1000-Mbps (or GigE) signals, as well as subrate signals for each of these interface types. Some Ethernet interface cards, such as the G-Series and CE-Series, provide transport (or Layer 1) services; other card types enable switching (Layer 2) as well as routing (Layer 3) functionality.

Transport (Layer 1) Ethernet Service Interfaces

Layer 1 Ethernet transport services, which are enabled using point-to-point ONS 15454 circuits, are provided using either the G-Series or CE-Series Ethernet interface cards. This type of service can also be provided using the E-Series cards when operated in what Cisco calls Port-Mapped mode. Table 6-5 provides a summary of these cards.

Table 6-5 *ONS 15454 Transport (Layer 1) Ethernet Service Interface Summary*

Card Name	Per-Port Maximum Bandwidth	Number of Ports	Connectors
G1000-4	1000 Mbps	4	4 GBIC (SC) receptacles
G1K-4	1000 Mbps	4	4 GBIC (SC) receptacles
E1000-2	1000 Mbps	2	2 GBIC (SC) receptacles
E1000-2-G	1000 Mbps	2	2 GBIC (SC) receptacles
E100T-12	100 Mbps	12	12 RJ-45 jacks
E100T-G	100 Mbps	12	12 RJ-45 jacks
CE-100T-8	100 Mbps	8	8 RJ-45 jacks

G-Series Ethernet Interface Cards

The G-Series interface cards are used to provide transport bandwidth for Ethernet frame forwarding between two locations in an MSPP network for point-to-point services. The bandwidth that can be allocated for linking G-Series ports is user-selectable, from a minimum of a single STS-1 up to full line-rate GigE. The G1000-4 and G1K-4 are hardware equivalents; the only difference between the two versions is XC card compatibility. The G1000-4 card is always limited to use with the XC10G cross-connect card and cannot be used in a chassis equipped with XCVT cross-connect cards. The G1K-4 card is a later version with additional flexibility. When ONS 15454 Software Release 4.0 or higher is used, the G1K-4 can be used with XCVT cross-connect cards; its use is limited to the high-speed slots (Slots 5, 6, 12, and 13).

G-Series cards have four Gigabit Interface Converter (GBIC) ports on the card faceplate. GBICs are similar in concept to SFPs, but in a larger physical package and with SC fiber connectors instead of an LC pair. Figure 6-11 shows a group of GBIC modules.

Figure 6-11 *GBIC Modules*

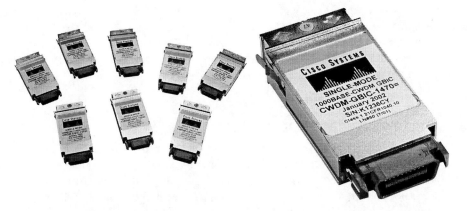

Each G-Series card GBIC receptacle can be independently equipped with SX (short reach 850 nm), LX (long reach 1300 nm), ZX (extended reach 1550 nm), Coarse DWDM, or DWDM GBICs. Each card has an LED labeled ACT, which indicates the card's status, and an LED labeled FAIL, which remains illuminated if the card's processor is not ready or if a failure has occurred. Additionally, each of the four ports has a status LED, labeled ACT/LINK. A solid green ACT/LINK lamp indicates that a link is not carrying traffic; a flashing ACT/LINK lamp means that the link is active and carrying traffic. A solid amber ACT/LINK lamp indicates that there is link but that traffic is inhibited, such as when a circuit has not been built to the port or when the port has not been enabled.

Circuits for carrying Ethernet frames are built from a single port on a G-Series card in one MSPP node to another single port on a G-Series card in another MSPP node. The G-Series cards permit contiguously concatenated circuits of certain sizes. The allowable sizes that can be accommodated in the SONET platform are STS-1, STS-3c, STS-6c, STS-9c, STS-12c, STS-24c, and STS-48c. The maximum bandwidth that can be provisioned to a single G-Series card is 48 STS-1 equivalents.

One restriction exists for the circuit sizes that can be used with the G-Series cards, which could preclude the network operator from using the full 48 STS-1 bandwidth. This restriction is applicable to instances in which a single line-rate circuit (STS-24c) and one or more subrate circuits are used on the same card. In this case, the total bandwidth of the subrate circuits is limited to a maximum of 12 STS-1s. To realize the full bandwidth capability (48 STS-1s) of the G-Series card, the possible combinations are either two line-rate circuits or a combination of up to four subrate circuits with total bandwidth of 48 STS-1s.

Besides Ethernet transport between two sites, there is one additional application for the G-Series Ethernet interface cards. These interfaces can also be configured for use as transponders. This would enable a conventional SX, LX, or ZX Gigabit interface to be converted to a dense wavelength-division multiplexing (DWDM) or coarse wavelength-division multiplexing (CWDM)–compatible wavelength signal, and to be directly connected to a DWDM/CWDM filter system. In this mode, the traffic passing through the G-Series card does not access the cross-connect fabric; this simply provides a method for conditioning Ethernet signals for transport across an xWDM network.

CE-Series Ethernet Interface Cards

Much like the G-Series, the CE-100T-8 interface card is used to provide transport bandwidth for Ethernet frame forwarding between two locations in an MSPP network for point-to-point services. The bandwidth that can be allocated for linking CE-Series ports is user-selectable, from a minimum of a single VT1.5, up to full line-rate 100Base-T Ethernet. The CE-100T-8 requires ONS 15454 software version 5.0.2 or higher and can be used in any of the 12 system traffic slots.

CE-100T-8 cards have eight 10/100 RJ-45 Ethernet ports on the card faceplate. Each card has an LED labeled ACT, which indicates the card's status, and an LED labeled FAIL, which remains illuminated if the card's processor is not ready or if a failure has occurred. Additionally, each of the eight ports has an ACT LED (amber) and a LINK LED (green). A solid green LINK lamp indicates that a link exists. A blinking amber ACT lamp indicates that traffic is being transmitted and received over the link.

Circuits built to connect ports on CE-100T-8 cards can be either contiguously concatenated or virtually concatenated, and either low order (VT1.5) or high order (STS-1 or STS-3c).

Combinations of the various circuit types and sizes are allowed. Some example circuits include these:

- **Line rate 100BASE-T**—To carry a full 100-Mbps signal, an STS-3c or an STS-1-3v can be provisioned.

- **Line rate 10BASE-T**—To carry a full 10-Mbps signal, an STS-1 (inefficient) can be used. A better option is a VT1.5-7v.

- **Subrate 100BASE-T**—An STS-1 can carry approximately 49 Mbps, or a VT1.5-14v can be used to provide 20-Mbps service.

Additional information on circuit sizing and CE-Series applications is provided in Chapter 7, "ONS 15454 Ethernet Applications and Provisioning."

E-Series Ethernet Interface Cards

The E-Series Ethernet cards can be used to provide Layer 1 transport or Layer 2 switched Ethernet services. Layer 1 services are provided when the cards on either end of a circuit (or circuits) are set up in what is known as port-mapped or linear-mapper mode. In this mode, Layer 2 features are disabled. The E-Series card works in a similar manner to the CE-100T-8.

Two types of E-Series cards exist, and there are two versions for each type, making a total of four cards in the set. The 10/100Base-T versions are the E100T-12 and the E100T-G. The difference between the two cards is that the E100T-12 can operate only with the XCVT cross-connect cards; the newer E100T-G does not have this restriction. The 1000-Mbps versions are the E1000-2 and the E1000-2-G. Like the 10/100 cards, the difference between these two cards is that the E1000-2 is supported only in systems equipped with the XCVTs; while the newer E1000-2-G is not limited to XCVT systems.

The E100T cards have 12 10/100 RJ-45 Ethernet ports on their faceplates; the E1000 cards have two GBIC receptacles. The E1000 cards can be equipped with either SX or LX GBICs. All E-Series cards have an LED called ACT, which indicates the card's status, and an LED labeled FAIL, which remains illuminated if the card's processor is not ready or if a failure has occurred. Additionally, each card has a port LED, which is green to indicate a link and amber to indicate that the port is active.

The E-Series cards are hardware-limited to a maximum circuit-termination size of STS-12c. Therefore, although a line-rate 100-Mbps Ethernet circuit can be provisioned to a port on the E100T cards, a GigE circuit provisioned to a port on the E1000 cards is bandwidth-limited to a maximum of 600 Mbps. The available circuit sizes for the E-Series cards in port-mapped mode are STS-1, STS-3c, STS-6c, and STS-12c. The E-Series cards do not support high-order or low-order virtual concatenation (VCAT) circuits.

Switching (Layer 2) and Routing (Layer 3) Ethernet Service Interfaces

Layer 2 and Layer 3 Ethernet services can be provisioned on an ONS 15454 network using the E-Series (Layer 2 only) and ML-Series (Layer 2/Layer 3) interface cards. These cards can be used to build multipoint, switched services over the SONET network, such as shared packet rings or resilient packet rings (RPR). Chapter 7 covers these services in detail. Table 6-6 provides a summary of these cards.

Table 6-6 *ONS 15454 Switching/Routing Ethernet Service Interface Summary*

Card Name	Per-Port Maximum Bandwidth	Number of Ports	Connectors
ML100T-12	100 Mbps	12	12 RJ-45 jacks
ML1000-2	1000 Mbps	2	2 SFP (LC) receptacles
ML100-FX	100 Mbps	8	8 SFP (LC) receptacles
E1000-2	1000 Mbps	2	2 GBIC (SC) receptacles
E1000-2-G	1000 Mbps	2	2 GBIC (SC) receptacles
E100T-12	100 Mbps	12	12 RJ-45 jacks
E100T-G	100 Mbps	12	12 RJ-45 jacks

E-Series Ethernet Interface Cards

The E-Series Ethernet interface cards, previously discussed within this chapter, can be used to provide Layer 2 services, such as virtual local-area network (VLAN) connectivity when provisioned in single-card EtherSwitch or multicard EtherSwitch mode. Chapter 7 covers these applications in detail.

ML-Series Ethernet Interface Cards

The ML-Series cards are Layer 2/Layer 3 switching cards that are integrated into the ONS 15454 MSPP system. The ML cards use a combination of the Cisco IOS command-line interface (CLI) and CTC for operational provisioning. ML-Series cards can support the RPR topology for bridging multiple LANs across a metro optical network. See Chapter 7 for a detailed explanation of RPR.

The ML-Series family consists of three different cards:

- **ML100T-12**—Provides 12 switched, autosensing, 10/100Base-T Ethernet ports for connecting to client equipment using RJ-45 faceplate connectors

- **ML100-FX**—Provides eight faceplate SFP receptacles supporting 10/100 SX and LX interfaces for connecting to client equipment

- **ML1000-2**—Provides two faceplate SFP receptacles supporting Gigabit SX and LX interfaces for connecting to client equipment

Each ML-Series card has two virtual Packet over SONET/SDH (POS) ports used to interconnect it to other Ethernet services interface cards (in the same or different ONS 15454 node), such as in a ring/RPR topology. These ports function similarly to OC-N card ports, and CTC is used to provision SONET/SDH circuits to these ports. The ML-Series cards can support contiguously concatenated (STS-1, STS-3c, STS-6c, STS-9c, STS-12c, STS-24c) and virtually concatenated (STS-1-2v, STS-3c-2v, STS-12c-2v) circuits. The ML-Series cards also support a software-based Link Capacity Adjustment Scheme (LCAS), which allows VCAT circuit group members to be added or removed from the circuit bandwidth in case of a failure or recovery from failure. Figure 6-12 shows an example of a four-node ONS 15454 MSPP ring equipped with seven total ML-Series cards being linked by SONET circuits over an OC-48 UPSR in an RPR configuration.

Figure 6-12 *RPR Application over ONS 15454 MSPP Using ML-Series Cards Example*

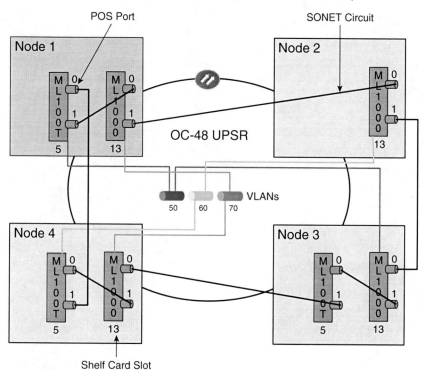

All ML-Series cards have an LED called ACT, which indicates the card's status, and an LED labeled FAIL, which remains illuminated if the card's processor is not ready or if a failure has occurred. Additionally, each card has a port LED, which is green to indicate a link and amber to indicate that the port is active.

Electrical Interface Cards

Electrical interface cards are used in the ONS 15454 to provide DS1, DS3, EC1, and DS3 transmux services. Table 6-7 provides a summary of the available electrical interface cards for the ONS 15454 MSPP, including the EIA types that can be used on a shelf side with each type of interface card.

Table 6-7 *ONS 15454 Electrical Interface Card Summary*

Card	Interface Type	Number of Ports	Bandwidth per Port	EIA Types Allowed
DS1-14/DS1N-14	DS1	14	1.544 Mbps	AMP Champ SMB UBIC-V UBIC-H
DS1-56	DS1	56	1.544 Mbps	UBIC-V UBIC-H
DS3-12/DS3N-12 DS3-12E/DS3N-12E	DS3	12	44.736 Mbps	BNC HD BNC HD Mini-BNC SMB UBIC-V UBIC-H
EC1-12	EC1	12	51.84 Mbps	BNC HD BNC HD Mini-BNC SMB UBIC-V UBIC-H
DS3/EC1-48	DS3/EC1	48	44.736 Mbps or 51.84 Mbps	HD Mini-BNC UBIC-V UBIC-H

Table 6-7 *ONS 15454 Electrical Interface Card Summary (Continued)*

Card	Interface Type	Number of Ports	Bandwidth per Port	EIA Types Allowed
DS3XM-6	DS3 transmux	6	44.736 Mbps	BNC HD BNC HD Mini-BNC SMB UBIC-V UBIC-H
DS3XM-12	DS3 transmux	12	44.736 Mbps	BNC HD BNC HD Mini-BNC SMB UBIC-V UBIC-H

All of the electrical interface cards are identical in appearance from their faceplate, with the exception of the card name marking. Each of the cards has three LEDs. The FAIL LED is an indication that the card is not yet ready, or that a card failure has occurred if it remains illuminated. The ACT/STBY LED is green for an active card and amber for a card in the standby state (protect card in a protection group). The SF lamp is illuminated to indicate a signal failure or a condition such as a loss of signal (LOS), a loss of frame (LOF), or a high bit error rate (BER) on one or more of the card's ports.

The sections that follow discuss each of the available electrical interface cards.

DS1-14 and DS1N-14 Interface Cards

DS1-14 and DS1N-14 cards each provide 14 DS1 ports, which operate at 1.544 Mbps. The difference between these two cards is that the DS1N-14 card contains additional circuitry, which allows it to act as the protection card in a 1:N (where N is less than or equal to 5) protection group (when installed in Slot 3 for Side A, or Slot 15 for Side B). The interface to these cards is through the shelf backplane EIA connectors. These cards can operate in any of the 12 traffic slots in the ONS 15454 chassis. A maximum of 12 DS1-14 or DS1N-14 cards can be active and providing service in a shelf. A typical shelf that is participating in a ring can have 112 working DS1 circuits using DS1-14 and DS1N-14 cards. This configuration would consist of ring OC-N cards in two slots, DS1N-14 standby cards in Slots 3 and 15, and working/active DS1-14 or DS1N-14 cards in the remaining eight slots, or 8 active cards $\times$ 14 ports per card = 112 working service ports.

DS1-56 Interface Card

The DS1-56 interface card provides 56 DS1 ports operating at 1.544 Mbps. This card can function as a working card in shelf Slots 1, 2, 16, or 17, or as a protection card in Slot 3 (protecting working cards in Slots 1 and 2) and Slot 15 (protecting working cards in Slots 16 and 17). The interface to these cards is through the shelf backplane EIA connectors. A maximum of four DS1-56 cards can be active and providing service in a shelf. A typical shelf that is participating in a ring can have 224 working DS1 circuits using active DS1-56 cards in Slots 1, 2, 16, and 17, with optional protection/standby cards in Slot 3 (protecting Slots 1 and 2) and Slot 15 (protecting Slots 16 and 17), or 4 active cards × 56 ports per card = 224 working service ports.

DS3-12, DS3N-12, DS3-12E, and DS3N-12E Interface Cards

DS3-12, DS3N-12, DS3-12E, and DS3N-12E cards each provide 12 DS3 ports, which operate at 44.736 Mbps. The distinction between the E versions (DS3-12E, DS3N-12E) and the non-E versions (DS3-12 and DS3N-12) is that the E versions have enhanced performance-monitoring capabilities that are not included in the non-E versions. This allows for earlier detection of transmission problems. In addition, each version has both regular (for example, DS3-12E) and N (for example, DS3N-12E) card types. The difference between these two cards is that the DS3N-12 and DS3N-12E cards contain additional circuitry, which allows them to act as the protection card in a 1:N (where N is less than or equal to 5) protection group (when installed in Slot 3 for Side A, or Slot 15 for Side B). The interface to these cards is through the shelf backplane EIA connectors. These cards can operate in any of the 12 traffic slots in the ONS 15454 chassis. A maximum of 12 DS3-12/DS3-12E and/or DS3N-12/DS3N-12E cards can be active and providing service in a shelf. A typical shelf that is participating in a ring can have 96 working DS3 circuits using DS3-12/DS3-12E and DS3N-12/DS3N-12E cards. This configuration would consist of ring OC-N cards in two slots, DS3N-12/DS3N-12E standby cards in Slots 3 and 15, and working/active DS3-12/DS3-12E or DS3N-12/DS3N-12E cards in the remaining 8 slots, or 8 active cards × 12 ports per card = 96 working service ports.

EC1-12 Interface Cards

EC1-12 cards each provide 12 EC-1 ports, which operate at 51.84 Mbps. The interface to these cards is through the shelf backplane EIA connectors. These cards can operate in any of the 12 traffic slots in the ONS 15454 chassis. A maximum of 12 EC1-12 cards can be active and providing service in a shelf. EC1-12 cards support 1:1 card protection only (1:N protection is not an available option with the EC1-12 cards; there is no N version of the card). For 1:1 protection, the working/active EC1-12 cards are installed in even-numbered slots (2, 4, 6, 12, 14, and 16), while the protection/standby EC1-12 cards are installed in the corresponding odd-numbered slots. For example, an active EC1-12 card in Slot 2 can be

1:1 protected by a standby EC1-12 card in Slot 1. Another example is an active EC1-12 card installed in Slot 14, protected by a standby EC1-12 card installed in Slot 15. One possible shelf configuration using 1:1 protection groups with the EC1-12 cards would provide for 60 protected EC-1 ports. One example of this type of configuration would consist of ring OC-N cards installed in Slots 5 and 6; working/active EC1-12 cards installed in Slots 2, 4, 12, 14, and 16; and protection/standby EC1-12 cards installed in Slots 1, 3, 13, 15, and 17, or 5 active cards × 12 ports per card = 60 working service ports.

DS3/EC1-48 Interface Cards

The DS3/EC1-48 interface card provides 48 DS3 (44.736 Mbps) or EC-1 (51.84 Mbps). With software releases 6 and higher, each port on the card can be user defined to operate as either a DS3 or an EC-1. This card can function as a working card in shelf Slots 1, 2, 16, or 17, or as a protection card in Slot 3 (protecting working cards in Slots 1 and 2) and Slot 15 (protecting working cards in Slots 16 and 17). The interface to these cards is through the shelf backplane EIA connectors. A maximum of four DS3/EC1-48 cards can be active and providing service in a shelf. A typical shelf that is participating in a ring can have 192 working DS3 and EC-1 circuits using active DS3/EC1-48 cards in Slots 1, 2, 16, and 17, with optional protection/standby cards in Slot 3 (protecting Slots 1 and 2) and Slot 15 (protecting Slots 16 and 17), or 4 active cards × 48 ports per card = 192 working service ports.

DS3XM-6 and DS3XM-12 Interface Card

The DS3XM-6 interface card provides six M13 multiplexing ports, each of which converts a framed DS3 into 28 VT1.5s for grooming and cross-connection. This card can function as a working card in any slot, or as a protection card in a 1:1 protection group (in an even-numbered slot, with a working card in the adjacent odd slot). The DS3XM-6 card does not support the 1:N protection scheme. The interface to these cards is through the shelf backplane EIA connectors.

The DS3XM-12 interface card provides 12 M13 multiplexing ports, each of which converts a framed DS3 into 28 VT1.5s for grooming and cross-connection. This card can function as a working card in any slot, as a protection card in a 1:1 protection group (with a working card in an adjacent slot), or as a protection card in a 1:N protection group if located in Slot 3 or Slot 15. The DS3XM-12 interface cards can operate in one of two modes: ported or portless. In ported mode, the interface to each of the 12 card ports is through the shelf backplane EIA connectors. In portless mode, the M13 DS3 is groomed to an OC-N port on an optical card in the ONS 15454 for optical connection to an external switch. A variety of configurations can be supported using this interface card, with engineering rules based on the type of cross-connect card as well as the mode of operation. Consult the Cisco ONS 15454 Reference Manual for detailed usage information.

Storage Networking Cards

The SL-Series Fibre Channel(FC)/FICON (Fiber Connection) interface card for the ONS 15454 MSPP, also referred to as the FC_MR-4 card, is a four-port card used to provide storage-area networking (SAN) extension services over a SONET/SDH ring. This card can be installed in Slots 5, 6, 12, or 13 in a shelf equipped with XCVT cross-connect cards, or in any slot in a shelf equipped with XC10G or XC-VXC-10G cross-connect cards. Each of the four client ports can be independently equipped with 1-Gbps single-rate or 1-Gbps/2-Gbps dual-rate GBICs. Each port can support 1.0625 Gbps or 2.125 Gbps of FC/FICON connections. A maximum bandwidth of STS-48 is supported per card.

The SL-Series card supports both contiguously concatenated (CCAT) and virtually concatenated (VCAT) SONET/SDH circuits, as follows:

1 Gbps FC/FICON can be mapped into this:

- STS-1, STS-3c, STS-6c, STS-9c, STS-12c, STS-18c, STS-24c, and STS-48c (minimum SONET CCAT size for line-rate service is STS-24c)
- VC4-1c, VC4-2c, VC4-3c, VC4-4c, VC4-6c, VC4-8c, and VC4-16c (minimum SDH CCAT size for line-rate service is VC4-8c)
- STS-1-nv, where $n=1$ to 24 ($n = 19$ for line-rate service)
- STS-3c-nv, where $n=1$ to 8 ($n = 6$ for line-rate service)
- VC4-nv, where $n=1$ to 8 ($n = 6$ for line-rate service)

2-Gbps FC/FICON can be mapped into:

- STS-1, STS-3c, STS-6c, STS-9c, STS-12c, STS-18c, STS-24c, and STS-48c (minimum SONET CCAT size for line-rate service is STS-48c)
- VC4-1c, VC4-2c, VC4-3c, VC4-4c, VC4-6c, VC4-8c, VC4-12c, and VC4-16c (minimum SDH CCAT size for line-rate service is VC4-16c)
- STS-1-nv, where $n = 1$ to 48 ($n = 37$ for line-rate service)
- STS-3c-nv, where $n = 1$ to 16 ($n = 12$ for line-rate service)
- VC4-nv, where $n = 1$ to 16 ($n = 12$ for line-rate service)

The SL-Series cards also support advanced SAN distance extension features, including the use of buffer-to-buffer (B2B) credits, which are supported by the connected FC switching devices, to overcome distance limitations in 1-Gbps and 2-Gbps line-rate SAN extension applications. Additionally, distance extensions functions enabled by Receiver Ready (R_RDY) spoofing enables the SL Series to serve as an integrated FC extension device, reducing the need for external equipment.

MSPP Network Design Case Study

A major healthcare services provider in a large metropolitan area is currently preparing to implement a major upgrade to its data and telecommunications networks. Backbone connectivity for University Healthcare System, Inc. (UHCS), in Brounsville is currently provided using multiple T1, DS3, and OC-3 links provided by a local exchange carrier, BrounTel. These leased lines are used for connectivity among various company locations, such as the corporate headquarters campus and various hospital locations throughout the metro area. These services are provisioned through a combination of BrounTel copper T1 span lines, legacy point-to-point asynchronous optical multiplexers, and first-generation SONET OC-3 and OC-12 systems. The company is seeking to upgrade its current network services for several reasons:

- **Network survivability**—To ensure business continuity, the UHCS IT staff wants to improve the reliability of its leased network services. In the past, the nonredundant portions of the network serving some of its locations have failed, causing unacceptable service outage times.

- **Flexibility**—A major driver in the upgrade decision is the capability to add bandwidth and services with the simple addition or upgrade of existing components, versus the delay associated with conditioning new T1 span lines or adding fiber facilities.

- **Scalability**—UHCS seeks to future-proof its infrastructure so that the network can grow as the business continues to expand.

- **Advanced network services**—Line-rate and sub-rate GigE connections are among the current network requirements, and storage networking, 10 GigE, and wavelength services could become requirements in the long term. These services cannot be provided using the existing BrounTel network facilities that serve the UHCS locations.

- **Cost reduction**—By reducing their network to a simpler, more advanced technology platform, the company plans to reduce recurring costs paid to BrounTel.

After discussing service requirements and contract terms with BrounTel, the IT managers have elected to contract with BrounTel to provide a leased dedicated SONET ring (DSR) service for connectivity between company sites, and for access to the public switched telephone network (PSTN). BrounTel will deploy a Cisco ONS 15454 MSPP solution for the DSR.

MSPP Ring Network Design

A total of seven sites in various parts of the metro area will need connectivity to the new network. BrounTel has existing standard single-mode fiber (SMF) optic cables serving some of the locations; it will use existing cable or build new optical cable facilities as required for diverse routing between the company locations and multiple BrounTel central offices. Three central office locations will have MSPP nodes on the SONET ring; the others will serve as fiber patch (or pass-through) locations. Table 6-8 gives a list of location names, addresses, and site types.

Table 6-8   *University Healthcare System DSR Locations and Requirements*

Node Number	Site Name	Site Address	Site Type
0	University Hospital North (UHCS)	4303 Thach Avenue	MSPP Node
-	Foy Central Office (BrounTel)	307 Duncan Drive	Fiber Patch
1	Magnolia Central Office (BrounTel)	41 Magnolia Avenue	MSPP Node
2	University Healthcare HQ (UHCS)	130 Donahue Drive	MSPP Node
-	Poplar Street Central Office (BrounTel)	8016 Poplar Street	Fiber Patch
3	Brounsville Main Central Office (BrounTel)	2004 Elm Street	MSPP Node
4	University Medical Center (UHCS)	34 South College Street	MSPP Node
-	Beech Street Central Office (BrounTel)	2311 Beech Street	Fiber Patch
5	University Hospital East (UHCS)	1442 Wire Road	MSPP Node
-	Roosevelt Drive Central Office (BrounTel)	1717 Roosevelt Drive	Fiber Patch
6	Samford Avenue Central Office (BrounTel and IXC POP)	940 Samford Avenue	MSPP Node
-	Mell Street Central Office (BrounTel)	1183 Mell Street	Fiber Patch
7	University Hospital South (UHCS)	9440 Parker Circle	MSPP Node
-	Haley Central Office (BrounTel)	1957 Concourse Way	Fiber Patch
-	Ross Central Office (BrounTel)	60 Wilmore Road	Fiber Patch
8	UHCS Data Center (UHCS)	1983 Draughon Trace	MSPP Node
-	Ramsay Central Office (BrounTel)	2322 Hemlock Drive	Fiber Patch
9	Jordan Memorial Hospital (UHCS)	1969 Goodwin Lane	MSPP Node
-	Morrison Central Office (BrounTel)	141 Morrison Drive	Fiber Patch

Table 6-9 shows the measured (for existing facilities) or calculated (for proposed facilities) fiber cable loss figures for the ring facilities. All loss figures include losses because of splices, connectors and patch panels, as well as the cable loss.

Table 6-9   *Fiber Cable Losses for UHCS Ring Network*

Fiber Span Number	From Location	To Location	Distance (km)	Loss at 1310 nm (dB)	Loss at 1550 nm (dB)
1	University Hospital North (UHCS)	Foy Central Office (BrounTel)	12.3	7.42	5.38
2	Foy Central Office (BrounTel)	Magnolia Central Office (BrounTel)	13.8	8.12	5.79
3	Magnolia Central Office (BrounTel)	University Healthcare HQ (UHCS)	17.1	9.53	6.73
4	University Healthcare HQ (UHCS)	Poplar Street Central Office (BrounTel)	6.2	3.73	2.76
5	Poplar Street Central Office (BrounTel)	Brounsville Main Central Office (BrounTel)	7.1	4.15	3.08
6	Brounsville Main Central Office (BrounTel)	University Medical Center (UHCS)	18.9	11.16	6.98
7	University Medical Center (UHCS)	Beech Street Central Office (BrounTel)	9.4	5.92	3.70
8	Beech Street Central Office (BrounTel)	University Hospital East (UHCS)	8.7	5.48	3.43
9	University Hospital East (UHCS)	Roosevelt Drive Central Office (BrounTel)	13.2	8.58	5.37
10	Roosevelt Drive Central Office (BrounTel)	Samford Avenue Central Office (BrounTel and IXC POP)	7.0	4.73	3.25
11	Samford Avenue Central Office (BrounTel and IXC POP)	Mell Street Central Office (BrounTel)	12.3	7.06	4.97

continues

Table 6-9 *Fiber Cable Losses for UHCS Ring Network (Continued)*

Fiber Span Number	From Location	To Location	Distance (km)	Loss at 1310 nm (dB)	Loss at 1550 nm (dB)
12	Mell Street Central Office (BrounTel)	University Hospital South (UHCS)	21.0	11.38	7.11
13	University Hospital South (UHCS)	Haley Central Office (BrounTel)	8.3	4.32	2.70
14	Haley Central Office (BrounTel)	Ross Central Office (BrounTel)	12.4	8.10	5.65
15	Ross Central Office (BrounTel)	UHCS Data Center (UHCS)	13.8	8.99	6.07
16	UHCS Data Center (UHCS)	Ramsay Central Office (BrounTel)	6.7	4.25	2.66
17	Ramsay Central Office (BrounTel)	Jordan Memorial Hospital (UHCS)	8.3	4.17	3.13
18	Jordan Memorial Hospital (UHCS)	Morrison Central Office (BrounTel)	10.3	7.72	5.33
19	Morrison Central Office (BrounTel)	University Hospital North (UHCS)	8.4	4.6	3.55

Because of the relatively short distances between MSPP node locations, polarization mode dispersion (PMD) will not be an issue in this deployment.

To determine the bandwidth requirements for the ring, the UHCS service demands must be considered. These services will be provided using the DSR:

- **Multipoint Switched Ethernet** will be used to connect the LANs at all the UHCS sites together using a resilient packet ring (RPR) with GigE links.

- **Private line Ethernet** connections, which are point-to-point Ethernet transport "pipes," will be used between a subset of the UHCS sites.

- **TDM services**, including DS1 and DS3 links, will be required for transport of voice traffic between UHCS Private Branch eXchange (PBX) systems, and between UHCS sites and BrounTel central offices.

Listing the requirements individually, take a look at Table 6-10, which shows the planned circuits for the ring.

Table 6-10   *DSR Circuit Requirements*

Circuit Type	Circuit Size	Quantity	Protection?	Locations	Purpose
GigE	STS-24c	1	None	Node 0 to Node 2	RPR for production data
GigE	STS-24c	1	None	Node 2 to Node 4	RPR for production data
GigE	STS-24c	1	None	Node 4 to Node 5	RPR for production data
GigE	STS-24c	1	None	Node 5 to Node 7	RPR for production data
GigE	STS-24c	1	None	Node 7 to Node 8	RPR for production data
GigE	STS-24c	1	None	Node 8 to Node 9	RPR for production data
GigE	STS-24c	1	None	Node 9 to Node 0	RPR for production data
GigE	STS-24c	1	UPSR	Node 0 to Node 2	Private line video and data application
GigE (Sub-Rate)	STS-12c	1	UPSR	Node 0 to Node 5	Private line video and data application
GigE (Sub-Rate)	STS-12c	1	UPSR	Node 0 to Node 7	Private line video and data application
DS3	STS-1	2	UPSR	Node 1 to Node 8	Data access
DS3	STS-1	3	UPSR	Node 2 to Node 6	Data access
DS3	STS-1	1	UPSR	Node 2 to Node 3	Data access
DS1	VT1.5	3	UPSR	Node 0 to Node 9	PBX voice trunks
DS1	VT1.5	4	UPSR	Node 0 to Node 2	PBX voice trunks
DS1	VT1.5	5	UPSR	Node 1 to Node 8	Voice access

continues

Table 6-10   *DSR Circuit Requirements (Continued)*

Circuit Type	Circuit Size	Quantity	Protection?	Locations	Purpose
DS1	VT1.5	2	UPSR	Node 2 to Node 9	PBX voice trunks
DS1	VT1.5	4	UPSR	Node 2 to Node 8	PBX voice trunks
DS1	VT1.5	3	UPSR	Node 2 to Node 7	PBX voice trunks
DS1	VT1.5	22	UPSR	Node 2 to Node 6	Voice access
DS1	VT1.5	3	UPSR	Node 2 to Node 5	PBX voice trunks
DS1	VT1.5	4	UPSR	Node 2 to Node 4	PBX voice trunks
DS1	VT1.5	2	UPSR	Node 2 to Node 3	Voice access
DS1	VT1.5	3	UPSR	Node 4 to Node 8	PBX voice trunks
DS1	VT1.5	2	UPSR	Node 5 to Node 7	PBX voice trunks
DS1	VT1.5	2	UPSR	Node 7 to Node 9	PBX voice trunks
DS1	VT1.5	2	UPSR	Node 7 to Node 8	PBX Voice Trunks

To calculate the bandwidth requirements for the ring, simply add each of the individual requirements to arrive at the total number of STS-1s needed. This calculation is shown in Table 6-11.

Table 6-11   *Ring Bandwidth Requirements*

Circuit(s)	Number of Ring STS-1s Required
(7) Unprotected GigE RPR links	24
(1) UPSR-Protected Line-Rate GigE	24
(2) UPSR-Protected Sub-Rate GigE (12 STS-1)	24
(6) UPSR-Protected DS3s	6
(61) UPSR-Protected DS1s	3
Total STS-1s Required	**81**

NOTE	The GigE links that form the RPR reuse the same bandwidth throughout the ring because they are built without SONET protection.

Because the initial requirements are 81 STS-1s, an OC-192 ring will be used. This allows sufficient capacity for the existing service requirements, as well as for future growth to the network.

OC-192 Ring Transmission Design

Having defined the network bandwidth requirements to be OC-192, you can now select the appropriate OC-192 interfaces to equip at each ONS 15454 MSPP node to link the ring sites. ONS 15454 OC-192 IR interfaces transmit at a nominal wavelength of 1550 nm and have an allowable link loss budget of about 13 dB. OC-192 LR interfaces also transmit at the 1550 nm wavelength and have an allowable link loss budget of 26 dB. There is also an available SR OC-192 interface, but the small allowable link loss budget is not suitable for the distances involved in the UHCS application. Therefore, either the IR or LR optics will be used, with 10 dB being the "breakpoint" between the two. This allows 3 dB of margin for future loss increases due to fiber cable degradation, future repair splicing, and component aging.

Based on the specifications of the various ONS 15454 OC-192 interfaces and the loss characteristics of each fiber section (outlined in Table 6-9), the node-to-node interface types can be determined for the ring. OC-192/10G operation is allowed in chassis Slots 5, 6, 12, and 13. You use a pair of these slots at each location for the East- and West-facing ring interfaces. Although any combination of two of the four available slots is acceptable, uniformly select Slots 5 and 6 at each of the nodes for operational simplicity. Table 6-12 shows the selection of OC-192 optics for each ring span.

Table 6-12 *OC-192 Ring Optics for UHCS DSR*

From East Node/Slot	To West Node/Slot	Loss at 1550 nm (dB)	OC-192 Interface Type
Node 0 Slot 6	Node 1 Slot 5	11.17	OC-192 1550 LR
Node 1 Slot 6	Node 2 Slot 5	6.73	OC-192 1550 IR
Node 2 Slot 6	Node 3 Slot 5	5.84	OC-192 1550 IR
Node 3 Slot 6	Node 4 Slot 5	6.98	OC-192 1550 IR
Node 4 Slot 6	Node 5 Slot 5	7.13	OC-192 1550 IR
Node 5 Slot 6	Node 6 Slot 5	8.62	OC-192 1550 IR
Node 6 Slot 6	Node 7 Slot 5	12.08	OC-192 1550 LR
Node 7 Slot 6	Node 8 Slot 5	14.42	OC-192 1550 LR
Node 8 Slot 6	Node 9 Slot 5	5.79	OC-192 1550 IR
Node 9 Slot 6	Node 0 Slot 5	8.88	OC-192 1550 IR

In addition to the optical trunk interface card selection, one of the BrounTel central office nodes will be designated as the Gateway Network Element (GNE) and will connect to BrounTel's Network Operations Center (NOC) using its IP-based interoffice management network. The Magnolia central office MSPP node will be chosen as the GNE.

For the purposes of network synchronization, each MSPP node located in a BrounTel central office will be connected to the office BITS and will be configured as externally timed. Line timing will be configured on the ONS 15454 systems located in the UHCS customer premises sites, with the OC-192 optical interface ports on the cards installed in Slots 5 and 6 serving as the primary and secondary reference sources.

With all the necessary parameters now defined, you can prepare all necessary engineering documentation, such as the network map, shelf card slot assignments, chassis EIA equipage, tributary protection group configuration, and cabling termination assignments.

Network Map

The network map or ring map is a key piece of documentation that assists in bringing an MSPP network online, as well as a future as-built reference for planning, troubleshooting, and performing upgrades or additions. Figure 6-13 shows the network map for the UHCS DSR. The following information has been included:

- Graphical representation of the network topology
- Location of all MSPP nodes and fiber pass-through offices (no equipment—just interconnection of outside plant fiber cables)
- Distance and loss/attenuation figures for each fiber link
- Slot assignments and card types for the interconnecting OC-192 interface cards
- Node names, IP addresses, and timing configurations
- GNE assignment and IP address of the default router
- Software version

The node numbers and Ring ID have also been provided as reference information; however, because this is not a BLSR network, this information is not required to provision the ring nodes.

Shelf Card Slot Assignments, EIA Equipage, and Tributary Protection Group Configuration

The versatility of the ONS 15454 MSPP gives the BrounTel engineers multiple options when selecting the chassis card slot assignments. Some assignments will be common for all ring nodes; others might vary to allow for maximum flexibility to add future services to the network.

Figure 6-13 *UCHS DSR Ring Map*

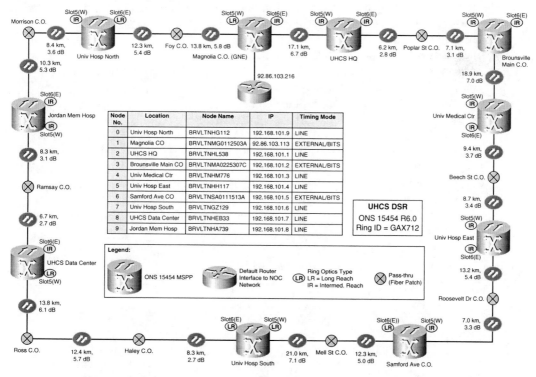

Node No.	Location	Node Name	IP	Timing Mode
0	Univ Hosp North	BRVLTNHG112	192.168.101.9	LINE
1	Magnolia CO	BRVLTNMG0112503A	92.86.103.113	EXTERNAL/BITS
2	UHCS HQ	BRVLTNHL538	192.168.101.1	LINE
3	Brounsville Main CO	BRVLTNMA0225307C	192.168.101.2	EXTERNAL/BITS
4	Univ Medical Ctr	BRVLTNHM776	192.168.101.3	LINE
5	Univ Hosp East	BRVLTNHH117	192.168.101.4	LINE
6	Samterd Ave CO	BRVLTNSA0111513A	192.168.101.5	EXTERNAL/BITS
7	Univ Hosp South	BRVLTNGZ129	192.168.101.6	LINE
8	UHCS Data Center	BRVLTNHEB33	192.168.101.7	LINE
9	Jordan Mem Hosp	BRVLTNHA739	192.168.101.8	LINE

UHCS DSR
ONS 15454 R6.0
Ring ID = GAX712

Legend:
ONS 15454 MSPP
Default Router Interface to NOC Network
Ring Optics Type   LR = Long Reach   IR = Intermed. Reach
Pass-thru (Fiber Patch)

An important factor in the determination of card slot assignments for electrical interface cards, such as the DS1 and DS3 cards in the UHCS ring, is the type of tributary card protection required. Additionally, these interface types and their locations will help to determine the type of EIA that must be ordered for the ONS 15454 chasses.

All 10 of the ring nodes have the TCCP2 cards in Slots 7 and 11. Recall that two TCC cards are required in every ONS 15454 MSPP node. Likewise, two XC-VXC-10G cross-connect cards will be placed in Slots 8 and 10 for each node. Finally, for standardization and operational simplicity, the OC-192 ring interface cards will be placed in Slots 5 (West) and 6 (East) in each of the ring nodes. For the other interface cards required in the ring nodes, customized slot assignments will be specified. According to the conditions of the DSR service contract, BrounTel will design all TDM service interfaces to be card-protected in either 1:1 or 1:*N* protection groups. Of course, the Ethernet service interfaces will be unprotected.

At the University Hospital–North node (Node 0), the initial service-termination requirements include seven DS1 circuits, three private-line GigE links (one line rate and two subrate), and the GigE RPR circuit connections. Figure 6-14 shows the shelf diagram with card locations for this node. The DS1s require a single working DS1-14 card. This card will be placed in Slot 4, with a DS1N-14 card installed in Slot 3 for protection. A 1:*N* protection group will be established for these cards, as indicated in the diagram. The use of Slots 3 and 4 will allow for future growth in DS1s; Slots 1 and 2 are left vacant for potential future DS3 requirements. Two G1K-4 cards are required because of the necessity of a line-rate (STS-24c) circuit, and two sub-rate circuits (STS-12c each), whose combined bandwidth exceeds 12 STS-1s. Slots 16 and 17 will be used for these cards. Note that the GBIC types for the ports to be equipped are indicated on the shelf diagram. The "SX" designation indicates that these ports will be equipped with 1000Base-SX (850-nm) GBICs. A single ML-Series card is required to connect this node to the RPR overlay. Because UHCS requires redundant GigE links from this interface, an ML1000-2 card with dual 1000Base-SX SFPs is specified, and this card will be installed in Slot 15. Also, an AIC-I card will be installed in Slot 9 so that, through the SONET overhead, the BrounTel network operations center can monitor the contact closure alarms from the associated direct current (DC) power plant. Card Slots 1, 2, 12, 13, and 14 will not be used for service interfaces initially and will be equipped with blanks.

Figure 6-14 *University Hospital North Node—Shelf Diagram*

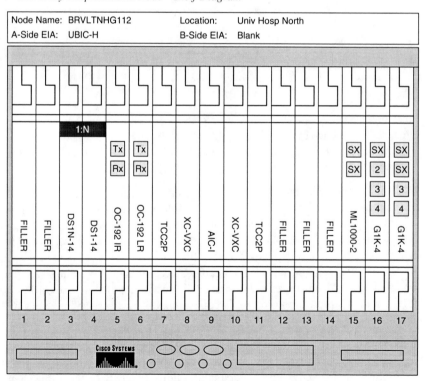

The DS1 interfaces in this node require backplane electrical connections. Because of the requirement for possible future DS3s, a UBIC-H EIA will be installed on Side A of the rear of the chassis. This enables both DS1 and DS3 interfaces to be cabled out from the rear of the shelf. Because all current and future requirements for Side B are front-cabled interface cards (Ethernet, storage, or optical), an EIA is not required to be installed on the rear of Side B, and the default blank cover can be used.

Each of the other MSPP nodes in the UHCS ring will be designed using similar interfaces, with an eye on future network-expansion requirements. A brief description of the requirements, interfaces, and protection groups for each is given in the next several sections, along with an accompanying shelf diagram.

Magnolia Central Office (Node 1)

Requirements are for five DS1s and two DS3s. Both interface types will be slotted on Side B, and a UBIC-H EIA will be equipped on the rear to accommodate the cabling for these cards. See Figure 6-15 for the shelf diagram.

Figure 6-15 *Magnolia Central Office Node—Shelf Diagram*

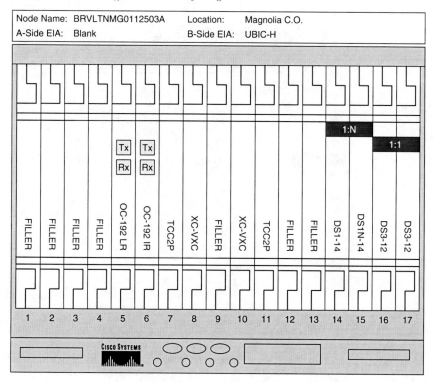

UCHS Headquarters (Node 2)

Initial requirements are 44 DS1s, 4 DS3s, 1 line-rate private line Ethernet circuit, and the RPR GigE links. The high-density (56-port) DS1 card will be used on Side B, with the DS3 cards on Side A. Both shelf sides will be equipped with the UBIC-H EIA. See Figure 6-16 for the shelf diagram.

Figure 6-16   *UCHS Headquarters Node—Shelf Diagram*

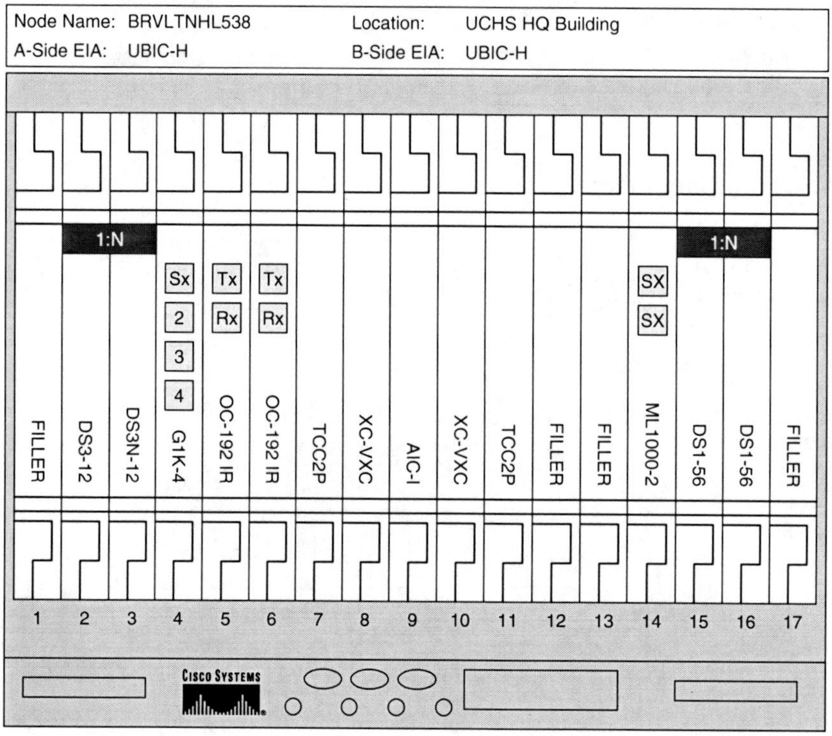

Brounsville Main Central Office (Node 3)

Requirements are for two DS1s and one DS3. Both interface types will be slotted on Side B, and a UBIC-H EIA will be equipped on the rear to accommodate the cabling for these cards. See Figure 6-17 for the shelf diagram.

University Medical Center (Node 4)

Initial requirements are for seven DS1s, as well as the RPR GigE links. DS1 interface cards and a UBIC-H EIA will be installed on Side B, with the ML1000-2 card and associated SX SFPs on Side A. See Figure 6-18 for the shelf diagram.

Figure 6-17 *Brounsville Main Central Office Node—Shelf Diagram*

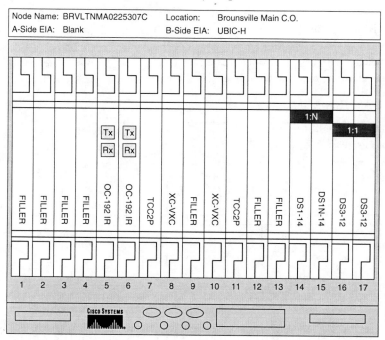

Figure 6-18 *University Medical Center Node—Shelf Diagram*

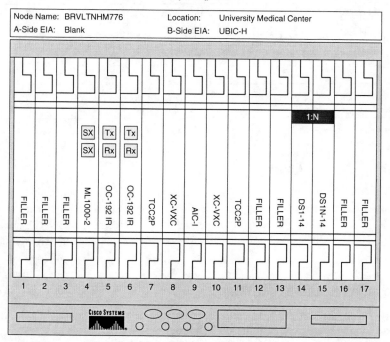

University Hospital—East (Node 5)

Five DS1s, a single subrate GigE private line circuit, and the RPR GigE links must be provisioned at this location. A DS1-14 card will be placed in Slot 14 and will be protected by a DS1N-14 card in Slot 15. The Ethernet cards will be installed on the A Side. Because electrical interfaces are required to be cabled only from the B side of the shelf, no EIA will be required to be installed on Side A. See Figure 6-19 for the shelf diagram.

Figure 6-19 *University Hospital–East Node—Shelf Diagram*

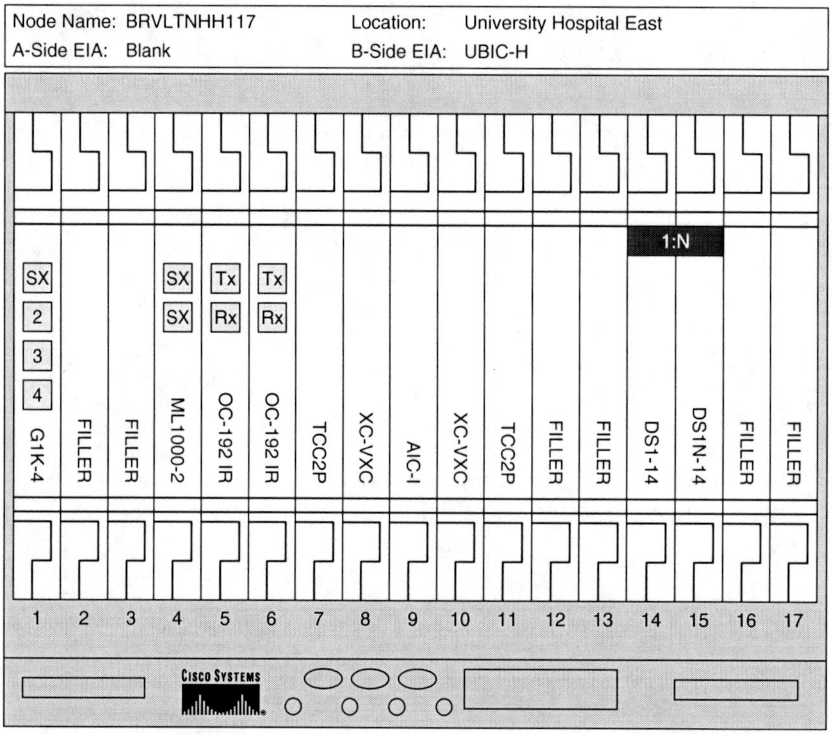

Samford Avenue Central Office (Node 6)

Initial requirements are for 22 DS1s and 3 DS3s. A pair of DS1-14 cards will be placed in Slots 16 and 17, with both cards being protected in a 1:*N* group by a DS1N-14 card in Slot 15. The DS3-12E card will be placed in Slot 2 and will be protected by a DS3N-12E card in Slot 3. UBIC-H EIAs will be installed on the rear for both sides. See Figure 6-20 for the shelf diagram.

University Hospital–South (Node 7)

Nine DS1s, a subrate (STS-12c) GigE private line connection, and the RPR GigE links are required. See Figure 6-21 for the shelf diagram.

Figure 6-20 *Samford Avenue Central Office Node—Shelf Diagram*

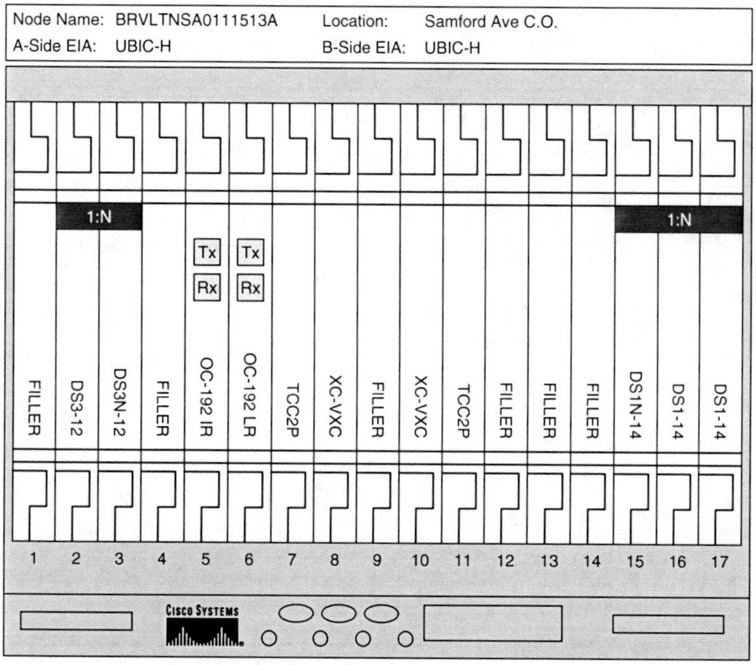

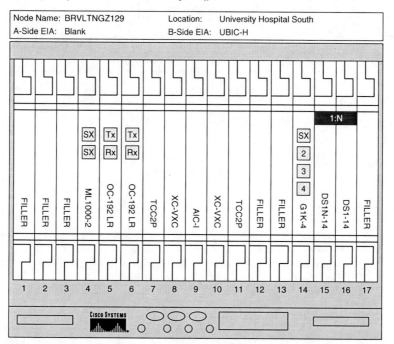

Figure 6-21 *University Hospital–South Node—Shelf Diagram*

| Node Name: BRVLTNGZ129 | Location: | University Hospital South |
| A-Side EIA: Blank | B-Side EIA: | UBIC-H |

UCHS Data Center (Node 8)

Initial requirements include 14 DS1s, 2 DS3s, and the RPR GigE links. Electrical interfaces will be installed only on Side B. See Figure 6-22 for the shelf diagram.

Figure 6-22 *UCHS Data Center Node—Shelf Diagram*

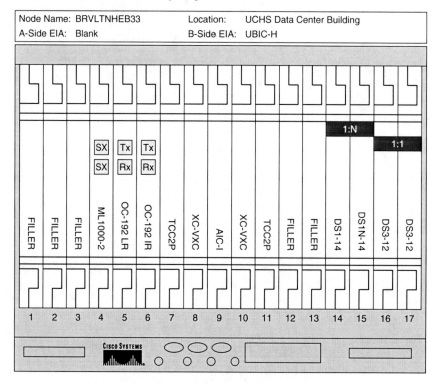

Jordan Memorial Hospital (Node 9)

Seven DS1s and the RPR GigE links are the initial requirements here. See Figure 6-23 for the shelf diagram.

Cabling Terminations

The UHCS DSR requires various interface types to be cabled out for interconnecting to the customer premises equipment (CPE) at UHCS locations, or for interfacing with the BrounTel or Interexchange Carrier (IXC) networks at the central office locations. Figure 6-24 shows a diagram with typical interface cabling for an ONS 15454 node location on the UHCS ring: the UHCS Headquarters node. The OC-192 ring optics will be cabled to the outside plant (OSP) fiber-termination panel using single-mode optical fibers from the SC faceplate connectors. DS1 and DS3 interface cards will be cabled to digital signal cross-connect (DSX) panels via the backplane UBIC EIA connectors. Ethernet interface cards, including the G1K-4 and ML1000-2, will be cabled to an optical splitter module panel using multimode fibers from the GBIC SC faceplate connectors (G1K-4) or the SFP LC faceplate connectors (ML1000-2).

Figure 6-23 *Jordan Memorial Hospital Node—Shelf Diagram*

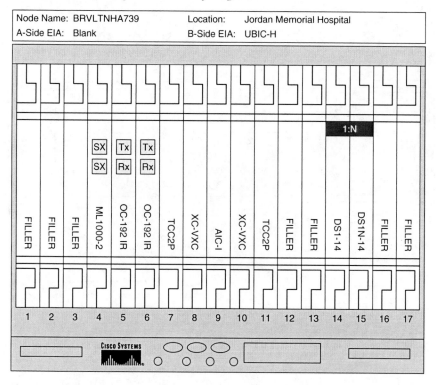

Table 6-13 shows an example cabling termination assignment chart for the UHCS Headquarters location.

Figure 6-24 *Cabling Termination Diagram, UCHS Headquarters Node*

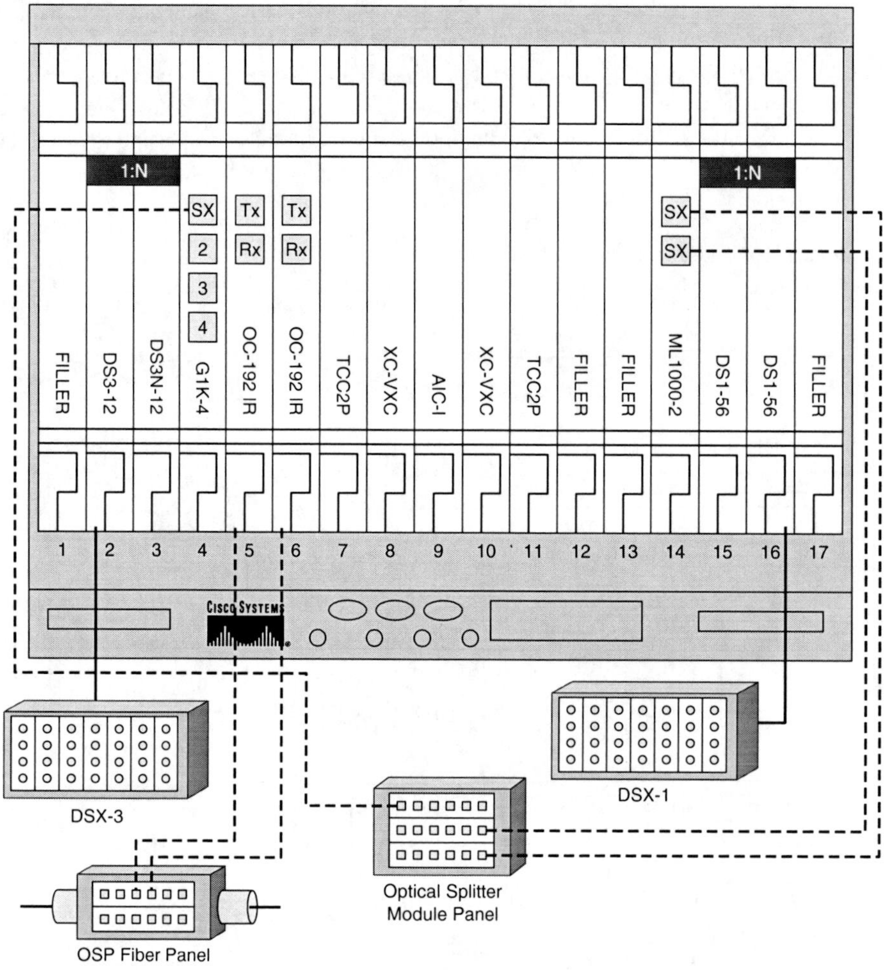

Table 6-13 *Customer Drop Cabling Terminations for UHCS HQ Node*

Panel Type	Panel Location	Connector/Jack	To Equipment	To Slot/Port
Optical	RR 101 PNL 4	1	ONS RR 101A	Slot 5 Tx
		2	ONS RR 101A	Slot 5 Rx
		3	ONS RR 101A	Slot 6 Tx
		4	ONS RR 101A	Slot 6 Rx
		5-24	FUTURE	FUTURE

Table 6-13 *Customer Drop Cabling Terminations for UHCS HQ Node (Continued)*

Panel Type	Panel Location	Connector/Jack	To Equipment	To Slot/Port
Optical/ Splitter	RR 101 PNL 3	Mod 1 SRC Tx	ONS RR 101A	Slot 4/1 Tx
		Mod 1 SRC Rx	ONS RR 101A	Slot 4/1 Rx
		Mod 1 Cus Tx	ONS RR 101A	CPE Tx
		Mod 1 Cus Rx	ONS RR 101A	CPE Rx
		Mod 2 SRC Tx	ONS RR 101A	Slot 14/1 Tx
		Mod 2 SRC Rx	ONS RR 101A	Slot 14/1 Rx
		Mod 2 Cus Tx	ONS RR 101A	CPE Tx
		Mod 2 Cus Rx	ONS RR 101A	CPE Rx
		Mod 3 SRC Tx	ONS RR 101A	Slot 14/2 Tx
		Mod 3 SRC Rx	ONS RR 101A	Slot 14/2 Rx
		Mod 3 Cus Tx	ONS RR 101A	CPE Tx
		Mod 3 Cus Rx	ONS RR 101A	CPE Rx
DSX-1	RR 101 PNL 2	1-56	ONS RR 101A	Slot 16/1-56
		57-84	FUTURE	FUTURE
DSX-3	RR 101 PNL 1	1-12	ONS RR101A	Slot 2/1-12

Summary

This chapter looked at the Cisco ONS 15454 MSPP, which is one of the most widely deployed MSPP systems worldwide. The system components, both hardware (shelf assembly, backplane interfaces, and EIAs) and interface modules (common/control, optical cards, Ethernet services cards, and electrical interface cards) were briefly described to provide an understanding of the basic functionality of the ONS 15454 MSPP.

After exploring the system, a network design example was presented. After identifying the requirements for the end customer, the entire network was designed, including network map, ring optics transmission design, interface slotting, EIA selection, protection group assignment, and interface cabling plan.

PART III

Deploying Ethernet and Storage Services on ONS 15454 MSPP Networks

This chapter covers the following topics:

- ONS 15454 E-Series Interface Cards
- ONS 15454 G-Series Interface Cards
- ONS 15454 CE-Series Interface Cards
- ONS 15454 ML-Series Interface Cards

CHAPTER 7

ONS 15454 Ethernet Applications and Provisioning

As noted in Chapter 6, "MSPP Network Design Example: Cisco ONS 15454," the ONS 15454 Multiservice Provisioning Platform (MSPP) provides multiple types of Ethernet interface cards for a variety of applications. These interface cards fall into two broad categories: transport (Layer 1) Ethernet interfaces and multilayer (Layer 2 and Layer 3) Ethernet interfaces. The following is a categorization and brief description of each of the Ethernet interface cards covered in this chapter:

- **E Series**—The E-Series cards can be considered members of both Ethernet interface card categories because they can be used in both transport applications and switched applications. The E-Series cards were the first Ethernet cards introduced for use in the ONS 15454 MSPP. This series includes an Ethernet/Fast Ethernet card and a Gigabit Ethernet (GigE) card. The E Series supports both point-to-point links and shared packet-ring topologies.

- **G Series**—The G-Series cards are four-port Layer 1 transport cards with GigE interfaces. These cards support subrate and line-rate GigE point-to-point links.

- **CE Series**—The CE Series currently includes an eight-port Layer 1 transport card with 10-/100-Mbps Ethernet/Fast Ethernet interfaces. CE-Series cards support subrate and line-rate Ethernet and Fast Ethernet point-to-point links.

- **ML Series**—Three card types comprise the "flagship" Ethernet interfaces for the ONS 15454, the ML Series. The ML Series can be thought of as Layer 2/Layer 3 switches embedded into the Synchronous Optical Network (SONET)/Synchronous Digital Hierarchy (SDH) network. These cards support a rich set of features and interface types.

Typically, each series (or family) of ONS 15454 Ethernet cards is used in a SONET/SDH network. For example, ML-Series cards are used to interoperate with other ML-Series cards within a SONET/SDH ring. However, some Ethernet cards from different series can interconnect with each other using circuits configured between their respective Packet over SONET/SDH (POS) interfaces. The limitation of mixing these types to interoperate in a network is that they are limited to frame-forwarding (Layer 1) functionality only. So, for example, if a circuit is built between the POS interface of an ML-Series card and the POS interface of a G-Series card, the switch features of the ML-Series card (such as VLANs) are not supported. To achieve Ethernet card POS interoperability between cards of dissimilar series, the POS encapsulation type (LEX, for example), the cyclic redundancy check (CRC) size, and the framing mode (HDLC, for example) must match at the POS endpoints.

It is possible to interoperate cards in the ML Series, G Series, and CE Series. However, the E-Series cards use proprietary E-Series encapsulation and do not support POS interoperability with any of the other Ethernet card series.

ONS 15454 E-Series Interface Cards

Cisco ONS 15454 E-Series Ethernet interface cards integrate a Layer 2 switch into the SONET/SDH platform, enabling multipoint Ethernet service delivery within the optical transport network. Additionally, through the use of the E Series "port-mapped" mode, point-to-point Ethernet services can be provisioned.

The two cards included in the E Series, the E100T-12/E100T-G and the E1000-2/E1000-2-G cards, were the initial Ethernet interface cards introduced with the ONS 15454 platform. (The E100T-G and E1000-2-G are later versions, with the additional capability of functioning with XC10G-equipped systems.) Although these cards are based on solid technologies, the later generations of Ethernet interfaces provide capabilities that the E-Series cards do not. For example, the CE Series and G Series provide for more efficient ways of delivering point-to-point services, while the ML-Series cards deliver a much richer feature set for multipoint services than the E-Series cards. For this reason, E-Series cards are most often found in existing networks and are typically not chosen for new network builds. The E-Series cards are covered only briefly in this chapter; you can find additional information in the Cisco ONS 15454 User Documentation, available at Cisco.com.

Table 7-1 provides a summary of the cards included in the E Series. Both 10-/100-Mbps Ethernet/Fast Ethernet and 1000-Mbps GigE versions of the E Series are available.

Table 7-1   *E-Series Ethernet Card Interface Characteristics*

E-Series Card	Number of User-Side Interfaces	Cross-Connect Card Compatibility	Supported Ethernet Interface Media Types	Port Bandwidth	Interface Connector Types	Port Lamps
E1000-2	2	XCVT	1000BASE-SX 1000BASE-LX	1000 Mbps	GBIC (SX or LX)	ACT/LINK
E1000-2-G	2	All	1000BASE-SX 1000BASE-LX	1000 Mbps	GBIC (SX or LX)	ACT/LINK
E100T-12	12	XCVT	10/100BASE-TX	10 Mbps 100 Mbps	RJ-45	ACT/LINK
E100T-G	12	All	10/100BASE-TX	10 Mbps 100 Mbps	RJ-45	ACT/LINK

ONS 15454 E-Series Card Modes and Circuit Sizes

E-Series cards can be placed in any of the ONS 15454 traffic card slots (1–6 and 12–17), and can operate in three different modes: single-card EtherSwitch, multicard EtherSwitch, and port-mapped mode. These modes are briefly described as follows:

- **Single-card EtherSwitch**—In this mode, each E-Series card in the ONS 15454 chassis acts as an independent Ethernet switching entity.

- **Multicard EtherSwitch**—In multicard mode, two or more E-Series cards act as a single Layer 2 Ethernet switch.

- **Port-mapped**—In port-mapped mode, the E-Series card Layer 2 features are disabled, and individual ports on the E-Series cards terminate point-to-point circuits for Layer 1 Ethernet frame transport.

Table 7-2 lists the circuit sizes that can be used to link E-Series cards for each mode.

Table 7-2   *ONS 15454 SONET/SDH Circuit Sizes for E-Series Cards*

ONS 15454 SONET E-Series Port-Mapped and Single-Card EtherSwitch Modes	ONS 15454 SONET E-Series Multicard EtherSwitch Mode	ONS 15454 SDH E-Series Port-Mapped and Single-Card EtherSwitch Modes	ONS 15454 SDH E-Series Multicard EtherSwitch Mode
STS-1	STS-1	VC4	VC4
STS-3c	STS-3c	VC4-2c	VC4-2c
STS-6c	STS-6c	VC4-4c	—
STS-12c	—	—	—

NOTE   GigE transport with the E-Series GigE cards is limited to 600 Mbps because Synchronous Transport Signal (STS)-12c is the maximum bandwidth that can be terminated to the card. In addition, 10-Mbps Ethernet service requires a minimum circuit size of STS-1 (approximately 50 Mbps), and 100-Mbps Fast Ethernet service requires a minimum circuit size of STS-3c (150 Mbps).

ONS 15454 E-Series Example Application and Provisioning

E-Series Ethernet circuits can be used to link E-Series cards in multiple ONS 15454 nodes in point-to-point (private line), hub-and-spoke, or shared packet ring (SPR) configurations. If a connection is needed between only two nodes, a point-to-point connection is typically used. If connectivity is needed among more than two sites, the hub-and-spoke or SPR topology can be used. Figure 7-1 shows an example of a network that uses the shared packet ring configuration. The ring will pass through E-Series cards in the four nodes (A, B, C, and D) on an OC-12 unidirectional path-switched ring (UPSR). In UPSR shared packet ring configurations, Layer 1 SONET protection is not used (Spanning Tree Protocol [STP] only).

Figure 7-1 *E-Series Ethernet Interface Card Shared Packet Ring Network Example*

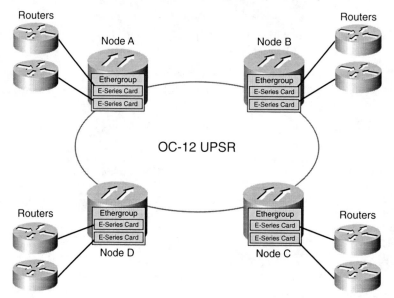

To illustrate an example application of the E-Series cards in an ONS 15454 MSPP network, consider this high-level overview of the steps in setting up a SPR:

Step 1   Verify that the Ethernet cards that will be in the SPR are provisioned for Multicard EtherSwitch mode.

Step 2   Provision the E-Series Ethernet ports. This can include assigning port names, performing autonegotiation and duplex, enabling the ports, assigning port queuing priorities, and enabling STP.

Step 3   Provision the E-Series ports for membership in one or more VLANs.

Step 4   Create the SPR Ethernet circuit. This circuit can be an STS-1, STS-3c, or STS-6c. For the circuit source and destination, select Ethergroup as the slot for the endpoint nodes. You can assign an existing virtual local-area network (VLAN) to the circuit or create a new VLAN.

Step 5   Manually route the SPR circuit through every node in the ring.

ONS 15454 G-Series Interface Cards

Cisco ONS 15454 G-Series (called G1K-4) Ethernet interface cards enable the provisioning of up to four point-to-point Layer 1–mapped Ethernet services, which are also known as Ethernet private line (EPL) services. Each of the card's four ports can be independently equipped with modular Gigabit interface converters (GBICs), as needed, for service requirements. "Private line" is good terminology for this type of service because these connections are provisioned and act much like traditional private line services, such as a DS3 provisioned over a SONET network. The only major differences are that the Ethernet frames are transported natively instead of being packaged into a DS1 or DS3 frame, and the bandwidth of the link is variable within certain STS-1 or STS-Nc circuit sizes (where N can be 1, 3, 6, 9, 12, 24, or 48). Recall that an STS-24c circuit is required to provide full line-rate GigE connectivity between the two circuit endpoints and that the maximum bandwidth that can be used across the four ports of the G-Series cards is equal to 48 STS-1s.

Figure 7-2 shows three examples of possible circuit terminations on the G-Series card.

Figure 7-2   *G-Series Ethernet Interface Card Circuit Termination Examples*

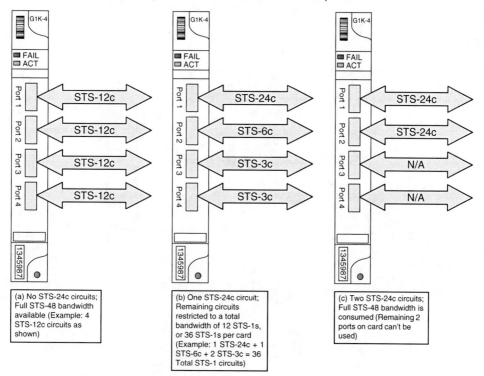

In Figure 7-2(a), each of the four ports is terminating a subrate (STS-12c or lower) circuit. For this type of configuration, there is no restriction on circuit terminations. In Figure 7-2(b), a single line-rate (STS-24c) circuit is terminated on port 1, and the other circuits terminated

on the card are subrate circuits. In this case, a hardware restriction limits the total
bandwidth provisioned across the combination of the other three ports to a maximum of
12 STS-1s. One possibility is shown in the example: An STS-6c circuit and two STS-3c
circuits are used in combination with the single STS-24c circuit. Finally, Figure 7-2(c)
shows an example of two line-rate STS-24c circuits terminated on ports 1 and 2 of the
G-Series card. In this case, the remaining two ports (ports 3 and 4 in the example)
cannot be used because the two line-rate circuits have reached the STS-48 total
card bandwidth limit.

ONS 15454 G-Series Card Example Application

Figure 7-3 shows an example application using the G-Series Ethernet interface card. In
the four-node ONS 15454 OC-48 MSPP ring shown, the G-Series card at Node A is
terminating a circuit from a corresponding G-Series card at each of the other three nodes.
A line-rate circuit is carried between Node A and Node B, while subrate circuits are carried
from Node A to Nodes C and D (an STS-9c and an STS-3c, respectively). This configuration
is known as a "hub-and-spoke" application; Node A is the hub, with spoke circuits
extending to each of the other sites.

Figure 7-3   *G-Series Card Example Application*

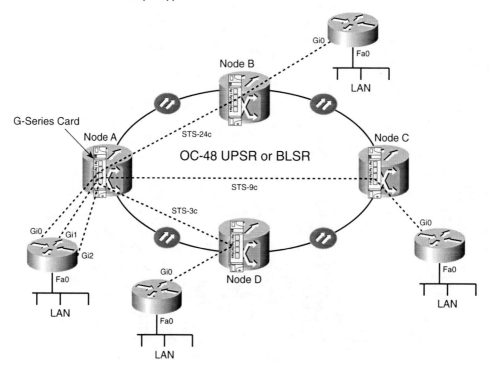

In this example, the router located at Node A is the hub router in the end user's data network. The interfaces of the router at the Node A location labeled Gi0 (GigE 0), Gi1, and Gi2 are equipped with GBICs or Small Form Factor Pluggables (SFPs) of the same wavelength as the GBICs installed in ports 1, 2, and 3 of the G-Series card installed in Node A. A fiber pair, either multimode (for 1000Base-SX GBICs/SFPs) or single-mode (for 1000Base-LX or 1000Base-ZX GBICs/SFPs) connects each G-Series card port to the corresponding router port. Similarly, the remote office routers located at Sites B, C, and D will have wavelength-matched interfaces connected to port 1 of their respective G-Series cards.

In addition to the 1000Base-SX, 1000Base-LX, and 1000Base-ZX GBICs, the G-Series Ethernet cards support a variety of coarse wavelength-division multiplexing (CWDM) and dense wavelength-division multiplexing (DWDM) GBICs.

The G-Series card supports any Layer 3 protocol that can be encapsulated and transported over GigE, such as Internet Protocol (IP) or Internet Packet Exchange (IPX).

Important Features of the ONS 15454 G-Series Card

The G-Series cards provide several important features to ensure smooth operation of the end-user overlay Layer 2/Layer 3 network. Among these features are the support of standard flow control and frame buffering, link aggregation, and end-to-end Ethernet link integrity. Each of these features is discussed briefly in the sections that follow.

Flow Control and Frame Buffering

G-Series cards support flow control and frame buffering as defined in the Institute of Electrical and Electronic Engineers (IEEE) 802.3z. Each port on the G-Series card port has 512 KB of buffer memory to help prevent frame loss in case the attached transmitting source exceeds the throughput bandwidth provisioned to the port. If this buffer memory nears capacity (approximately 95 percent, by default), the G-Series port transmits a pause frame to the attached station instructing it to stop sending frames for a specified amount of time until the memory buffer is depleted and capable of storing additional information.

The G-Series flow control is symmetrical, meaning that it can both send pause frames to the attached device and respond to pause frames received from the attached device. Flow control is set up on the G-Series card based on the provisioning of autonegotiation and flow control settings for the card port. The provisioning of these parameters is done separately in software releases 5.0 and later. If both the autonegotiation and flow control are enabled, the G-Series port attempts to negotiate symmetrical flow control with the attached equipment, and it might or might not be used. If autonegotiation is enabled but flow control is disabled, the G Series proposes no flow control. If autonegotiation is disabled, the G-Series card flow control is enabled or disabled based only on the card provisioning, regardless of the provisioning of the attached port.

The G-Series card ports have provisionable thresholds, or watermarks, for pause/resume signaling to the attached port. These watermarks define the maximum data for the buffer to hold before the transmission of a pause frame, as well as the amount of data at which the card will signal to resume transmission to the attached device. The default value for the transmission of a pause frame is 485 KB; the default value for transmission of a signal to resume transmission is 25 KB.

Link Aggregation

The G-Series card supports point-to-point link aggregation between attached devices. This enables the bundling of multiple GigE links for greater bandwidth or higher levels of equipment redundancy to ensure high availability. G-Series cards support all types of link aggregation, including the IEEE 802.3ad implementation and the Cisco-proprietary Gigabit EtherChannel (GEC). The GEC protocol does not run on the G-Series cards, but it allows the SONET ring to be transparent to GEC-enabled devices routed through the network. STP running on these devices operates as though the multiple links were a single link, and does not block the redundant paths.

The total throughput of the aggregated link is the sum of the bandwidth of the circuits provisioned. Any combination of circuit sizes can be used to form the parallel combination. These circuits can be distributed over G-Series card ports to provide varying levels of redundancy:

- **Ports on the same card**—SONET protection, but no card- or node-level protection. A card or node failure disrupts the link.

- **Ports on different cards, in same chassis**—SONET and card-level protection, but no node-level protection. A card failure results in bandwidth degradation, but service is maintained. A node failure still results in complete link loss.

- **Ports on different cards, in different chassis**—SONET protection, card-level protection, and node-level protection.

Ethernet Link Integrity

The G-Series cards also support Ethernet link integrity, which helps to ensure correct operation of the Layer 2 and Layer 3 protocols running on the attached Ethernet devices, such as customer premises equipment (CPE). This feature operates by shutting down the transmit laser on the G-Series port at each circuit-termination point if any part of the Ethernet link experiences a failure. This allows the end devices to detect a loss of carrier and mark the link as out of service. If one of the endpoint ports is administratively disabled (through Cisco Transport Controller [CTC], for example) or has been set in loopback, it is considered to be in a failed condition for the purposes of the operation of link integrity, and the transmit laser at each end is disabled.

ONS 15454 CE-Series Interface Cards

The ONS 15454 CE-Series (CE-100T-8) card is similar in concept to the G-Series card, in that both cards provide Layer 1 point-to-point (mapped) Ethernet circuits within the bandwidth of a SONET/SDH network. However, several major differences exist, including the number of ports, bandwidth, and support of advanced features such as virtual concatenation (VCAT) and Link Capacity Adjustment Scheme (LCAS). These features are covered later in this chapter.

The CE-Series cards provide eight client interfaces that support line-rate or subrate 10-/100-Mbps Ethernet using standard RJ-45 receptacles on the card's faceplate. Each client port has a corresponding virtual network interface Packet over SONET (POS) port that it uses to encapsulate and transmit Ethernet frames over the SONET/SDH ring. The client ports are statically mapped to the POS ports. Figure 7-4 illustrates the relationship between the front Ethernet ports and their corresponding virtual POS ports. As with the G-Series Ethernet cards, the CE-Series cards can transport any Layer 3 protocol that can be encapsulated and carried over Ethernet, such as IP or IPX.

ONS 15454 CE-Series Queuing

The CE-Series cards also support priority queuing based on IEEE 802.1Q class of service (CoS) and IP type of service (ToS) values that the end user can mark traffic with before inserting it into the CE-Series interface on the SONET/SDH ring. This allows for preferential treatment of latency-sensitive traffic (such as Voice over IP [VoIP]) when buffering occurs. The CE-Series card uses a "normal" queue and a "priority" queue to implement this feature. When buffering takes place, the priority traffic supersedes the normal traffic in transiting the SONET/SDH ring.

The CE-Series cards use thresholds to place traffic into the separate queues. These thresholds are set independently for the CoS and ToS values in the card view for the CE-Series card in CTC. There are 256 values provisionable for the ToS threshold, ranging from 0 to 255. By default, the ToS threshold is set at 255 so that all traffic is treated equally. Similarly, eight values are provisionable for the CoS threshold, ranging from 0 to 7. Again, the default CoS threshold value is set at 7, giving all traffic an equal weight. Priority queuing occurs when one or both of these values are set to below the default. For example, if the ToS threshold value is set to 150, any traffic tagged with a ToS value of 151 to 255 is sent to the priority queue. As another example, if the CoS threshold value is set to 4, any traffic tagged with a CoS value of 5, 6, or 7 is placed in the priority queue.

If the default settings are left in place (255 for ToS and 7 for CoS), no traffic is sent to the priority queue. One or both of these parameters must be configured at a value of lower than the default for priority queuing to occur. If both parameters are configured at less than the default on a CE-Series port, priority queuing is done based on whether the Ethernet frame has a VLAN tag.

If the Ethernet frame is VLAN tagged, priority queuing works as follows:

- If the CoS value is set at 7 (default) and the ToS value is set at 255 (default), priority queuing is disabled and all traffic is sent to the normal queue.

- If the CoS value is set at a value less than 7, the CoS value is used for queuing. The ToS value is ignored.

- If the CoS value is set at 7 and the ToS value is set at less than 255, the ToS value is used for queuing.

If the Ethernet frame is not VLAN tagged, priority queuing is done as follows:

- If the CoS value is set at 7 and the ToS value is set at 255, priority queuing is disabled and all traffic is sent to the normal queue.

- If the ToS value is set at a value less than 255, the ToS value is used for queuing. The CoS value is ignored.

- If the ToS value is set at 255 and the CoS value is set at less than 7, the CoS value is used for queuing.

Priority queuing on the CE Series has no effect when one of the following two conditions exists:

- The CE-100T-8 is configured with STS-3c circuits. Because the STS-3c connection bandwidth exceeds the 100 Mbps bandwidth of the Fast Ethernet port, no congestion occurs and no buffering is needed. Priority queuing is used only when buffering occurs.

- The CE-100T-8 port has flow control enabled. This is also done in the card view of CTC. When flow control is enabled, the prioritization function must be handled by the externally attached device (end-user switch or router), which performs buffering in response to receiving the pause frames as a result of the flow control implementation.

Figure 7-4 *CE-Series Ethernet Card Client-to-POS Port Mappings*

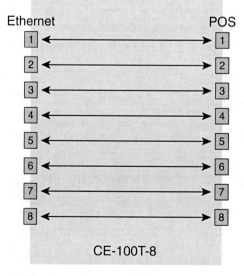

ONS 15454 CE-Series SONET/SDH Circuit Provisioning

The eight POS ports on the CE-Series card can terminate either a single contiguously concatenated circuit or a virtually concatenated circuit. A variety of circuit sizes is available for use with the CE-100T-8. Tables 7-3 and 7-4 show these available sizes for both the SONET and SDH ONS 15454 platforms. High-order VCAT circuits refer to STS-1 level and higher circuits; low-order VCAT circuits are lower than STS-1 (such as VT1.5).

Table 7-3   *SONET Circuit Sizes for the CE-100T-8*

Contiguously Concatenated (CCAT) Circuit Sizes	High-Order Virtually Concatenated (VCAT) Circuit Sizes	Low-Order Virtually Concatenated (VCAT) Circuit Sizes
STS-1	STS-1-1v	VT1.5-nV (n = 1 to 64)
STS-3c	STS-1-2v	—
—	STS-1-3v	—

Table 7-4   *SDH Circuit Sizes for the CE-100T-8*

Contiguously Concatenated (CCAT) Circuit Sizes	VC-3 Virtually Concatenated (VCAT) Circuit Sizes	VC-12 Virtually Concatenated (VCAT) Circuit Sizes
VC-3	VC-3-1v	VC-12-nV (n = 1 to 63)
VC-4	VC-3-2v	—
—	VC-3-3v	—

Although some of the circuit sizes shown represent greater than 100 Mbps of bandwidth, such as the STS-3c or STS-1-3v, the total bandwidth of the client port is limited to 100 Mbps because of the hardware limitation of the front Fast Ethernet port.

The number of circuits that can be terminated on a single CE-100T-8 card, as well as the total combined bandwidth, is determined by the combination of circuit sizes that are configured on the card. The total maximum card bandwidth is limited to STS-12/STM-4. Table 7-5 shows the different available combinations for high-order CCAT and VCAT circuit sizes for both SONET and SDH.

Table 7-5   *CCAT High-Order Circuit Size Available Combinations for SONET/SDH*

CCAT SONET Circuits	
Number of STS-3c Circuits	Maximum Number of STS-1 Circuits
0	8
1	7
2	6
3	3
4	0

continues

Table 7-5 *CCAT High-Order Circuit Size Available Combinations for SONET/SDH (Continued)*

CCAT SDH Circuits	
Number of VC-4 Circuits	**Maximum Number of VC-3 Circuits**
0	8
1	7
2	6
3	3
4	0
VCAT SONET Circuits	
Number of STS-1-3v Circuits	**Maximum Number of STS-1-2v Circuits**
0	4
1	3
2	2
3	1
4	0
VCAT SDH Circuits	
Number of VC-3-3v Circuits	**Maximum Number of VC-3-2v Circuits**
0	4
1	3
2	2
3	1
4	0

Circuits created on the CE-100T-8 card are assigned to one of four hardware-based "pools." Each pool has a capacity of up to three STS-1s (SONET) or three VC-3s (SDH). You can see the utilization of each of the four pools by accessing the Maintenance tab in the card-level view for a CE-100T-8 card in CTC. Figure 7-5 shows an example of this view for a CE-100T-8 with multiple circuit terminations. The Pool Utilization table displays the pool circuit type, usage, and pool usage for each pool. For each of the POS ports, the POS Port Map table shows the circuit size and type, the LCAS type for VCAT circuits, and the number (from 1 to 4) of the pool to which the circuit is assigned.

Certain rules govern the manner in which circuits are allocated in the ONS 15454 to each of the pools. For example, all VCAT circuit members must be from the same pool.

The Cisco ONS 15454 documentation available at Cisco.com has a detailed description of how this works.

Figure 7-5 *CE-Series CTC Maintenance Tab View: Pool Utilization*

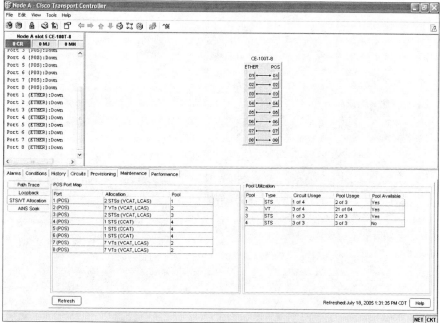

ONS 15454 CE-Series Card Example Application

Figure 7-6 shows an example application of the CE-Series card. Node A is the "hub" site in the end-user's data network, with Ethernet connections required from Node A to each of the other locations (Nodes B, C, and D) in the SONET OC-12 network. Node B has a connection bandwidth requirement of 50 Mbps, Node C requires 100 Mbps, and Node D requires 10 Mbps. To implement these service requirements, a single CE-100T-8 card is installed in each MSPP node, with the card in Node A terminating the three circuit connections from the other nodes, and each card in Nodes B, C, and D terminating a single circuit from hub Node A.

As mentioned, Node B has a bandwidth requirement of 50 Mbps, so a single STS-1 circuit is provisioned between a port on the CE card at Node A and the CE card at Node B. Of course, this circuit will be Layer 1–protected by the SONET protection scheme implemented for this network, which can be either UPSR or bidirectional line switch ring (BLSR).

Figure 7-6 *CE-Series Ethernet Interface Card Application Example*

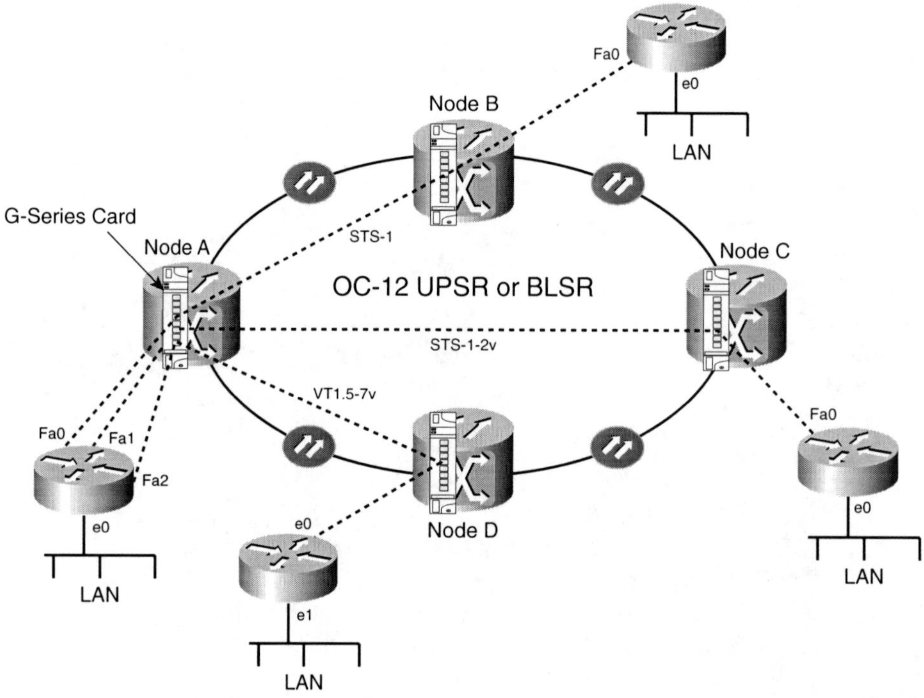

NOTE	The connection between the customer's routers at each end is made using Fast Ethernet ports, which have an available bandwidth of 100 Mbps. However, the total data throughput is limited to the STS-1 bandwidth provided by the ONS 15454 CE-Series circuit.

Node C has a larger bandwidth requirement, 100 Mbps. This connection can be provided using two virtually concatenated STS-1 circuits, referred to as an STS-1-2v. These STS-1 links can be contiguous within the OC-12 bandwidth of the ring, such as STS-1 #1 and #2, or they can be separate, such as STS-1 #3 and #6. This circuit can be terminated on any port of the CE card at Node C and are terminated on any available port on the card at Node A. Finally, to provide the 10-Mbps requirement for the customer location at Node D, seven virtually concatenated VT1.5 circuits can be used (VT1.5-7v). This circuit can be terminated on any port on the CE card at Node D, with the other circuit endpoint terminating on an available port on the card at Node A.

ONS 15454 ML-Series Interface Cards

The ONS 15454 ML (Multilayer)-Series cards are essentially Layer 3 switches embedded in the SONET/SDH transport system. These cards are used to efficiently transport Ethernet frames between two or more locations on the network by using nondedicated SONET/SDH circuit links. Unlike the G-Series and CE-Series Ethernet cards, which use dedicated or "nailed-up" point-to-point connections over the MSPP network, the ML Series uses shared pathways to link two or more locations, providing a VLAN across a metro or regional area.

The ML family consists of three cards. The ML1000-2 card is a GigE interface card; the ML100T-12 and ML100X-8 are Fast Ethernet interface cards. Each of the three ML-Series cards uses the same basic hardware and software, and has similar features. The primary difference between the cards is in their Layer 1 faceplate user-side interface types. Table 7-6 summarizes the Layer 1 interface features for each of the three card types.

Table 7-6   *ML-Series Ethernet Card Interface Characteristics*

ML-Series Card	Number of User-Side Interfaces	Interface Port Numbers (CTC)	Supported Ethernet Interface Media Types	Port Bandwidth	Interface Connector Types	Port Lamps
ML1000-2	2	0,1	1000Base-SX 1000Base-LX	1000 Mbps	SFP (SX or LX)	ACT LINK
ML100T-12	12	0–11	10/100Base-TX	10 Mbps 100 Mbps	RJ-45	ACT LINK
ML100X-8	8	0–7	100Base-FX 100Base-LX10	100 Mbps	SFP (FX or LX 10)	ACT LINK

As shown in Table 7-6, each ML-Series card has both an ACT (Active) and a LINK LED for each interface port. The ACT and LINK LED states have the following meanings for all cards:

- **ACT LED (Amber)**—A blinking ACT LED indicates that traffic is flowing (normal operation). A steady ACT LED indicates that a link is detected, but an issue is preventing traffic from transiting properly.

- **LINK LED (Green)**—A blinking LINK LED indicates that traffic is flowing. The rate at which this LED flashes is proportional to the rate at which traffic is being sent and received over the port. A solid LINK LED indicates that a link is present and detected but that traffic is not being sent and received over the port.

Each ML-Series card has two virtual Packet over SONET/SDH (POS) ports, which interface the faceplate Ethernet ports to the SONET/SDH ring. These POS ports are administered through the Cisco Internetwork Operating System and can be thought of as the "trunk" or network-side ports for the ML-Series cards. They function in a manner comparable to the way that OC-N/STM-N ports operate in a SONET/SDH interface card. These POS ports support CCAT or VCAT circuit types, as well as a software-based LCAS. The POS ports are called POS0 and POS1 in the Internet Operating System (IOS) software interface. They can be provisioned in a point-to-point manner with POS ports on other Ethernet cards in the network, or as East- and West-facing interfaces in a metro-/regional-area Ethernet overlay ring configuration such as resilient packet ring (RPR). RPR is discussed in greater detail later in this chapter. Figure 7-7 illustrates the POS ports and Ethernet ports for an ML100T-12 card.

Figure 7-7 *ML100T-12 Ethernet Interface Card: POS and Fast Ethernet Ports*

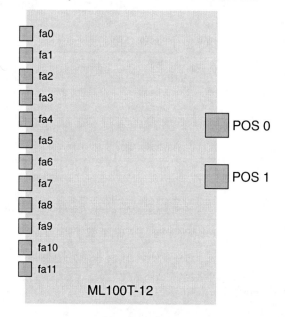

The ML-Series cards are provisioned and managed by a combination of CTC and Cisco IOS software. Cisco Transport Manager (CTM), which is the element management system (EMS) for multiple Cisco networking equipment (including the ONS 15454), can also be used for provisioning and managing the ML-Series cards. IOS software is the primary user interface for the ML-Series cards and is used for tasks such as Ethernet interface configuration, bridging configuration, VLAN configuration, and RPR configuration.

CTC is also used for configuring and managing the ML Series, and is required for SONET/SDH circuit provisioning to link the ML-Series card(s) with other Ethernet cards in the

SONET/SDH ring. This cannot be done using the IOS command-line interface (CLI). CTC can be used for the following actions related to the ML-Series cards:

- **Display of port-level Ethernet statistics for the Ethernet and POS ports**—These statistics can be displayed using the card-level view in CTC. Examples of the statistics available include link status, total packets and bytes received, total packets and bytes transmitted, and total received errors.

- **Display of Ethernet and POS ports provisioning information**—Although the Ethernet and POS ports must be provisioned using IOS and cannot be provisioned through CTC (other than port name), the current provisioning of the ports can be displayed in card-level view in CTC. Examples of provisioning information that is displayed in CTC include the administrative state, link state, and Maximum Transmission Unit (MTU) size. Provisioning information displayed varies by card type.

- **Provisioning of framing mode**—The provisioning mode for the POS ports can be set in CTC. The options are High-Level Data Link Control (HDLC) or Frame-Mapped Generic Framing Protocol (GFP-F). A connected POS port must have the same framing type provisioned as the far-end peer port.

- **Managing SONET/SDH alarms**—CTC is used to manage and report alarms from the ML-Series cards in the same manner as it does for other ONS 15454 cards. These can be viewed in the Alarms tab in CTC. Examples of ML-Series alarms include Carrier Loss, Encapsulation Mismatch, RPR Wrapped, and IOS Configuration alarms.

- **Display of maintenance information**—The Maintenance tab in card view of the ML-Series card displays the card's Field Programmable Gate Array (FPGA) version, and also displays the shelf type (SONET or SDH) in which the card is installed. Two FPGA versions exist: one for HDLC framing and one for GFP-F. In ONS 15454 software Release 5.0 and later, the correct version is automatically loaded to the card when the user modifies the framing type.

- **J1 path trace**—This is a mechanism for monitoring problems or changes in the SONET/SDH traffic terminating on the card. The J1 path trace uses a repeated, fixed number of bytes for testing purposes. If the received string on a circuit drop port does not match the expected information, an alarm is raised. ML-Series cards support the J1 path trace function, and it is managed in the CTC card-level view.

ML-Series Card Transport Architecture Examples

From a transport network perspective, the ML-Series cards can be used several different ways in a SONET/SDH network. In this section, you will learn three different options for ML-based network topologies, including point-to-point, hub and spoke, and shared resilient packet ring (RPR). The type of network topology implemented depends on several factors, including the traffic pattern required and the availability of fiber facilities between ONS 15454 locations.

Point-to-Point Transport Architecture

ML-Series cards can be connected in a point-to-point design to transport data between two MSPP nodes in a SONET/SDH ring. Figure 7-8 shows this type of network.

Figure 7-8 *ML-Series Point-to-Point Network Application Example*

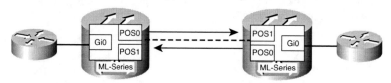

RPR Transport Architecture

RPR is a network architecture designed to reliably transport Ethernet frames at high speeds over local, metro, or regional optical-fiber ring networks. RPR is a Layer 2 technology. It was standardized by the IEEE in 2004 and is known as IEEE 802.17. The current generation of Cisco ONS 15454 ML Series uses a prestandard implementation of RPR and does not fully comply with the IEEE standard. A discussion of differences between IEEE 802.17 and ML-Series RPR comes later in this chapter.

RPR is used in lieu of STP or Rapid Spanning Tree Protocol (RSTP) to ensure loop-free transmission with faster network convergence time in the event of a failure. Unlike STP or RSTP, RPR uses spatial reuse of bandwidth on the SONET/SDH network and does not need to block redundant paths.

A detailed description of RPR follows in this section, but first review the following key benefits of ML-Series RPR:

- **Dual rotating rings**—Traffic is sent both clockwise and counterclockwise simultaneously, to fully use the bandwidth allocated to the RPR on the SONET/SDH ring.

- **High availability**—RPR offers restoration from failures, such as a fiber cut or node failure, in less than 50 ms, and can accomplish this without using Layer 1 SONET/SDH-protected circuits.

- **Spatial reuse**—The same bandwidth time slots can be used in different sections in the RPR simultaneously, creating excellent bandwidth efficiency for Ethernet transport.

- **Granular RPR capacity**—Various levels of bandwidth, up to line-rate 1 Gbps transmission, can be supported over ML-Series RPR topologies by provisioning the appropriate SONET/SDH CCAT or VCAT circuit size.

RPR Operation in the ONS 15454 ML-Series Cards

To explore the application and operation of RPR using the ONS 15454 ML-Series interface cards, consider the example shown in Figure 7-9. In this example, you want to connect four locations with an ONS 15454 SONET MSPP OC-48 UPSR ring, and you want to allow any-to-any Ethernet connectivity among all locations with carrier-class (less than 50 ms) failover in the event of a facility failure. This can be provisioned using ML-Series cards in an RPR network architecture. Each node will contain an ML1000-2 card installed in Slot 5. See Figure 7-10 for a typical node configuration in this network.

Figure 7-9   *ML-Series RPR Network Application Example*

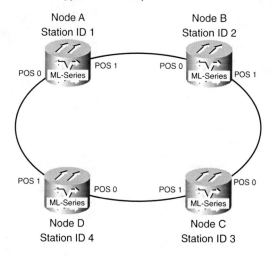

Figure 7-10   *Typical Node Configuration—ML-Series RPR Network Application Example*

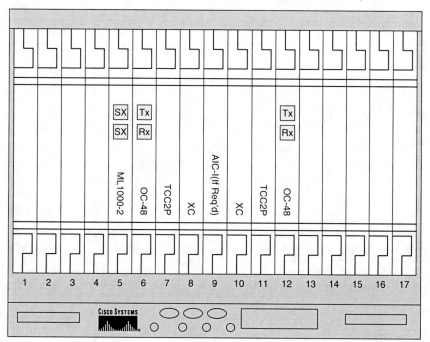

As shown in Figure 7-9, you connect the ML-Series cards installed in each of the nodes with a point-to-point, unprotected (from a Layer 1 ring standpoint) SONET circuit, connecting POS port 1 at one node to POS port 0 at the next adjacent node. For example, Node A POS port 1 is connected to Node B POS port 0. Again, these circuits are provisioned as unprotected because the failover protection is provided through the Layer 2 RPR instead of the Layer 1 SONET protocol. As previously mentioned, multiple circuit sizes can be used based on the bandwidth required between ML-Series cards participating in the RPR. Table 7-7 shows the available circuit sizes that can be provisioned between POS ports on ML-Series cards.

Table 7-7   *Available SONET and SDH Circuits for ML-Series POS Ports*

SONET Circuit Size	SDH Circuit Size	Circuit Type	Approximate Circuit Link Bandwidth
STS-1	VC-3	Nonconcatenated	50 Mbps
STS-1-2v	VC-3-2v	Virtually concatenated	100 Mbps
STS-3c	STM-1 (VC4)	Contiguously concatenated	150 Mbps
STS-6c	STM-2 (VC4-2c)	Contiguously concatenated	300 Mbps
STS-3c-2v	VC-4-2v	Virtually concatenated	300 Mbps
STS-9c	STM-3 (VC4-3c)	Contiguously concatenated	450 Mbps
STS-12c	STM-4 (VC4-4c)	Contiguously concatenated	600 Mbps
STS-24c	STM-8 (VC4-8c)	Contiguously concatenated	1200 Mbps (1.2 Gbps)
STS-12c-2v	VC-4-4c-2v	Virtually concatenated	1200 Mbps (1.2 Gbps)

To provide the maximum amount of bandwidth between ML-Series cards, STS-24c circuits are provisioned between network-adjacent POS ports in our example network. These circuits can be provisioned using CTC. Table 7-8 provides a summary of the circuits provisioned for the RPR.

Table 7-8   *Circuits Provisioned for Example RPR Network*

Source	Destination	Circuit Size
Node A/Slot 5/POS1	Node B/Slot 5/POS0	STS-24c
Node B/Slot 5/POS1	Node C/Slot 5/POS0	STS-24c
Node C/Slot 5/POS1	Node D/Slot 5/POS0	STS-24c
Node D/Slot 5/POS1	Node A/Slot 5/POS0	STS-24c

When virtually concatenated circuits are used with the ML-Series cards, a software-based LCAS can be used. ML-Series LCAS allows for the automatic addition or removal of a VCAT circuit member in case of the failure or recovery of a SONET/SDH facility. For example, suppose that you are using STS-12c-2v (two virtually concatenated STS-12c circuits) for the RPR instead of single STS-24c circuits in a 2-Fiber SONET BLSR. You could provision one of the two members of the STS-12c-2v as a Protection Channel Access (PCA) circuit. This PCA circuit would be dropped in the event of a line protection switch, but LCAS would allow this member to be dropped from the VCAT group and the link would continue to function with half the normal bandwidth. Software-based LCAS differs from the hardware-based version primarily because it is not errorless in operation.

Each of the ML-Series cards installed in the ONS 15454 nodes is designated with a unique RPR Station ID: The Node A ML-Series card is assigned to be RPR Station 1, Node B is RPR Station 2, Node C is RPR Station 3, and Node D is RPR Station 4.

NOTE   In some cases, multiple ML-Series cards might be installed in the same ONS 15454 node and might be participating in the same RPR. This could be the case if card redundancy is required to provide a very high level of network availability at one or more MSPP locations. In this scenario, the two ML-Series cards still have unique RPR Station IDs, even though they are located in the same node.

Each ML-Series card can insert traffic onto the RPR on either the POS0 or POS1 interfaces, and Ethernet frames are stripped from the RPR network at either the source or destination RPR station, depending on the traffic type. Frames are forwarded onto the RPR using either the POS0 or POS1 based on the following criteria:

- For unicast Ethernet traffic, a calculation involving the source and destination addresses of the unicast frame is performed to determine which POS interface to forward the traffic out.

- For multicast and broadcast Ethernet traffic, or for Ethernet frames with unknown destination addresses, the POS interface to forward the frame is determined by the Ethernet interface through which the frame is received. Half of the ML-Series card's Ethernet interfaces are associated with POS0, and the other half are associated with POS1.

ONS 15454 ML-Series RPR Frame Format

The RPR framing is added to each Ethernet frame for transport on the RPR, and is removed at the egress ML-Series port. The RPR frame format on the current ML Series is Cisco proprietary. Figure 7-11 shows the RPR frame format and encapsulation of an Ethernet frame.

Figure 7-11 *ML-Series RPR Frame Format and Encapsulation of an Ethernet Frame*

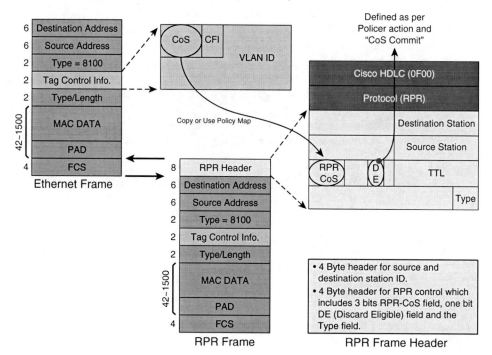

The RPR header includes the following:

- **Destination station ID**—An 8-bit field that contains the station ID of the frame destination. Two well-known destination station IDs are 0xFF (multicast destination address MAC) and 0x00 (unknown destination address MAC flood)

- **Source station ID**—Also an 8-bit field.

- **RPR control**—4 bytes, including the following:

  — CoS—3 bits

  — Discard Eligible Flag—1 bit

  — Time to Live (TTL)—9 bits

  — Type—The packet type (data or control)

ML-Series Bridge Groups

Unlike many Layer 2 switches, such as the Cisco 6500 series, the ML-Series cards forwards no traffic by default, including untagged frames and frames tagged as VLAN 1. For any traffic to be bridged by the ML Series over the RPR, one or more bridge groups must be created. The bridge group controls which ports traffic can be forwarded to and which VLANs can be forwarded on those ports. The bridge group can be compared to the cross-connect matrix of the SONET/SDH network node, in that each bridge group on the ML Series is locally significant (like the cross-connect matrix on an individual ONS 15454 node) and provides local bridging between the selected interfaces on the ML-Series card. Configuration of bridge groups can be accomplished using the IOS CLI to the ML Series.

Figure 7-12 shows the relationship between the ML-Series interfaces, including the physical client interfaces, the virtual POS interfaces, and the RPR through the bridge group. Each bridge group is configured with a bridge group number in the range of 1 to 255. Ethernet or GigE interfaces (or subinterfaces) are associated with bridge groups as required. The ML-Series card bridges traffic among the interfaces configured with the same bridge group. Any interface that is not configured with a bridge group cannot forward bridged traffic. For bridge groups that need to be connected to the RPR ring, the bridge group is configured to a subinterface on the RPR interface.

Figure 7-12 *Ethernet Bridging over RPR (Bridge Group and VLAN)*

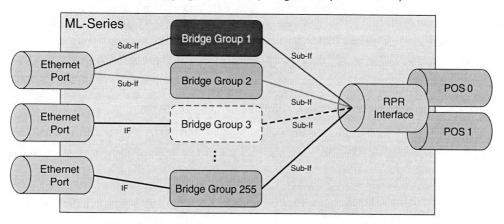

Ethernet Bridging over RPR (Bridge Group and VLAN)

RPR Operation

ML-Series cards with the RPR implementation prevent loops by using source and destination station frame stripping, and by using the RPR header information. If a facility failure occurs, such as a fiber cable cut, frame loss is prevented through a wrapping procedure executed within the RPR protocol. The forwarding table in the ML Series is implemented with high-speed content addressable memory (CAM). When a frame is received and a database lookup is done from SONET/SDH, a typical Layer 2 learn occurs for the source MAC address; the source station ID is also saved.

Each ML-Series RPR card can handle an Ethernet frame in three ways: with bridging, pass-through, and stripping. Each of these is briefly described as follows:

- **Bridging**—A packet from a client Ethernet port is inserted into the RPR ring, with the addition of the source station ID (which is the local ML-Series card's station ID) and either the well-known destination station ID (which is set to 0x00 for packets with an unknown unicast MAC destination address, and 0xFF for packets with a multicast/broadcast MAC destination address), or the particular destination station ID, if it is determined from the table lookup (such as a previously learned MAC). The transmit direction on the RPR ring is determined based on the previously discussed criteria. Before transmitting the frame in either direction on the RPR ring, the ML-Series card sets the direction bit based on whether the frame is being sent out on interface POS0 or interface POS1. If the packet is being sent out on POS0, the direction bit is set to 0. If it is sent out on POS1, the bit is set to 1.

- **Pass-through**—The pass-through stations forward the RPR frame around the ring as transit or express traffic without performing the normal forwarding lookups, such as bridging table destination address MAC and source address MAC lookups. The transit or pass-through decision is made on the ML-Series cards using the Destination Station field and the bypass flag of each RPR frame. When the destination station does not match the ML-Series card's station ID or one of the well-known destination station values, the bypass flag is set in the received RPR frame and is treated as transit or pass-through traffic.

- **Stripping**—When an ML-Series card receives a multicast or broadcast RPR frame and the Source Station field matches the ML-Series card's station ID (under normal, nonwrapped conditions), that frame is stripped off the ring.

To further understand each of these operations, examine how some frames are handled in your example RPR network.

In Figure 7-13, you can see the sequence of events involved in transporting a unicast packet from a router connected to the ML-Series card in Node A (RPR Station 1) to a router connected to the ML-Series card in Node C (RPR Station 3). Assume that this is the initial Ethernet frame added from Station 1.

The frame arrives at the ML1000-2 card on client interface GigE 0 (gi0). The Station 1 ML Series adds the RPR header to the Ethernet frame and sets its station ID (1) in the Source Station field of the RPR header. Station 1 also sets the destination station address in the packet to the well-known 0x00 address to flood the packet onto the ring. Station 1 then determines the direction of transit on the RPR ring, based on the previously mentioned algorithm. In this example, assume that the packet is transmitted in the clockwise direction, which is the link from Node A to Node B, using the POS1 interface.

Figure 7-13 *RPR Operation: Unicast Frame Transport from Node A to Node C*

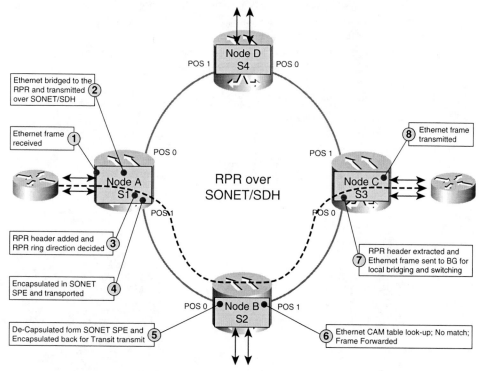

When Station 2 (which is the ML-Series card located in Slot 5 at Node B) receives the packet on its POS0 interface, it checks the source station ID and determines that it did not originate this packet. Station 2 then examines the destination station ID and finds that it is not a transit packet because the destination station is set for 0x00. Therefore, Station 2 must perform a lookup in its bridging table. Station 2 performs a MAC lookup for the source address MAC and completes a standard MAC learning. It also learns the source station from the packet and saves it in the CAM table. Because Station 2 finds no match in the bridging table lookup, it floods the packet forward on the Node B to Node C link using its POS1 interface.

Any subsequent station on the RPR would perform the same operation as Station 2 until the packet arrives at the POS0 interface of the proper destination station, which, in this case, is Station 3. If none of the other stations participating in the RPR can find a match for the destination address MAC, the packet eventually reaches the originating station (in this case, Station 1 on its POS0) and is dropped because the source station ID matches and the bypass flag is not set.

Figure 7-14 illustrates the RPR operation for the initial frame transmitted from Station 3 to Station 1. Because this frame is added at one of the client interfaces at Station 3, it is not flooded as in the previous example because the bridging table lookup finds the destination address MAC in the table along with the destination station ID. This information was learned from the initial packet sent from Station 1 in the previous example. The frame is encapsulated with the RPR header, Station 3's ID is set as the source station, and Station 1's ID is set as the destination station. Station 3 then performs the calculation to determine the direction of transmission onto the RPR. In this case, it is transmitted clockwise on the link between Node C and Node D using its POS1 interface.

Figure 7-14 *RPR Operation: Unicast Frame Transport from Node C to Node A*

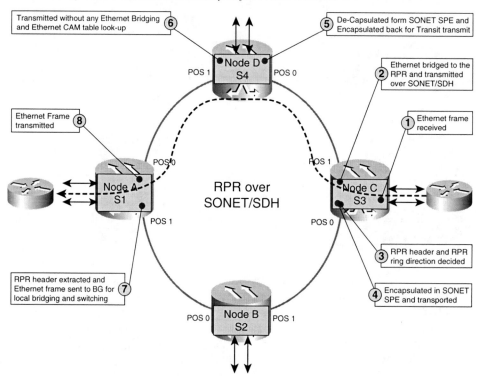

Station 4, which is the ML-Series card located in Slot 5 of Node D, checks the source station in the frame header and determines that it did not originate the frame. Station 4 then can determine that this is a pass-through frame by looking at the destination station ID, which is set for Station 1. Therefore, it forwards the frame onto the ring using its POS1 interface.

When Station 1 receives the frame on its POS0 interface, it determines that the destination station is a match. A bridging table lookup and learn are performed, with two possible outcomes. In the normal case, a match is found in the table and the frame is forwarded out one of the ML-Series card's front client ports. In this case, the frame is not transmitted back out onto the RPR. However, if the destination address MAC is not found, the frame is flooded back onto the RPR using the POS1 interface. In the second scenario, the frame will

eventually be removed from the RPR by the originating station, which is Station 3 in this
example. Station 3 would also invalidate its CAM entry used to send the destination address
MAC to Station 1.

RPR Operation in Failure Scenarios

The ML-Series RPR implementation efficiently handles failure scenarios, such as fiber
facility failures or node/station failures. To illustrate the operation of the RPR in a fiber cut,
refer to the example network shown in Figure 7-15.

Figure 7-15 *RPR Operation During Fiber Facility Failure*

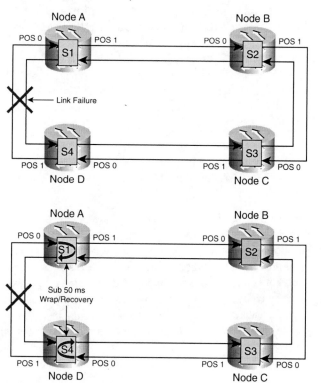

When the fiber facility between Nodes A and D is cut, the failure is detected in less than
50 ms, and the RPR stations (in this case, stations 1 and 4) transition into "wrap" mode and
can be said to be "wrapped." When in wrapped mode, the RPR stations wrap packets back
into the direction opposite the facility failure. Figure 7-15 shows this wrapping operation.
To understand how packets are handled under the wrap failure-recovery system, examine
a unicast packet sent from Node B to Node D.

After the adjacent stations 1 and 4 detect the cut, a unicast packet is added at Station 2 (Node B) with a destination station of 4 (Node D). Station 2 encapsulates the packet with the RPR header, adds its station ID as the source station, and performs a calculation to determine the direction that the packet will traverse the ring. In this case, the packet will be forwarded over the link from Node B to Node A in the counterclockwise direction on the ring, using the Node B POS0 interface. After receiving the packet on its POS1 interface, Station 1 would normally inspect the packet's destination station ID, see that it is intended for Station 4, and forward it out using its POS0 interface. However, because of the fiber cut, Station 1 is in the wrapped condition. Because Station 1 is wrapping, it sets the wrap flag (equal to 1), sets the bypass flag (equal to 1), assigns its station ID of 1 to the Wrap Station field, and transmits the packet back out the incoming port (POS1).

When Station 2 receives the packet (on the POS0 interface), it detects that it is the originating station by finding its station ID as the source station. However, because the bypass flag is set, it resets the bypass flag (equal to 0) and performs a pass-through operation, sending the packet back out onto the ring using its POS1 interface. When Station 3 receives the packet on its POS0 interface, it determines that it is not the originating station (because the source station is set as Station 2) and that it is not the intended destination (because the destination station is set as Station 4), so the packet is treated as transit and forwarded out onto the ring using its POS1 interface.

Station 4, which is the ML-Series card in Slot 5 at Node D, receives the packet on its POS0 interface. After determining that it is the destination station, it performs a lookup and forwards the packet out of a client port. The packet is then not forwarded back onto the RPR (dropped).

Another possible failure scenario is the loss of a station in the RPR because of facility isolation or catastrophic node failure (such as a loss of power to the MSPP node). In this case, the adjacent nodes to the failed station perform a wrap and packets are handled much as described in the previous fiber cut scenario. A special case of this scenario occurs when the station that fails (or is removed) is the intended destination of a packet. In this situation, when the packet arrives at the first wrapping node, it sets the wrap and bypass flags (equal to 1) and wraps the packet. All other nodes then perform a pass-through operation as the packet circles the ring in the opposite direction until it arrives at the other wrapping node, where it is dropped. Figure 7-16 shows this situation.

RPR Spatial Reuse

Now that we have examined the operation of the RPR, including the capability to avoid loops and recover from facility and node failures, it is easier to understand the capability of the RPR network to simultaneously transmit packets among multiple station sets on the ring. This is shown in Figure 7-17: Packet flows are simultaneously occurring between Station 1 and Station 3 (Frame A), Station 4 and Station 5 (Frame B), and Station 7 and Station 10 (Frame C).

Figure 7-16 *RPR Operation During Single-Node Isolation/Failure*

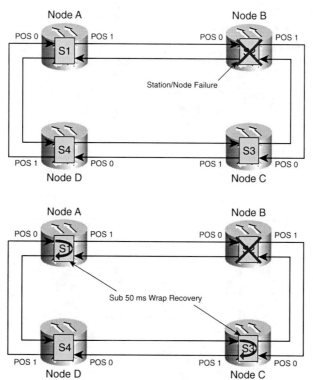

Figure 7-17 *RPR Spatial Reuse*

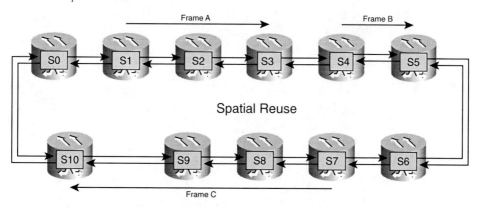

Provisioning RPR Using the ONS 15454 ML-Series Cards

Setting up an RPR topology with the ML-Series cards can be accomplished using CTC and the Cisco IOS CLI for the ML Series. Configuration can also be accomplished using CTM, which is Cisco's integrated optical element-management system. Chapter 15, "Large-Scale Management," covers CTM in greater detail. This section concentrates on RPR provisioning using CTC. The network in Figure 7-18 shows the process of creating the RPR. In this example, a group of seven ML-Series cards installed in four different ONS 15454 (SONET) MSPP nodes (participating in an OC-48 UPSR) will be placed into the RPR, with three VLANs, as shown in the figure.

Figure 7-18 *ML-Series RPR Network Example Configuration*

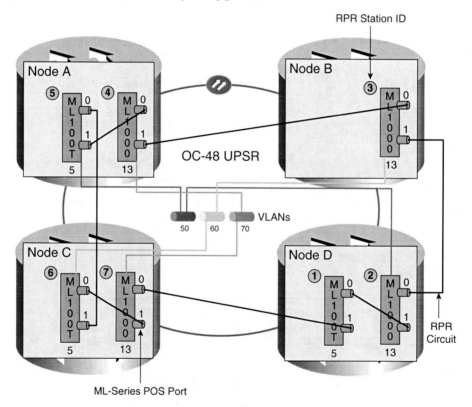

CTC RPR Circuit and Framing Mode Provisioning

Provisioning the "backbone" circuits between the RPR stations in CTC is a fairly straightforward process. This is essentially the same process as provisioning a normal TDM circuit, such as a DS1 or DS3; the primary difference is that the RPR circuits are typically provisioned as unprotected. This is because failover is provided using the RPR station wrap functionality instead of SONET protection switching.

A high-level overview of this circuit-provisioning process in CTC is provided in the following steps:

Step 1   In the Circuit Type window, select a circuit type of either STS (for contiguously concatenated circuits) or STS-V (for virtually concatenated circuits). For this example, select STS.

Step 2   In the Circuit Attributes window, select the circuit size. For this example, an STS-12c is used.

Step 3   For the circuit source, select one of the nodes to be placed in the RPR, the appropriate ML-Series slot, and one of the two POS ports (POS0 or POS1). In this example, select Node A, the ML100T in Slot 5, and the POS1 port as the source.

Step 4   For the circuit destination, select the next adjacent node, the correct ML-Series card, and the appropriate POS port (opposite of the POS number for the source). For the purposes of this example, select Node A (again), the ML1000-2 in Slot 13, and POS0 as the destination.

Step 5   Next, in the Circuit Routing Preferences window, uncheck the Fully Protected Path option so that the circuit will not be created with SONET UPSR Layer 1 protection.

Step 6   To complete this circuit, in the Route Review/Edit window, map the circuit from the source node to the destination node and select the STS path to use. If two or more ML-Series cards are in the same node, as is the case with some nodes in this example, this step is unnecessary because the circuit exists only internally to the particular MSPP node.

Step 7   Similar circuits should then be built linking the POS ports for all the other ML-Series cards that participate in the RPR until the initial source card is reached. So, for this example, this process would be repeated for the following links:

— Node A/Slot 13/POS1 to Node B/Slot 13/POS0

— Node B/Slot 13/POS1 to Node C/Slot 13/POS0

— Node C/Slot 13/POS1 to Node C/Slot 5/POS0

— Node C/Slot 5/POS1 to Node D/Slot 13/POS0

— Node D/Slot 13/POS1 to Node D/Slot 5/POS0

— Node D/Slot 5/POS1 to Node A/Slot 5/POS0

Figure 7-19 shows an example of the CTC Detailed Circuit View for an RPR circuit.

Figure 7-19 *CTC Detailed Circuit View for an Unprotected RPR Circuit*

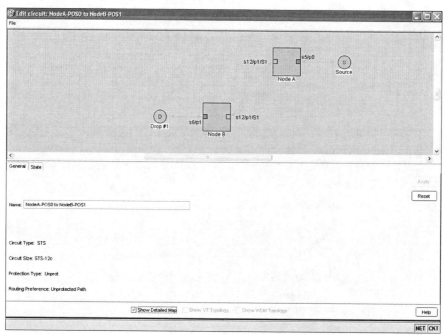

In addition to the SONET/SDH circuit provisioning, the framing mode for the ML-Series card is configured using CTC. This is accomplished in card view for the ML-Series card, under the Provisioning tab and Card subtab. Framing mode options are HDLC or GFP-F. HDLC is the default. Peer POS ports must have the same framing mode.

NOTE   The ML-Series card will reboot if the configured framing mode is changed.

ML-Series IOS Configuration File Management

After provisioning the circuits in CTC, the remaining provisioning required for RPR functionality can be performed in the ML-Series IOS CLI. Each ML-Series card is configured independently. The ML-Series CLI is typically accessed either by using the console port located on the faceplate of the card (using the RJ-11 to RJ-45 adapter cable that ships with the ML-Series cards) or by opening an IOS CLI session in CTC. It can also be accessed using a Telnet session.

You can load the startup configuration file that the ML Series requires using CTC. Cisco supplies a sample IOS configuration file, called Basic-IOS-startup-config.txt on the ONS 15454 software CDs. CTC can also be used to load a user-supplied configuration file. If the configuration file is uploaded to the Timing, Communications, and Control Cards

(TCC+, TCC2, or TCC2P) before the ML-Series card is installed, the ML Series is download and applies the file when it is installed. This allows the card to immediately begin functioning as a fully configured card. If the configuration file is uploaded after the ML Series is installed, the ML-Series card will need to be reset to use the file.

The **copy-run-start** command executed in the CLI copies the current running configuration on the ML-Series card to the Flash memory of the ML Series. This is also saved to the node's database-restoration file after approximately 30 seconds. If the user performs a database backup, this IOS configuration file is stored in the backup file along with all of the other provisioning in the node. The ML-Series card's IOS configuration file can be managed by right-clicking the ML-Series card in CTC shelf view, which opens a shortcut menu. This is also one way to open a Telnet IOS CLI session with the card.

RPR Provisioning and Verification in the IOS CLI

In the example to demonstrate the IOS RPR configuration process, place all seven of the ML-Series cards in the four-node ring into the same RPR. Two of the cards, the ML100T-12 in Slot 5 in Node A and the ML1000-2 in Slot 13 of Node C, will participate in VLAN 50. Likewise, VLANs 60 and 70 will also be used between two stations (each) in the RPR, as shown in Figure 7-18.

Table 7-9 shows a step-by-step configuration of the ML-Series card in Node D, Slot 5. Assume that an IOS session has been established with the ML-Series card using the Open IOS Session shortcut option in CTC shelf view. A similar configuration is performed for each of the other ML-Series cards that will be placed into the RPR.

Table 7-9   *IOS RPR Configuration for ML-Series Card in Node A, Slot 5*

Step	Command	Purpose
1	Router>**enable**	Enter privileged EXEC mode.
2	Router#**config terminal**	Enter terminal configuration mode.
3	Router(config)#**bridge irb**	Turn on Integrated Routing and Bridging (IRB). Enables the Cisco IOS software to both route and bridge a given protocol on separate interfaces within a single ML-Series card.
4	Router(config)#**hostname ML100NodeDSlot5**	Assigns a hostname to the ML-Series card. Use a name that is descriptive of the card type and location.
5	ML100NodeDSlot5(config)#**interface spr1**	Configures the shared packet ring interface on the ML-Series card. The only valid SPR number is 1.
6	ML100NodeDSlot5(config-if)#**spr station-id 6**	Configures the RPR station ID. The valid range is from 1 to 254. Use 6 for this ML-Series card.

continues

Table 7-9 *IOS RPR Configuration for ML-Series Card in Node A, Slot 5 (Continued)*

Step	Command	Purpose
7	ML100NodeDSlot5(config-if)#**interface pos 0**	Enters interface configuration mode for the POS 0 interface. Each of the ML-Series card's two POS ports must be assigned to the SPR interface.
8	ML100NodeDSlot5(config-if)#**spr-intf-id 1**	Assigns the POS 0 interface to the SPR interface. The SPR number must match the number that was assigned to the SPR interface.
9	ML100NodeDSlot5(config-if)#**interface pos 1**	Enters interface configuration mode for the POS 1 interface.
10	ML100NodeDSlot5(config-if)#**spr-intf-id 1**	Assigns the POS 1 interface to the SPR interface. The SPR number must match the number that was assigned to the SPR interface.
11	ML100NodeDSlot5(config-if)#**interface fa0**	Enter interface configuration mode for the Fast Ethernet 0 (front/client) interface. Configure this ML-Series interface for VLAN membership.
12	ML100NodeDSlot5(config-if)#**no shut**	Enables the interface.
13	ML100NodeDSlot5(config-if)#**bridge-group 60**	Assigns bridge group/VLAN 60 to the interface. Refer to Figure 7-12 to see the relationship between the bridge group and the interfaces.
14	ML100NodeDSlot5(config-if)#**int spr1.1**	Creates a subinterface for the SPR interface, and enters subinterface configuration mode. Refer to Figure 7-12 to see the relationship between the bridge group and the interfaces.
15	ML100NodeDSlot5(config-subif)#**bridge-group 60**	Assigns the subinterface to bridge group/VLAN 60. Refer to Figure 7-12 to see the relationship between the bridge group and the interfaces.
16	ML100NodeDSlot5(config-subif)#**encap dot1q 60**	Sets the encapsulation on the VLAN to IEEE 802.1Q.
17	ML100NodeDSlot5(config-subif)#**end**	Returns to privileged EXEC mode.
18	ML100NodeDSlot5#**copy running-config startup-config**	If desired, saves the changes to the running configuration to the nonvolatile random-access memory (NVRAM).

To verify the configuration, the **show running-config** command can be entered in privileged mode. Example 7-1 shows sample output for this command.

NOTE   The other Fast Ethernet interfaces (1 through 11) have been omitted for clarity.

Example 7-1   *Verifying RPR Configuration for ML-Series Card in Node A, Slot 5*

```
ML100NodeDSlot5#show running-config
Building configuration...
Current configuration : 1392 bytes
!
! Last configuration change at 14:43:12 PST Fri Jun 25 2004
! NVRAM config last updated at 14:43:19 PST Fri Jun 25 2004
!
version 12.1
no service pad
service timestamps debug uptime
service timestamps log uptime
no service password-encryption
!
hostname ML100NodeDslot5
!
enable password CISCO15
!
clock timezone PST -8
ip subnet-zero
!
!
bridge irb
!
!
interface SPR1
 no ip address
 no keepalive
 spr station-id 6
 hold-queue 150 in
!
interface SPR1.1
 encapsulation dot1Q 60
 bridge-group 60
 bridge-group 60 spanning-disabled
!
interface FastEthernet0
 no ip address
 bridge-group 60
 bridge-group 60 spanning-disabled
!
interface POS0
 no ip address
 spr-intf-id 1
 crc 32
!
interface POS1
 no ip address
```

continues

Example 7-1 *Verifying RPR Configuration for ML-Series Card in Node A, Slot 5 (Continued)*

```
 spr-intf-id 1
 crc 32
!
ip classless
no ip http server
!
!
!
line con 0
line vty 0 4
 password CISCO15
 login
!
!
end
```

You can use the **show vlans** command to view the interface VLAN configuration, as shown in Example 7-2.

Example 7-2 *Displaying Interface VLAN Configuration*

```
ML100NodeDslot5#sh vlans
Virtual LAN ID:  1 (IEEE 802.1Q Encapsulation)

vLAN Trunk Interfaces:  SPR1
 This is configured as native Vlan for the following interface(s) :

SPR1

Protocols Configured:    Address:              Received:          Transmitted:

Virtual LAN ID:  60 (IEEE 802.1Q Encapsulation)

   vLAN Trunk Interface:    SPR1.1

   Protocols Configured:    Address:             Received:         Transmitted:

        Bridging         Bridge Group 60            0                  0
```

Now configure all the other ML-Series cards in the RPR in a similar manner. Table 7-10 shows the parameters used in the command series to configure the network to match the functionality shown in Figure 7-18.

Table 7-10 *RPR Provisioning Parameters for Example Network Stations 1–5 and 7*

ML-Series Card	Hostname	RPR Station ID	Configured Bridge Group
Node A Slot 5	ML100NodeAslot5	5	50
Node A Slot 13	ML1000NodeAslot13	4	70
Node B Slot 13	ML1000NodeBslot13	3	60

Table 7-10   *RPR Provisioning Parameters for Example Network Stations 1–5 and 7 (Continued)*

ML-Series Card	Hostname	RPR Station ID	Configured Bridge Group
Node C Slot 13	ML1000NodeCslot13	2	50
Node C Slot 5	ML100NodeCslot5	1	None
Node D Slot 13	ML1000NodeDslot13	7	70

When all other ML-Series cards in the RPR are configured, you can use the **show cdp neighbors** command to see the other RPR stations, as shown in Example 7-3.

Example 7-3   *Displaying Neighboring RPR Stations*

```
JohnJohn#show cdp neighbors
Capability Codes: R - Router, T - Trans Bridge, B - Source Route Bridge
                  S - Switch, H - Host, I - IGMP, r - Repeater, P - Phone
Device ID          Local Intrfce    Holdtme   Capability  Platform  Port ID
ML1000NodeAslot13  SPR1             170       R           ONS-ML1000SPR1
ML1000NodeBslot13  SPR1             160       R           ONS-ML1000SPR1
ML1000NodeCslot13  SPR1             121       R           ONS-ML1000SPR1
ML100NodeAslot5    SPR1             168       R           ONS-ML100TSPR1
ML100NodeCslot5    SPR1             131       R           ONS-ML100TSPR1
ML1000NodeDslot13  SPR1             152       R           ONS-ML1000SPR1
```

Summary

This chapter discussed the various ONS 15454 Ethernet interface cards, including the E-Series, G-Series, CE-Series, and ML-Series cards. These interfaces are used to provide highly reliable point-to-point or multipoint Ethernet transport services over SONET/SDH rings.

The E-Series cards, which were the original ONS 15454 Ethernet interfaces, can be used for Layer 1 or Layer 2 services. However, these cards have been made largely obsolete by the newer card types and are typically found only in embedded networks. The E Series includes the E1000-2 and E1000-2-G for GigE connections, and the E100T-12 and E100T-G for 10-Mbps and 100-Mbps connections.

The G-Series and CE-Series cards are used to provide Layer 1 point-to-point private line services for subrate and line-rate forwarding of Ethernet frames for 10-Mbps, 100-Mbps, and 1000-Mbps services. The G Series contains a single card, the G1K-4, and provides up to four ports of subrate and line-rate GigE services. The CE Series includes the CE-100T-8 card and is used to provision 10/100 Mbps services. The CE-100T-8 also supports priority queuing through inspection of the ToS and CoS bits.

The ML-Series cards are used to provide Layer 2 and Layer 3 services. The ML Series includes the ML100T-12 (10/100Base-TX), ML100X-8 (100Base-FX and 100Base-LX10), and ML1000-2 (1000Base-SX and 1000Base-LX). The RPR network application using the ML-Series cards provides an efficient and resilient method of multipoint Ethernet transport in an optical network. The operation and configuration of RPR was discussed in detail in this chapter.

This chapter covers the following topics:

- SAN Review
- SAN Protocols
- SONET or DWDM?
- Data Storage Mirroring
- A Single-Chassis SAN Extension Solution: ONS 15454
- Storage over Wavelength
- Storage over SONET
- Fibre Channel Multirate 4-Port (FC-MR-4) Card 1G and 2G FC
- Overcoming the Round-Trip Delay Limitation in SAN Networks
- Using VCAT and LCAS
- SAN Protection

CHAPTER 8

ONS 15454 Storage-Area Networking

Because applications have a great need for storage capacity, the limited internal storage for servers has not been enough over the past few years. As a result, storage capacity has been added externally to the server, and storage-networking protocols have been developed to allow the server to access the external storage devices. With the growth of servers and storage devices, managing this network became difficult. Many companies consolidated their servers and storage to help overcome this management problem. The Small Computer System Interface (SCSI) protocol has been used to connect servers to storage devices because most servers and storage devices supported this protocol. However, the cabling distance for SCSI is severely limited, and other protocols, such as Fibre Channel, were developed to overcome this distance limitation.

This chapter focuses on the storage-area networking (SAN) extension using the Cisco ONS 15454. SAN extensions provide the capability to use the different locations of storage devices. This is commonly referred to as having "islands" of storage devices. You can link these islands using SAN extension solutions.

More specifically, you can use 15454 networks to connect SANs in different geographical locations. This is important because of the need to consolidate data center resources and to create architecture for disaster recovery and high availability. Service providers can offer managed SAN services for businesses that do not have the resources to extend their SANs on their own.

This chapter introduces you to the SAN extension capabilities of the ONS 15454 and addresses the following areas:

- Using the Multiservice Provisioning Platform (MSPP) for storage transport using Synchronous Optical Network (SONET) and wavelengths
- Using the FC-MR-4-series card
- Overcoming long distances by using buffer credits

SAN Review

In simple terms, a SAN is a network infrastructure deployed between servers and storage resources. For example, one solution is to provide a storage network where Fibre Channel (FC) switches, such as the Cisco MDS 9000, can be interconnected over a 15454 network. The 15454 network interfaces with the FC switches, as shown in Figure 8-1.

Figure 8-1 *SAN Extension Example*

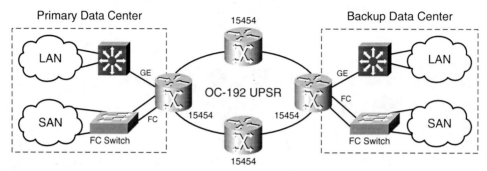

A SAN is normally a dedicated network optimized for the exchange of large amounts of data within the data center to support better utilization of available storage and to provide for disaster-recovery solutions. Companies typically have two data centers, one that is active and one in standby, in case the active data center fails. The SAN enables the two data centers to be synchronized so that the standby center has up-to-date information. A SAN creates a dedicated link between the server and storage devices that must follow certain requirements, such as low delay, high bandwidth, and high reliability. FC and Enterprise Systems Connection (ESCON) switches are used at the local data centers to provide access to storage discs.

NOTE   Network attached storage (NAS) is similar to a SAN but uses files instead of blocks. Blocks of data represent a more granular representation of the data, compared to files. It's similar to the concept of Microsoft Windows, in which files represent data. In addition, the data is stored on your hard disk in a much more granular format, which you can think of as blocks. SAN replicates the data between the servers and storage devices at the block level instead of the file level. Databases, such as Oracle, store data as blocks. This method of replicating data at the block level makes more efficient use of the storage devices than the file replication method used by NAS.

SAN Protocols

The most commonly used SAN protocols include these:

- **Small Computer System Interface (SCSI)**—SCSI is the most common method for connecting storage devices to servers. Multiple versions of serial SCSI are in use today. For example, SCSI can be mapped on top of FC, which enables SCSI to overcome its distance limitation and achieve much greater speeds and access to more devices. SCSI is widely supported by servers and storage devices.

- **FC**—FC is capable of transferring large blocks of data over longer distances at very high data rates, such as 4 Gbps. Not all SAN protocols are capable of connecting over long distances.

- **Fiber Connection (FICON)**—FICON is the IBM version of FC. FICON connects mainframes directly with control units or ESCON aggregation switches. The transport requirements for FICON are the same as those for FC.

- **Fibre Channel over IP (FCIP)** is another option of extending storage data. FCIP tunnels Fibre Channel over Transmission Control Protocol/Internet Protocol (TCP/IP). FCIP encapsulates the FC protocol within the Internet Protocol (IP) packet. As a result, existing IP-based networks can be used to interconnect FC-based storage networks.

- **ESCON**—Similar to FC, ESCON is commonly used by many businesses. ESCON has a large installed base because it has been used for many years. Unlike FC, ESCON is a low-speed IBM protocol that has distance limitations of up to 40 km.

You can use different network technologies, such as Dense Wave Division Multiplexing (DWDM), SONET, and IP, to transport these SAN protocols.

SONET or DWDM?

Multiple methods exist for transporting SAN protocols between geographical locations. You can use a SONET-based infrastructure in which you can add a storage extension card in the existing equipment located at the data centers. This is possible when the SONET rings have been built out using the ONS 15454. In this scenario, you can use the FC-MR-4 card to extend the FC or FICON between the locations that contain the servers and the storage devices.

Alternatively, you can use a DWDM-based infrastructure. This is applicable when high density and high bandwidth are needed between the server and storage device locations. The ONS 15454 supports this environment when the ONS 15454 is used as the DWDM transmission network and is used to add the wavelength service, such as SAN extension, on the DWDM network. In this scenario, you select the appropriate Cisco 15454 transponder card that supports the SAN protocol, the number of SAN ports, and the bandwidth required between the desired locations. The section "A Single-Chassis SAN Extension Solution: ONS 15454," later in this chapter, lists the available transponder cards used for SAN extension on the 15454.

NOTE   Both solutions, SONET and DWDM, offer quick protection on the order of 50 ms.

Data Storage Mirroring

The time it takes to back up storage data and the time it takes to recover from a failure are critical components to many companies, especially ones in the financial marketplace. In the past, tapes were used to replicate the storage data and were physically moved off-site. This process of backing up data is unacceptable to many companies today because of the slow recovery time. Today, large companies such as banks require an almost immediate recovery of their services. This can be accomplished by mirroring the data between locations, known as data replication or data mirroring, and it is the primary reason businesses want to transport storage data between data centers.

Two main types of data replication exist: synchronous and asynchronous.

- In synchronous data replication, data is written to the local storage device and the remote storage device in real time, ensuring that a remote copy of the data is identical to the primary copy.

- In asynchronous data replication, data is written to primary storage and the next operation continues; there is no pause for confirmation that the data has been successfully written to the secondary site, as with synchronous replication.

Synchronous Data Replication

For synchronous data replication, the data is referred to as being "mirrored" between these two sites. When you want to achieve high availability, you must immediately switch over to a backup storage device that contains the exact data as the primary data device. You use synchronous data replication to achieve this. However, it is important to understand that latency between the data centers is critical to the data-replication application. Synchronous data replication is sensitive to end-to-end latency because an application running on the server must wait until it receives an acknowledgment that the data has been replicated on the primary and the backup storage devices. The data-replication application will not work properly if this time is too long. Figure 8-2 shows an example of synchronous data replication.

Figure 8-2 *Example of Synchronous Data Replication Using SONET or DWDM*

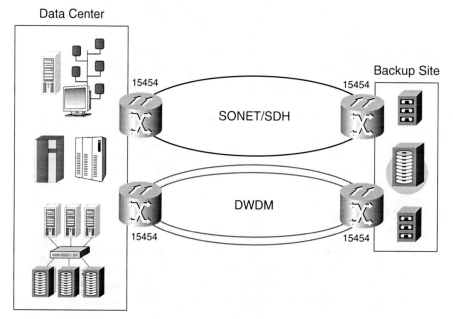

Another type of synchronous mirroring occurs when you mirror the application in addition to the storage device. GDPS, a protocol developed by IBM, implements this capability. GDPS requires that special control and timer links be established between locations, in addition to transporting the storage protocol. Cisco transponder cards support this environment.

For synchronous replication, the following events occur:

1 The server pauses application processing while writing data to the local storage device.

2 The local storage system sends data and writes the request to the backup storage device.

3 The backup storage device writes data and sends an acknowledgment to the local storage device.

4 An I/O complete signal is sent to the server, and application processing resumes.

Figure 8-3 shows how these four steps are carried out.

Figure 8-3   *Synchronous Data Replication Events*

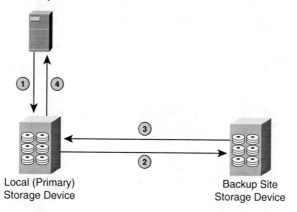

Local (Primary)
Storage Device

Backup Site
Storage Device

Asynchronous Data Replication

Asynchronous data replication disconnects the primary write operation from the remote write operation. In other words, the application writes the data to primary storage and continues with the next operation without pausing to wait for confirmation that data has been successfully written to the secondary site. Asynchronous replication is typically used where synchronous data replication is not practical, especially in long-haul applications in which the end-to-end delay is too much for synchronous mirroring. Unlike synchronous data replication, asynchronous is not latency sensitive and can accommodate a certain amount of delay in the network and data loss. Therefore, you can use asynchronous data replication across longer distances.

You must determine the distances that the storage devices can support using synchronous or asynchronous applications. This helps determine the type of network that can support SAN extension.

For asynchronous replication, the following events occur:

1 The server pauses application processing while writing data to the local disk.

2 The local storage device queues a copy of the data for transmission to the remote storage device and sends an I/O complete signal to the file server so that application processing can resume.

3 The local storage device sends data to the backup storage device.

4 The backup storage device sends a write acknowledge back to the local storage device.

Figure 8-4 shows how these four steps are carried out.

Figure 8-4   *Asynchronous Data Replication Events*

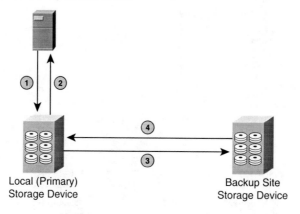

Local (Primary)
Storage Device

Backup Site
Storage Device

A Single-Chassis SAN Extension Solution: ONS 15454

The ONS 15454 is capable of supporting both synchronous and asynchronous data replication. The ONS 15454 provides storage extension over SONET and DWDM networks. Some characteristics of using the ONS 15454 MSPP solution for SAN extension follow:

• Commonly used for asynchronous applications but can be used for synchronous applications as well

• Achieve connections over large distances up to 10,000 km

• Introduces higher latency than ONS 15454 DWDM

• Has a lower storage protocol port density versus ONS 15454 DWDM

Some characteristics of using an ONS 15454 DWDM solution for SAN extension follow:

- Commonly used for synchronous data replication to up 200 km
- Achieves high density of SAN extension links over fiber

When selecting the right SAN extension solution, it is important to answer the following questions:

- How much bandwidth is needed?
- What distance is required?
- What is the latency tolerance of the SAN edge devices and applications?
- How many connections are required between data centers?
- Is an existing infrastructure (SONET or DWDM) being used?
- What is the availability of dark fibers?
- What managed services is the service provider offering?
- What are the regulatory requirements for your environment? For example, some regulations require certain financial institutions to recover critical functions the same business day a disruption occurs.

Storage over Wavelength

Most FC used today is at 1 Gbps; however, 2 Gbps FC is commonly used as well. Two recent options, 4G and 10G, have been introduced into the market. The 10G FC enables customers to better use the available wavelengths between their data centers. These 10G FC on the ONS 15454 transponder card interconnect to FC switches that support 10G trunk interfaces, such as the Cisco MDS 9000 switch.

The ONS 15454 DWDM solution provides the capability for SAN extension using DWDM transponder and muxponder cards. These DWDM line cards support SAN extension:

- **10G MR transponder (1 port)**—Supports 10G FC.
- **10G data muxponder (8 ports)**—Supports 1G, 2G, and 4G FC. Also supports IBM GDPS protocol enabling application mirroring, and storage mirroring. Supports buffer-to-buffer credits, enabling a 1-Gbps FC signal to extend about 1200 km.
- **2.5G MR transponder (1 port)**—Supports 1G and 2G FC. Two versions of this card exist: the splitter protection version and an unprotected trunk version.
- **2.5G data muxponder (8 ports)**—Supports 1G and 2G FC, as well as ESCON. Two versions of this card exist: a splitter protection version and an unprotected trunk version. Supports buffer-to-buffer credits, enabling a 1-Gbps FC signal to extend about 1600 km.

You can use these cards to translate the incoming traffic, such as FC, FICON, ESCON, or GigE, into a standard International Telecommunications Union (ITU) wavelength. The ports on the 15454 transponder and muxponder cards can support other traffic types in addition to storage interfaces. For example, if the appropriate small form factor pluggable (SFP) is chosen, the 10G data muxponder card can have any of its eight ports as GigE. Because the muxponder cards support multiple traffic types based on SFP type, these cards provide the flexibility for you to multiplex FC, GigE, and IBM InterSystem Channel (ISC) interfaces over a single lambda. This capability enables you to use the 15454 DWDM network efficiently by allowing different traffic types to run over it. This is practical when a service provider offers a private customer ring because multiple services can be sold to the end customer using a single muxponder card in the 15454. Figure 8-5 shows a managed private DWDM ring supporting SAN extension.

Figure 8-5   *Example of a Managed DWDM Ring Supporting SAN Extension*

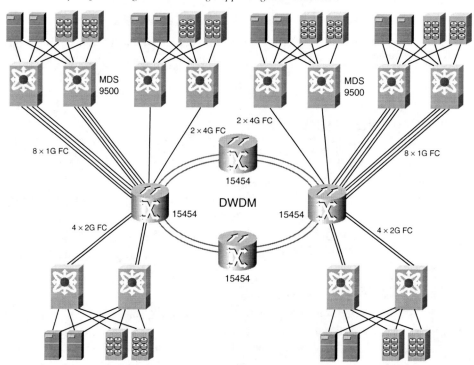

Table 8-1 identifies the different SFPs and 10-Gbps small form factor pluggables (XFPs) that you can use to support SAN extension on the ONS 15454.

Table 8.1 *ONS 15454 SFPs and XFPs Available for SAN Extension*

15454 Card	Client Ports	SFP/XFP Product ID	SAN Interfaces Supported on SFP/XFP*	Fiber	Connector	Tx Power Min/Max (dBm)	Rx Power Min/Max (dBm)
10G Multirate Transponder	1 (XFP)	ONS-XC-10G-S1	10G FC	1310 nm single mode	LC	–8.2 to +0.5	–14.4 to +0.5
		ONS-XC-10G-L2	10G FC	1550 nm single mode	LC	0 to +4	–24 to –7
10G Data Muxponder	8 (SFP)	ONS-SE-G2F-SX	1G-FC, 2G-FC	850 nm multimode	LC	–9.5 to 0	–17 to 0
		ONS-SE-G2F-LX	1G-FC, 2G-FC	1310 nm single mode	LC	–9.5 to –3	–19 to –3
		ONS-SE-4G-MM	4G FC	850 nm multimode	LC	–9 to –2.5	–15
		ONS-SE-4G-SM	4G FC	1310 nm single mode	LC	Average power of –7.3 dBm at an ER of 9 dB	Average power of –17.3 dBm at an ER of 9 dB
2.5G MR Transponder	1 (SFP)	ONS-SE-200-MM	ESCON	1310 nm multimode	LC	–20.5 to –15	–29 to –14
		ONS-SE-G2F-SX	1G-FC, 2G-FC	850 nm multimode	LC	–9.5 to 0	–17 to 0
		ONS-SE-G2F-LX	1G-FC, 2G-FC	1310 nm single mode	LC	–9.5 to –3	–19 to –3
2.5G Data Muxponder (SFP)	8 (SFP)	ONS-SE-G2F-SX	1G-FC, 2G-FC	850 nm multimode	LC	–9.5 to 0	–17 to 0
		ONS-SE-G2F-LX	1G-FC, 2G-FC	1310 nm single mode	LC	–9.5 to –3	–19 to –3
		ONS-SE-200-MM	ESCON	1310 nm multimode	LC	–20.5 to –15	–29 to –14

NOTE	Other interfaces, such as SONET and Ethernet, are supported by the same SFP and XFP listed in Table 8-1. Refer to the Cisco ONS 15454 documentation for a list of non-SAN interfaces supported by the SFPs and XFPs just shown. XFPs are 10G optical transceiver modules that combine transmitter and receiver functions in one module. XFP modules are protocol independent and can be used to support OC-192/STM-64, 10-Gigabit Ethernet, 10-Gigabit Fibre Channel, and G.709 data streams in routers and switches.

Storage over SONET

The ONS 15454 MSPP can support 1G and 2G line rate FC and FICON over SONET using Generic Framing Procedure (GFP). As with other services cards in the ONS 15454 MSPP, the FC-MR-4 card terminates the SONET signal on the cross-connect card, which switches the signal to the trunk optical card, such as the OC192 card.

Fibre Channel Multirate 4-Port (FC-MR-4) Card

The ONS 15454 MSPP uses the FC-MR-4 card to extend the storage networks over a SONET network. Service providers can offer their customers extended storage. For example, Figure 8-6 shows a customer's private SONET ring using existing services, point-to-point DS3, SONET, and Ethernet purchased from the service provider. You can simply add a four-port FC-MR-4 card into the existing 15454 MSPP chassis to the customer's data center MSPPs to extend the FC network between data centers.

Figure 8-6 *Using ONS 15454 to Extend FC*

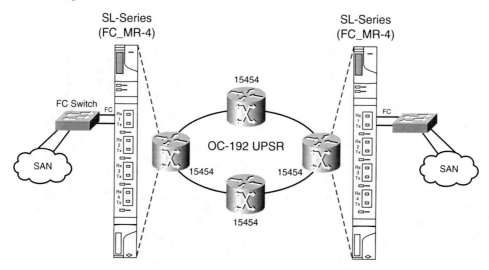

1G and 2G FC

The ONS 15454 can support 40 subrate Fibre Channel terminations. Each FC-MR-4 series card supports four point-to-point SAN extension circuits. This is accomplished by using four pluggable optical interfaces on the front of the FC-MR-4 card. These pluggable interfaces, called gigabit interface converters (GBICs), are used to interface to the SAN device. This GBIC is dual rate and can be set to FC or FICON at either 1 Gbps or 2 Gbps using the Cisco Transport Controller (CTC), the graphical-based interface. Because the FC-MR-4 card has 48 STS-1s going out of the back of the card, the card supports two 1-Gbps FC or one 2-Gbps Fibre Channel.

NOTE   The GBIC comes in two varieties: 850-nm and 1310-nm wavelengths.

The distance of the FC-MR-4 card depends upon the speed of the interface and whether buffer-to-buffer credit spoofing is used. You can achieve long distances when using the buffer-to-buffer credit spoofing on the FC-MR-4 card. For example, you can achieve a distance of 2300 km at 1 Gbps and 1150 km at 2 Gbps.

Overcoming the Round-Trip Delay Limitation in SAN Networks

FC uses flow control to avoid data loss. The transmitting device waits for an acknowledgment (called R_RDY signal) for every data frame that is sent. The number of frames that the device can send cannot exceed the buffer capacity of the receiving device. The sending device, therefore, has a certain number of buffer credits, with each buffer credit allowing the transmitting node to send one data frame. The amount of time that the acknowledgment takes to return to the sending device increases as the end-to-end delay increases. Increasing the distance is the major contributor to end-to-end delay. The buffer capacity dictates the amount of distance that you can go while still sending at line rate. You can estimate the approximate distance by knowing that one buffer credit is needed for every 2 km for a 1-Gbps FC link.

You can achieve much longer distances between the data centers with buffer-to-buffer credit spoofing when using FC. Buffer-to-buffer credit spoofing is supported in the FC-MR-4 card. This means that the FC-MR-4 card terminates the "receiver readies" messages from the attached SAN device. This is commonly referred to as spoofing. In addition, the idle frames are terminated at the local FC-MR-4 card and regenerated at the other FC-MR-4 card. This results in increased distance and bandwidth between the data centers.

Using VCAT and LCAS

The FC-MR-4 card supports subrates. In other words, you are not restricted to set up point-to-point 1 Gbps (STS24c) or 2 Gbps (STS48c); you can set up a variety of other SONET contiguous concatenated (CCAT) rates and high-order virtual concatenated (VCAT) rates. With high-order VCAT, you can group individual STS1s and STS3cs to better match the bandwidth required between the data centers—or perhaps you need to capture stranded bandwidth on the SONET ring. The Synchronous Transfer Signals (STSs) that are grouped together are referred to as a VCAT Group (VCG). It is important to understand that the VCG does not have to contain STSs that are contiguous within the SONET signal. You can select from the rates in Table 8-2.

Table 8-2 *FC-MR-4 Series Available CCAT and VCAT Rates*

CCAT Rates	1-Gbps FC/FICON	2-Gbps FC/FICON
STS-1	√	√
STS-3c	√	√
STS-6c	√	√
STS-9c	√	√
STS-12c	√	√
STS-18c	√	√
STS-24c	√	√
STS-36c	N/A	√
STS-48c	√	√
VCAT Rates		
STS-1-Xv (X = 1 to 24)	√	—
STS-3c-Xv (X = 1 to 8)	√	—
STS-1-Xv (X = 1 to 48)	—	√
STS-3c-Xv (X = 1 to 16)	—	√

Subrate support on the FC-MR-4 card allows the service provider to offer multiple tiers of SAN extension service. This enables smaller businesses that don't require full-line rate or that can't afford this service level to use SAN extension service from their service provider.

The VCG is reconfigurable with the use of the Link Capacity Adjustment Scheme (LCAS). LCAS is a capacity-control mechanism that adds and removes circuits into and out of a VCG in a hitless manner. This means that STS1 or STS3c circuits can be added as part of an STS1-based VCG or STS3c-based VCG, respectively, without affecting the SAN extension service. LCAS also can dynamically add a member back to the VCG when that member becomes available after a failure occurred. The individual circuits are referred to as members of the VCG. For example, an STS1 within the VCG can be removed if the circuit has an operational state of Out-of-service (OOS).

It is important to understand that when using VCAT, different path lengths among the individual member circuits of a VCG can exist. This can make the traffic arrive at the destination out of phase with other members of the VCG. The FC-MR-4 card solves this problem by using a buffer to compensate for the differential delay among members.

SAN Protection

The ONS 15454 supports a variety of protection options when performing SAN extension. These protection options depend on which type of SAN extension card you use. Table 8-3 defines the path-protection options for the SAN extension cards that are available on the 15454 SONET platform and the 15454 DWDM platform.

Table 8-3   *Path-Protection Options for ONS 15454 SAN Extension*

SAN Extension Card	Path-Protection Options on ONS 15454 MSPP	Path-Protection Options on ONS 15454 DWDM
FC-MR-4 (4 ports)	*Trunk protection:* Unidirectional path-switched ring (UPSR), bidirectional line-switched ring (BLSR), Path Protected Mesh Networking (PPMN), protection channel access (PCA), unprotected	—
10G MR Transponder (1 port)	—	*Client side:* Y cable *Trunk side:* Unprotected (1-XFP client to one trunk interface)
2.5G MR Transponder (1 port) (2 card versions: unprotected and splitter protected)	—	*Client side:* Y cable for unprotected trunk option *Trunk side:* Splitter protection or unprotected (1-SFP client to one or two trunk interfaces)

continues

Table 8-3 *Path-Protection Options for ONS 15454 SAN Extension (Continued)*

SAN Extension Card	Path-Protection Options on ONS 15454 MSPP	Path-Protection Options on ONS 15454 DWDM
2.5G Data Muxponder (8 ports) (2 card versions: unprotected and splitter protected)	—	*Client Side:* Y cable for unprotected trunk option *Trunk side:* Splitter protection or unprotected (8-SFP client to one or two trunk interfaces)
10G Data Muxponder (8 ports)	—	*Client side:* Y cable *Trunk side:* Unprotected (8-SFP clients to one trunk interface)

Protection for the transponder and muxponder cards includes client-side protection using Y cables and wavelength splitter protection. Automatic protection switching (APS) is supported when you select wavelength splitter protection on the 2.5G transponder and 2.5G muxponder cards. This enables you to protect the traffic on a DWDM ring because the transponder and muxponder card transmits the traffic from the storage device (such as MDS 9000) in both the east and west directions on the ring. Less than 50 ms switchover to the protect wavelength occurs when the working path fails or the performance degrades.

The 10G transponder and 10G muxponder do not support wavelength splitter protection. However, you can provide trunk-side protection by using two transponder cards or two muxponder cards. A Y-cable can be used to connect to a single trunk interface on the storage device. Keep in mind that this same configuration applies when using the unprotected card version of the 2.5G transponder and 2.5G muxponder.

Y-cable protection is supported on all the DWDM transponder and muxponder cards and is an optional feature. This means that there are two identical client-side cards in the ONS 15454. These two cards are configured as a protection group, which means that one card is designated as the working card and the other card is designated as the protecting card. Only one of the two client transmitters is on at any given time. If the working path fails, the transmitter on the protecting card is turned on, and the transmitter on the working card is turned off.

Summary

This chapter gives an overview of different SAN extension solutions that the ONS 15454 supports. SAN extension is supported over SONET and DWDM using the ONS 15454. You use the FC-MR-4 card to extend FC and FICON over SONET. You use the transponder and muxponder DWDM cards to extend FC, FICON, and ESCON over DWDM. The decision of whether to use SONET or DWDM is influenced by the distance between the data centers, the bandwidth required between the data centers, and the number of connections needed between the data centers. In addition, synchronous data replication affects the allowable distance between the primary and backup data centers.

The FC-MR-4 card is typically used when an existing SONET infrastructure is already in place, such as a service provider's private SONET rings sold to customers, when the bandwidth between data centers is less than 10G, and when a high density of SAN connections is not required between locations.

PART IV

Building DWDM Networks Using the ONS 15454

This chapter covers the following topics:

- ONS 15454 Shelf Assembly
- Timing, Communications, and Control Cards
- Optical Service Channel Module
- Optical Service Channel—Combiner/Separator Module
- Alarm Interface Controller Card
- ONS 15454 MSTP DWDM ITU-T Channel Plan
- 32-Channel Multiplexer Cards
- 32-Channel Demultiplexer Cards
- Four-Channel Multiplexer/Demultiplexer Cards
- Four-Band OADM Filters
- One-Band OADM Filters
- Channel OADM Cards
- ONS 15454 MSTP ROADM
- ONS 15454 MSTP Transponder/Muxponder Interfaces
- ONS 15454 MSTP Optical Amplifiers
- ONS 15454 MSTP Dispersion Compensation Unit
- ONS 15454 MSTP Supported Network Configurations

CHAPTER 9

Using the ONS 15454 Platform to Support DWDM Transport: MSTP

As noted in earlier chapters, the flexible, scalable design of the Cisco ONS 15454 lends itself to multiple uses within the carrier transport environment. Carriers can take advantage of its time-division multiplexing (TDM) and Ethernet solutions, and its support of dense wavelength-division multiplexing (DWDM) as well. The ONS 15454 is transformed into a Multiservice Transport Platform (MSTP) by slotting its common control and interface cards with DWDM-specific interface cards. Although the ONS 15454 can be used for "hybrid" TDM/DWDM applications, this chapter and others focus solely on the DWDM operation. This chapter highlights the basic building blocks of the MSTP. You will examine the following features and functions associated with each MSTP component:

- Shelf assembly
- Common control cards
- Multiplexer/demultiplexer cards
- Optical add/drop interface cards
- Transponder/muxponder interface cards
- Amplifier interface cards
- Dispersion compensation unit (DCU)

After a thorough examination of each interface type, the chapter concludes with network topology and shelf configuration examples. Each shelf configuration example is intended to show the most common equipment configurations applicable to today's networks.

ONS 15454 Shelf Assembly

The shelf assembly used for ONS 15454 MSTP operation is the same shelf assembly used for ONS 15454 Multiservice Provisioning Platform (MSPP) configurations. Chapter 6, "MSPP Network Design Example: Cisco ONS 15454," covered in detail the features and functions of this shelf assembly; the information is repeated here for convenience. The ONS 15454 shelf assembly is a 17-slot chassis with an integrated fan tray, rear electrical terminations, and front optical, Ethernet, and management connections. Slots 1–6 and 12–17 are used for traffic interface cards; Slots 7–11 are reserved for common control cards. Slots 1–6, on the left side as you face the front of the shelf, are considered Side A;

while Slots 12–17 are considered Side B. Table 9-1 summarizes the card slot functions for the ONS 15454 shelf assembly when it is used for MSTP configurations.

Table 9-1 *ONS 15454 Shelf Assembly Slot Functions*

Slot Number	Shelf Side	MSTP Interface Card Function(s)
1	A	OSC-CSM,32 MUX-0, 32 WSS, 32 DMX-0, 32 DMX, 4MD-xx.x, AD-1B.xx.x, AD-4B.xx.x, AD-1C-xx.x, AD-2C-xx.x, AD-4C-xx.x, MR/P-L1-xx.x, 10E-L1-xx.x, 10ME-xx.x, DM/P-L1-xx.x, OPT-PRE, and OPT-BST/E
2	A	OSC-CSM,32 MUX-0, 32 WSS, 32 DMX-0, 32 DMX, 4MD-xx.x, AD-1B.xx.x, AD-4B.xx.x, AD-1C-xx.x, AD-2C-xx.x, AD-4C-xx.x, MR/P-L1-xx.x, 10E-L1-xx.x, 10ME-xx.x, DM/P-L1-xx.x, OPT-PRE, and OPT-BST/E
3	A	OSC-CSM,32 MUX-0, 32 WSS, 32 DMX-0, 32 DMX, 4MD-xx.x, AD-1B.xx.x, AD-4B.xx.x, AD-1C-xx.x, AD-2C-xx.x, AD-4C-xx.x, MR/P-L1-xx.x, 10E-L1-xx.x, 10ME-xx.x, DM/P-L1-xx.x, OPT-PRE, and OPT-BST/E
4	A	OSC-CSM,32 MUX-0, 32 WSS, 32 DMX-0, 32 DMX, 4MD-xx.x, AD-1B.xx.x, AD-4B.xx.x, AD-1C-xx.x, AD-2C-xx.x, AD-4C-xx.x, MR/P-L1-xx.x, 10E-L1-xx.x, 10ME-xx.x, DM/P-L1-xx.x, OPT-PRE, and OPT-BST/E
5	A	OSC-CSM,32 MUX-0, 32 WSS, 32 DMX-0, 32 DMX, 4MD-xx.x, AD-1B.xx.x, AD-4B.xx.x, AD-1C-xx.x, AD-2C-xx.x, AD-4C-xx.x, MR/P-L1-xx.x, 10E-L1-xx.x, 10ME-xx.x, DM/P-L1-xx.x, OPT-PRE, and OPT-BST/E
6	A	OSC-CSM, 32 DMX, 4MD-xx.x, AD-1B.xx.x, AD-4B.xx.x, AD-1C-xx.x, AD-2C-xx.x, AD-4C-xx.x, MR/P-L1-xx.x, 10E-L1-xx.x, 10ME-xx.x, DM/P-L1-xx.x, OPT-PRE, and OPT-BST/E
7	—	TCC
8	—	OSCM
9	—	AIC
10	—	OSCM
11	—	TCC

Table 9-1   *ONS 15454 Shelf Assembly Slot Functions (Continued)*

Slot Number	Shelf Side	MSTP Interface Card Function(s)
12	B	OSC-CSM,32 MUX-0, 32 WSS, 32 DMX-0, 32 DMX, 4MD-xx.x, AD-1B.xx.x, AD-4B.xx.x, AD-1C-xx.x, AD-2C-xx.x, AD-4C-xx.x, MR/P-L1-xx.x, 10E-L1-xx.x, 10ME-xx.x, DM/P-L1-xx.x, OPT-PRE, and OPT-BST/E
13	B	OSC-CSM,32 MUX-0, 32 WSS, 32 DMX-0, 32 DMX, 4MD-xx.x, AD-1B.xx.x, AD-4B.xx.x, AD-1C-xx.x, AD-2C-xx.x, AD-4C-xx.x, MR/P-L1-xx.x, 10E-L1-xx.x, 10ME-xx.x, DM/P-L1-xx.x, OPT-PRE, and OPT-BST/E
14	B	OSC-CSM,32 MUX-0, 32 WSS, 32 DMX-0, 32 DMX, 4MD-xx.x, AD-1B.xx.x, AD-4B.xx.x, AD-1C-xx.x, AD-2C-xx.x, AD-4C-xx.x, MR/P-L1-xx.x, 10E-L1-xx.x, 10ME-xx.x, DM/P-L1-xx.x, OPT-PRE, and OPT-BST/E
15	B	OSC-CSM,32 MUX-0, 32 WSS, 32 DMX-0, 32 DMX, 4MD-xx.x, AD-1B.xx.x, AD-4B.xx.x, AD-1C-xx.x, AD-2C-xx.x, AD-4C-xx.x, MR/P-L1-xx.x, 10E-L1-xx.x, 10ME-xx.x, DM/P-L1-xx.x, OPT-PRE, and OPT-BST/E
16	B	OSC-CSM,32 MUX-0, 32 WSS, 32 DMX-0, 32 DMX, 4MD-xx.x, AD-1B.xx.x, AD-4B.xx.x, AD-1C-xx.x, AD-2C-xx.x, AD-4C-xx.x, MR/P-L1-xx.x, 10E-L1-xx.x, 10ME-xx.x, DM/P-L1-xx.x, OPT-PRE, and OPT-BST/E
17	B	OSC-CSM, 32 DMX, 4MD-xx.x, AD-1B.xx.x, AD-4B.xx.x, AD-1C-xx.x, AD-2C-xx.x, AD-4C-xx.x, MR/P-L1-xx.x, 10E-L1-xx.x, 10ME-xx.x, DM/P-L1-xx.x, OPT-PRE, and OPT-BST/E

ONS 15454 Shelf Assembly Backplane Interfaces

Backplane interfaces of the ONS 15454 chassis can be divided into four general areas:

- Power terminal area
- Alarm and chassis ground area
- Side A Electrical Interface Assembly
- Side B Electrical Interface Assembly

Figure 9-1 shows the location of each of these areas as you face the rear of the shelf assembly. Electrical Interface Assemblies (EIAs) are required for terminating electrical traffic signals, such as DS1s and DS3s, on the shelf. EIAs are not required for MSTP operation.

Figure 9-1   *ONS 15454 Backplane Interfaces*

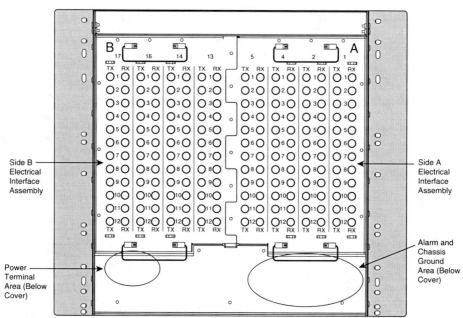

The power terminal area consists of four power terminal screws on the lower left side. The RET1/BAT1 terminals are for the A power connection; the RET2/BAT2 terminals are for the B connection. These connections are redundant, and either can power the entire shelf. The A and B designations do not refer to the A and B sides of the shelf.

The alarm and chassis ground area is located in the rear of the chassis on the lower right side and has the following terminations:

- **Frame ground terminals**—Two terminals with kepnuts are provided for ground-wire lug connection. This connection ensures that the shelf assembly is at the same electrical potential with the office ground.

- **BITS**—(Building Integrated Timing Supply) This consists of four wire-wrap pin pairs for connection to a BITS or for wiring out to external equipment when the BITS-Out feature is used to supply timing from the ONS 15454.

- **LAN**—(Local-area network) A LAN connects the MSPP node to a management work-station or network. It consists of four wire-wrap pin pairs; typically, two pairs are used.

- **Environmental alarms**—Sixteen wire-wrap pin pairs are provided for external alarms and controls. The Alarm Interface Controller card, covered later in this chapter, is required to use these connections.

- **ACO**—Alarm cutoff, a wire-wrap pin pair that is used to deactivate the audible alarms caused by the contact closures on the shelf backplane. This operation is described in the "Timing, Communications, and Control Cards" section, later in this chapter.

- **Modem**—Four pairs of wire-wrap pins are provided for connecting the ONS 15454 to a modem for remote management.

- **Craft**—Two wire-wrap pin pairs are provided for a TL1 craft-management connection. VT100 emulation software is used to communicate with the system by way of this connection.

- **Local alarms**—Eight wire-wrap pin pairs are used for Critical, Major, Minor, and Remote audible and visual alarms.

Timing, Communications, and Control Cards

The Timing, Communications, and Control (TCC) cards used for ONS 15454 MSTP operation are the same TCC used for ONS 15454 MSPP configurations. Chapter 6 covered the features and functions of the TCC interface in detail. The information is repeated here for convenience.

The Timing, Communications, and Control (TCC) cards are required for operation of the ONS 15454 MSPP system and are installed in a redundant pair in shelf Slots 7 and 11. Two current versions are available from Cisco: the Advanced Timing, Communications, and Control card (TCC2), and the Enhanced Advanced Timing, Communications, and Control card (TCC2P). Both perform the same basic functions, but the TCC2P is an updated version of the TCC2 and includes some security enhancements and additional synchronization options that are not available in the TCC2. Both cards have a purple square symbol on their faceplates, which corresponds to matching symbols on the front of the ONS 15454 shelf assembly. This serves as an aid in easily identifying the correct location to install the card. TCC cards are the only card type allowed in Slots 7 and 11, and both slots should always be equipped. Cisco does not support the operation of the ONS 15454 MSPP system with only a single TCC card installed. Although the system technically can function with only a single card, the second card is necessary for redundancy and to allow for continuity of system traffic in case of a failure or reset of the primary card. The system raises the "Protection Unit Not Available" (PROTNA) alarm if the secondary TCC card is not installed.

Two earlier versions of the TCC2 and TCC2P cards exist, called simply the Timing, Communications, and Control (TCC) card and the TCC Plus (TCC+). These older-version TCC cards provide similar functionality to the current cards, but they are much more limited in processing power. Although they can may be installed in some existing systems, Cisco no longer produces the TCC or TCC+ versions.

The TCC2 and TCC2P cards perform a variety of critical system functions, including the following:

- **System initialization**—The TCC2s/TCC2Ps are the first cards initially installed in the system and are required to initialize system operation.

- **Software, database, and Internet Protocol (IP) address storage**—The node database, system software, and assigned system IP address (or addresses) are stored in nonvolatile memory on the TCC2/TCC2P card, which allows for quick restoration of service in case of a complete power outage.

- **Alarm reporting**—The TCC2/TCC2P monitors all system elements for alarm conditions and reports their status using the faceplate and fan tray light-emitting diodes (LEDs). It also reports to the management software systems.

- **System timing**—The TCC card monitors timing from all sources (both optical and BITS inputs) for accuracy. The TCC selects the timing source, which is recovered clocking from an optical port, a BITS source, or the internal Stratum level 3 clock.

- **Cell bus origination/termination**—The TCC cards originate and terminate a cell bus, which allows for communication between any two cards in the node and facilitates peer-to-peer communication. These links are important to ensure fast protection switching from a failed card to a redundant-protection card.

- **Diagnostics**—System performance testing is enabled by the TCC cards. This includes the system LED test, which can be run from the faceplate test button on the active TCC or using Cisco Transport Controller (CTC).

- **Power supply voltage monitoring**—An alarm is generated if one or both of the power-supply connections is operating at a voltage outside the specified range. Allowable power supply voltage thresholds are provisionable in CTC.

Figure 9-2 shows a diagram of the faceplate of the TCC2/TCC2P card.

The two card types are identical in appearance, with the exception of the card name labeling. The cards have 10 LEDs on the faceplate, including the following:

- **FAIL**—This red LED that is illuminated during the initialization process. This LED flashes as the card boots up. If the LED does not extinguish, a card failure is indicated.

- **ACT/STBY**—Because the TCC cards are always installed as a redundant pair, one card is always active while the other is in standby state. The active TCC2/TCC2P has a green illuminated ACT/STBY LED; the standby card is amber/yellow.

- **PWR A and B**—These indicate the current state of the A- and B-side power-supply connections. A voltage that is out of range causes the corresponding LED to illuminate red; an acceptable level is indicated with green.

- **CRIT, MAJ, and MIN**—These indicate the presence of a critical (red), major (red), or minor (amber) alarm (respectively) in the local ONS 15454 node.

- **REM**—This LED turns red if an alarm is present in one or more remote DCC-connected systems.

- **SYNC**—This green SYNC lamp indicates that the node is synchronized to an external reference.

- **ACO**—The Alarm Cutoff lamp is illuminated in green if the ACO button on the faceplate is depressed. The ACO button deactivates the audible alarm closure on the shelf backplane. ACO stops if a new alarm occurs. If the alarm that originated the ACO is cleared, the ACO LED and audible alarm control are reset.

The faceplate also has two push-button controls. The LAMP TEST button initiates a brief system LED test, which lights every LED on each installed card and the fan tray LEDs (with the exception of the FAN FAIL LED, which does not participate in the test).

Figure 9-2   *TCC2/TCC2P Faceplate Diagram*

The RS-232 and Transmission Control Protocol/Internet Protocol (TCP/IP) connectors allow for management connection to the front of the ONS 15454 shelf. TCP/IP is an RJ-45 that allows for a 10Base-T connection to a PC or workstation that uses the CTC management system. A redundant local-area network (LAN) connection is provided via the backplane LAN pins on the rear of the shelf. The RS-232 is an EIA/TIA-232 DB9-type connector used for TL1 management access to the system. The CRAFT wire-wrap pins on the shelf backplane duplicate the functionality of this port.

Optical Service Channel Module

The optical service channel module (OSCM)) provides a point-to-point, bidirectional communications link between DWDM nodes. This communications link—the optical service channel (OSC)—is similar in function to the data communications channel (DCC) used in Synchronous Optical Network (SONET) topologies. However, in addition to providing an out-of-band DWDM communications link, the OSCM interface card provides other functions in an MSTP network, such as clock synchronization and orderwire channel communications. The OC-3 formatted optical service channel is considered out-of-band because it operates at the 1510 nm wavelength, which is typically outside of the erbium-doped fiber amplifier (EDFA) pass-band of 1528 to 1565 nm.

Because the optical service channel is a point-to-point link, two OSCM interface cards are required for most configurations—one for the East-to-West span, and one for the West-to-East span. The OSCM interface card operates solely in Slots 8 or 10 on the ONS 15454 shelf assembly. It is important to note that the OSCM is used in amplified MSTP configurations only. For amplified MSTP systems, the optical service channel is split from the composite DWDM fiber signal at the booster amplifier before being distributed to the OSCM card. For unamplified (passive) DWDM configurations, the optical service channel–combiner/separator module (OSC-CSM) interface card (detailed in the next section) is used.

In addition to the OC-3 formatted optical service channel, the OSCM card provides a 100-Mbps user data channel (UDC). The UDC uses the OC-3 payload bandwidth of the optical service channel and can be accessed on the front faceplate of the OSCM using an RJ45 connector.

Figure 9-3 shows the faceplate layout of the OSCM.

Figure 9-3 *OSCM Card Faceplate Diagram*

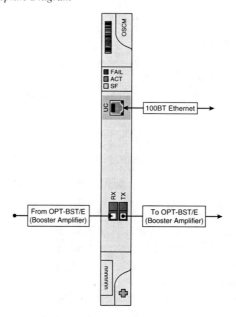

OSC-CSM

OSC-CSM provides the equivalent functionality of the OSCM interface card. In contrast to the OSCM interface, the OSC-CSM provides on-board circuitry to separate or combine the optical service channel from the composite DWDM fiber signal. Therefore, it is used for MSTP configurations that do not employ the booster amplifier—that is, passive systems.

After the OSC is extracted from the composite DWDM fiber signal and distributed to the OSC interface, the remaining composite signal is distributed to optical add/drop multiplexer (OADM) units for customer channel access.

Unlike the OSCM, the OSC-CSM can be operationally inserted in any of the service interface slots of the ONS 15454 shelf assembly—Slots 1–6 or 12–17. Figure 9-4 shows the faceplate layout of the OSCM.

Figure 9-4   *OSC-CSM Card Faceplate Diagram*

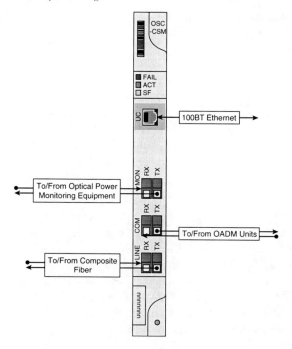

Alarm Interface Controller Card

The Alarm Interface Controller (AIC-I) card used for ONS 15454 MSTP operation is the exact same AIC-I card used for ONS 15454 MSPP configurations. The features/functions of the AIC-I interface were covered in Chapter 6. The information is repeated here for convenience.

The AIC-I card is an optional circuit pack that is installed in shelf Slot 9. The faceplate of the card is marked with a red diamond, corresponding to the symbol marked on the front of the ONS 15454 shelf assembly. This serves as an aid in easily identifying the correct location to install the card. For MSPP sites where the AIC-I is not required, a BLANK/FILLER is required to maintain proper airflow through the system while operating without the front door, and also to allow the system to meet Network Equipment Building Standards (NEBS), electromagnetic interference (EMI) standards, and electrostatic discharge (ESD) standards.

When is the AIC-I card required? The card provides four main capabilities to the network operator:

- Environmental alarm connection and monitoring
- Embedded voice-communication channels, known as orderwires
- A-Side and B-Side power supply input voltage monitoring
- Access to embedded user data channels

Each of these major functions, along with the associated card faceplate LEDs and cabling connectors, is covered in the next section. Figure 9-5 shows the faceplate layout of the AIC-I.

An earlier version of the Alarm Interface Controller is called the AIC (no -*I*). This older version provides a more limited environmental alarm-monitoring capacity and does not provide user data channel access or input voltage monitoring. Although they may be installed in some existing systems, Cisco no longer produces the AIC version.

Similarly to all ONS 15454 common control cards, the AIC-I has a FAIL LED and ACT LED on the upper part of the card faceplate, just below the top latch. The FAIL LED is red and indicates that the card's processor is not ready for operation. This LED is normally illuminated during a card reset, and it flashes during the card boot-up process. If the FAIL LED continues to be illuminated, this is an indication that the card hardware has experienced a failure and should be replaced. The ACT (Active) LED is green and illuminates to indicate that the card is in an operational state. Unlike the XC cards and TCC cards, the ACT LED does not have a standby (STBY) state because there is no secondary or back-up card to protect the active AIC-I card. If the card fails, the system can continue to operate normally, with the exception of the functionality provided by the AIC-I.

Figure 9-5 *AIC-I Card Faceplate Diagram*

Environmental Alarms

Environmental alarms are associated with events that affect the operation of the system and are specific to the surrounding environment and external support systems at an MSPP node location. These alarms are usually provisioned and monitored at locations other than those staffed and maintained by a carrier (for example, a central office). This can include an end-user customer's telecom equipment room or an outside plant location, such as a controlled-environment vault (CEV) or concrete hut. Some examples of these alarms include power system performance degradation or failure, hazardous condition alarms (for example, smoke, heat, rising water, and so on), and intrusion alarms (for example, unauthorized entry into a secured area). The ONS 15454 can use the alarm-monitoring capability of the AIC-I to report alarms via the SONET overhead back to the network operations center for trouble resolution or dispatch of maintenance personnel. Figure 9-6 shows an example of this application. A pair of LEDs, labeled as INPUT and OUTPUT, are included on the card faceplate, and illuminate when any input alarm or output control are active.

Figure 9-6 *Environmental Alarms Reported Using AIC-I Card Interfaces*

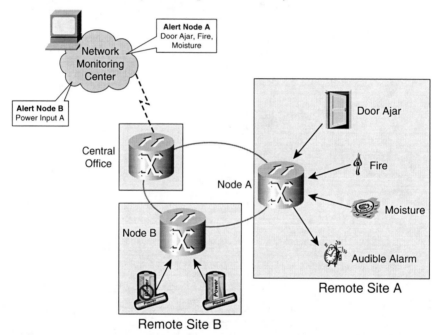

CTC enables the user to provision several parameters related to the operation of the environmental alarms, including an assigned severity (Critical, Major, Minor, or Not Reported), an alphanumeric alarm description, and the capability to set the alarm to be raised upon detection of an "open" or "closed" condition across the alarm contacts. The AIC-I card provides 12 alarm input connections and 4 additional connections that are provisionable as either inputs or outputs. An output is used to control operation of an external device, such as an alarm-indication lamp or a water pump. The backplane of the ONS 15454 chassis has 16 wire-wrap pin pairs for connection to the external equipment to be monitored or controlled. Figure 9-7 shows these connections.

Figure 9-7   *Backplane Environmental Alarm Connections*

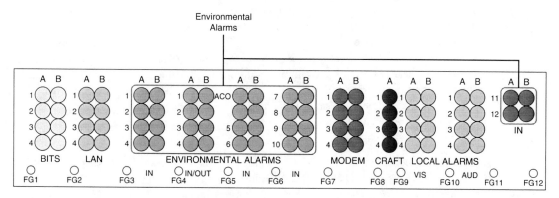

By using an additional piece of hardware, called the Alarm Expansion Panel (AEP), the AIC-I can actually be used to provide up to 32 alarm inputs and 16 outputs, for a total of 48 connections. The AEP is a connector panel that is wired to a subset of the environmental alarm wire-wrap pins and attached to the backplane. Cables can then be installed from the AEP to an external terminal strip for connecting alarm contacts to the system.

One interesting application that involves the use of both an environmental alarm and a control is referred to as a "virtual wire." A virtual wire enables the user to consider the activation of an incoming environmental alarm as triggering the activation of a control. One such scenario is shown in Figure 9-8, where the activation of an alarm at the remote location of Node A causes a control to activate an audible alarm at the staffed location of Node B. A virtual wire is used to associate the alarm with the control activation.

Figure 9-8   *Virtual Wire Operation*

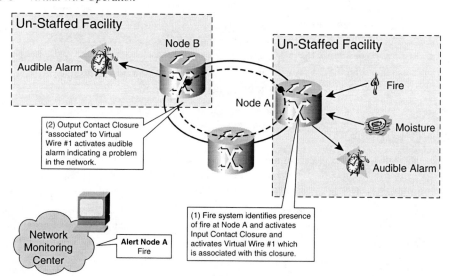

Orderwires

Orderwires allow technicians to attach a phone to the faceplate of the AIC-I card and communicate with personnel at other ONS 15454 MSPP sites. The AIC-I provides two separate orderwires, known as local and express. These can be used simultaneously, if desired. The local orderwire uses the E1 byte in the Section overhead to provide a 64-kbps voice channel between section-terminating equipment, while the express orderwire uses the E2 byte in the Line overhead to provide a channel between line-terminating equipment. Both orderwires operate as broadcast channels, which means that they essentially behave as party lines. Anyone who connects to an orderwire channel can communicate with everyone else on the channel.

Phone sets are connected to the AIC-I using the two standard RJ-11 jacks marked "LOW" (Local Orderwire) and "EOW" (Express Orderwire). A green LED labeled "RING" is provided for each jack. The LED lights and a buzzer/ringer sounds when an incoming call is detected on the orderwire channel.

Phone sets are connected to the AIC-I using the two standard RJ-11 jacks marked LOW (Local Orderwire) and EOW (Express Orderwire). A green LED labeled RING is provided for each jack. The LED lights and a buzzer/ringer sounds when the orderwire channel detects an incoming call.

Power Supply Voltage Monitoring

The AIC-I monitors the A and B power supply connections to the ONS 15454 for the presence of voltage, under-voltage, and over-voltage. Two bicolor LEDs are provided on the AIC-I faceplate for visual indication of either normal (green) or out-of-range (red) power levels. These LEDs are marked as PWR A and PWR B, and are located on the upper portion of the faceplate between the FAIL and ACT LEDs. The TCC2 and TCC2P controller cards also monitor the A and B power supplies for the chassis, and will override this feature of the AIC-I if installed in the same shelf. The TCC2/TCC2P force the power monitor LEDs on the AIC-I faceplate to match the state of their power-monitor LEDs. Because the older TCC+ controller cards do not include the power-monitoring feature, this feature of the AIC-I is more useful when installed with them.

User Data Channels

Four point-to-point data communications channels are provided for possible network operator use by the AIC-I, with two user data channels (UDC-A and UDC-B) and two data communications channels (DCC-A and DCC-B). These channels enable networking between MSPP locations over embedded overhead channels that are otherwise typically unused. The two UDCs are accessed using a pair of RJ-11 faceplate connectors; the two DCCs use a pair of RJ-45 connectors.

The UDC-A and UDC-B channels use the F1 Section overhead byte to form a pair of 64-kbps data links, each of which can be routed to an individual optical interface for connection to another node site. The DCC-A and DCC-B use the D4-D12 line-overhead bytes to form a pair of 576-kbps data links, which are also individually routed to an optical interface.

ONS 15454 MSTP DWDM ITU-T Channel Plan

The ONS 15454 MSTP interfaces comply with the ITU-T 100-GHz, C-band DWDM grid. All DWDM interfaces used on the ONS 15454 MSTP support the C-band wavelengths and are listed in Table 9-2.

Table 9-2   *DWDM C-Band Wavelengths Supported by the ONS 15454 MSTP*

100-GHz Channel Number	λ(nm)	Used by Cisco
59	1530.33	√
58	1531.12	√
57	1531.90	√
56	1532.68	√
55	1533.47	—
54	1534.25	√
53	1535.04	√
52	1535.82	√
51	1536.61	√
50	1537.40	—
49	1538.19	√
48	1538.98	√
47	1539.77	√
46	1540.56	√
45	1541.35	—
44	1542.14	√
43	1542.94	√
42	1543.73	√
41	1544.53	√
40	1545.32	—
39	1546.12	√

continues

Table 9-2 *DWDM C-Band Wavelengths Supported by the ONS 15454 MSTP (Continued)*

100-GHz Channel Number	λ(nm)	Used by Cisco
38	1546.92	√
37	1547.72	√
36	1548.51	√
35	1549.32	—
34	1550.12	√
33	1550.92	√
32	1551.72	√
31	1552.52	√
30	1553.33	—
29	1554.13	√
28	1554.94	√
27	1555.75	√
26	1556.55	√
25	1557.36	—
24	1558.17	√
23	1558.98	√
22	1559.79	√
21	1560.61	√
20	1561.40	—

For future system scalability, some ONS 15454 MSTP DWDM interfaces support wavelength channels on the ITU-T 50-GHz, C-band DWDM grid. This interface capability is noted in this chapter, where applicable.

32-Channel Multiplexer Cards

Two types of 32-channel multiplexer cards exist: the 32 MUX-O and the 32 wavelength selective switch (WSS). Each has the general function of accepting 32, 100-GHz-spaced ITU-T wavelength channels and multiplexing them into a common fiber output.

32 MUX-O Multiplexer Card

Channel multiplexing is accomplished using an on-board arrayed waveguide grating (AWG). Each of the 32 channel inputs can be sourced from a variety of optical products, each of which might have varied channel output power. For this reason, the optical inputs to the 32-channel multiplexer are equipped with an automatic variable optical attenuator (VOA) to equalize channel power among the 32 channel inputs. This ensures uniform gain distribution for each channel in the case of channel amplification, and also minimizes the risk of interchannel cross-talk. The maximum insertion loss for each input channel, assuming the minimum VOA attenuation, is 8.5 dB. ITU-T wavelength channels are input into the 32 MUX-O card using four MPO ribbon connectors on the front faceplate. Each ribbon connector contains eight transmit pairs for interfacing to a fiber cross-connect or termination panel.

The 32 MUX-O is a double-slotted interface card. It can be inserted in the ONS 15454 shelf assembly Slots 1–5 or 12–16.

NOTE   The card cannot be inserted in Slots 6 or 17 because it must occupy dual adjacent universal interface slot positions. Figure 9-9 shows the faceplate layout of the 32 MUX-O.

Figure 9-9   *32 MUX-O Card Faceplate Diagram*

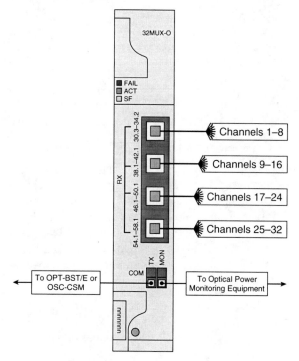

32 WSS Multiplexer Card

The 32 WSS interface unit is used to provide reconfigurable optical add/drop multiplexing (ROADM). With ROADM, any channel(s) within the 32-channel MSTP wavelength system can be added, dropped, or passed through to another DWDM site using software-controlled optical switches. The 32 WSS employs an integrated 80/20 optical splitter to initially drop wavelength channels to the 32-channel demultiplexer unit (covered later in this chapter). In addition to channel multiplexing, the 32 WSS provides the capability to add or pass through any of the 32 ITU-T channels in the MSTP system. This is accomplished using an on-board 2×32 AWG, coupled with 32 1×2 optical switches. Figure 9-10 details a block diagram of the 32 WSS interface module.

Figure 9-10 *The 32 WSS Card Block Diagram*

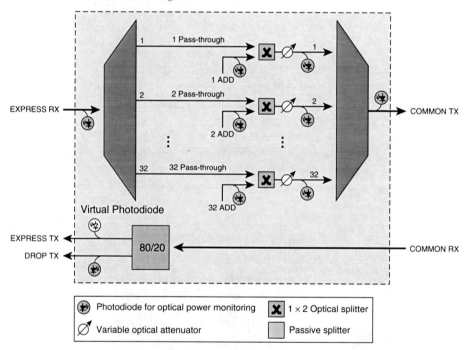

Each of the 32 channel inputs can be sourced from a variety of optical products, each of which can have varied channel output power. For this reason, the optical inputs to the 32-channel multiplexer are equipped with an automatic VOA to equalize channel power among the 32 channel inputs. This ensures uniform gain distribution for each channel (in the case of channel amplification) and also minimizes the risk of interchannel cross-talk. The maximum insertion loss for each input channel, assuming the minimum VOA attenuation, is 11.3 dB.

ITU-T wavelength channels are input into the 32 WSS card using four MPO ribbon connectors on the front faceplate. Each ribbon connector contains eight transmit pairs for interfacing to a fiber cross-connect or termination panel. In addition, six LC-PC-II optical connectors are used on the front faceplate to interface the 32 WSS to other interface cards within the MSTP system. Figure 9-11 details the fiber connections for the 32 WSS.

The 32 WSS is a double-slotted interface card. It can be inserted in the ONS 15454 shelf assembly Slots 1–5 or 12–16.

NOTE   The card cannot be inserted in Slots 6 or 17 because it must occupy dual adjacent universal interface slot positions.

Figure 9-11 shows the faceplate layout of the 32 WSS.

Figure 9-11 *The 32 WSS Card Faceplate Diagram*

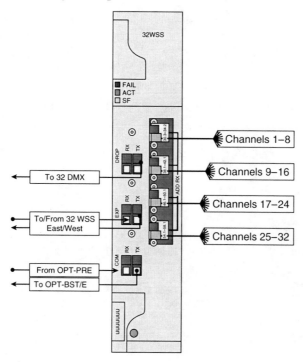

32-Channel Demultiplexer Cards

Two types of 32-channel demultiplexer (DMX) cards exist: 32 DMX-O and 32 DMX. Each has the general function of demultiplexing a composite fiber signal containing 32 100 Ghz–spaced ITU-T wavelength channels.

32 DMX-O Demultiplexer Card

Channel demultiplexing is accomplished using an on-board AWG. The optical outputs of the 32-channel demultiplexer are equipped with a VOA. ITU-T wavelength channels are extracted from the 32 DMX-O card using four MPO ribbon connectors on the front faceplate. Each ribbon connector contains eight receive pairs for interfacing to a fiber cross-connect or termination panel.

The 32 DMX-O is a double-slotted interface card. It can be inserted in the ONS 15454 shelf assembly Slots 1–5 or 12–16.

NOTE   The card cannot be inserted in Slots 6 or 17 because it must occupy dual adjacent universal interface slot positions.

Figure 9-12 shows the faceplate layout of the 32 DMX-O.

Figure 9-12   *The 32 DMX-O Card Faceplate Diagram*

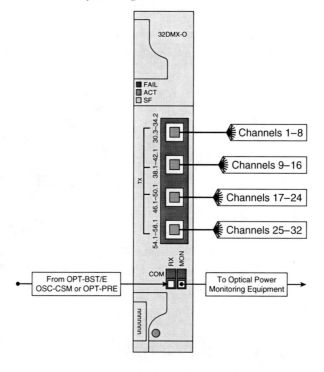

32 DMX Demultiplexer Card

The 32 DMX performs the same functions as the 32 DMX-O; however, the 32 DMX uses a single software-controlled VOA for all the channels on the common receive port. There are no individual software-controlled VOAs on each drop port, as with the 32 DMX-O.

The 32DMX is a single-slotted interface card. It can be inserted in the ONS 15454 shelf assembly Slots 1–6 or 12–17. Figure 9-13 shows a faceplate layout of the 32 DMX interface module.

Figure 9-13 *The 32 DMX Card Faceplate Diagram*

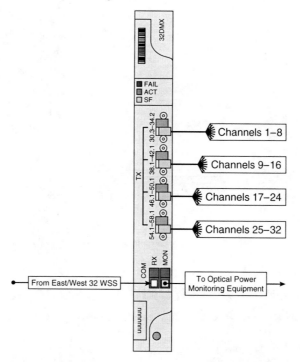

The 32 DMX and 32 WSS work in concert to provide ROADM capability. As noted earlier, the composite input signal is interfaced at the WSS, where an 80/20 splitter drops the 32 channels to the 32 DMX card and also distributes the channels to the corresponding East/West 32 WSS card as pass-through channels. A software-controlled switch in the 32 WSS determines whether the wavelength channel is added to the composite line span—and, thus, dropped at the 32 DMX—or whether the channel is passed through. Channels can be either added or passed through, but not both. Figure 9-14 illustrates the ROADM operation.

Figure 9-14 *ROADM Block Diagram*

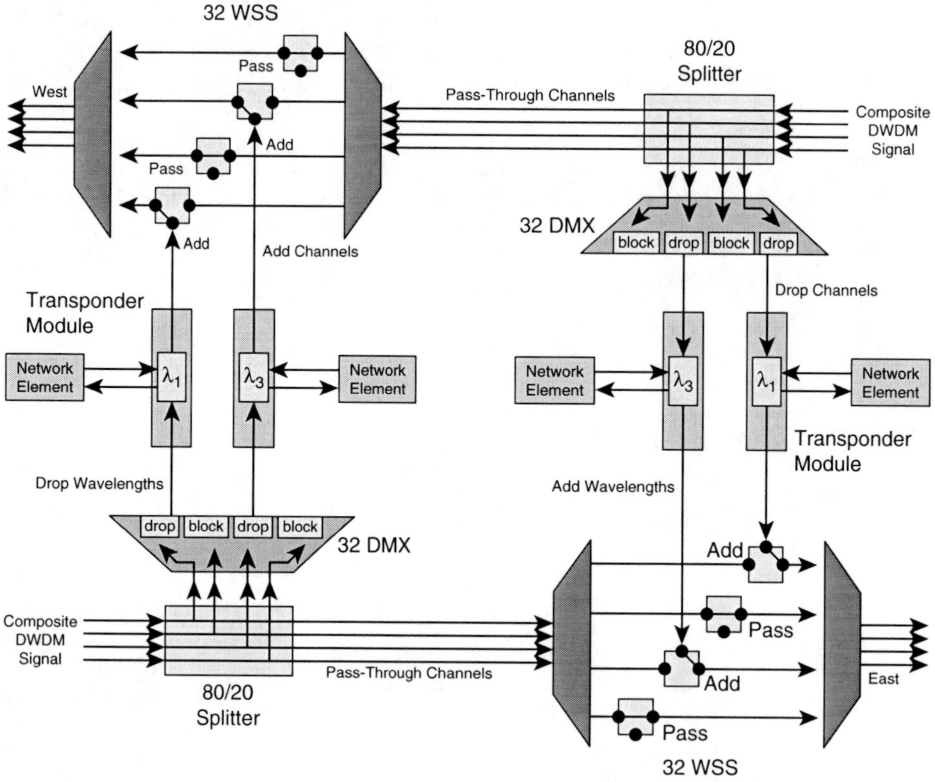

Four-Channel Multiplexer/Demultiplexer Cards

The four-channel multiplexer/demultiplexer card (4MD-xx.x) provides add/drop access to four 100 GHz–spaced ITU-T wavelength channels. Channels are multiplexed and demultiplexed using a passive cascade of interferential filters.

Each of the four optical client input ports on the 4MD-xx.x is equipped with an automatic VOA. The maximum insertion loss for each input channel, assuming the minimum VOA attenuation, is 3.6 dB on input and 3.2 dB on output. ITU-T wavelength channels are accessed on the 4MD-xx.x card using eight LC-PC-II connectors on the front faceplate. The 4MD-xx.xx card is a single-slotted interface card. It can be inserted in the ONS 15454 shelf assembly Slots 1–6 or 12–17. Figure 9-15 shows the faceplate layout of the 4MD-xx.x card.

Figure 9-15 *The 4MD-xx.x Card Faceplate Diagram*

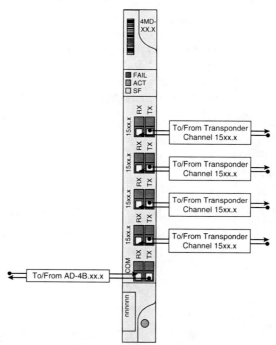

The 4MD-xx.xx adds/drops four adjacent 100 Ghz–spaced ITU-T wavelength channels. It is used in concert with the four-channel banded OADM (covered in the next section). Eight discreet product numbers are associated with the 4MD-xx.xx interface card, covering the entire 32-channel MSTP C-band spectrum. Table 9-3 summarizes the product numbers and wavelength assignments for each 4MD-xx.x interface.

Table 9-3   *4MD-xx.x Product Numbers*

4-Channel Multiplexer/Demultiplexer	ITU-T Wavelength Channels
4MD-30.3	1530.3, 1531.2, 1531.9, and 1532.6
4MD-34.2	1534.2, 1535.0, 1535.8, and 1536.6
4MD-38.1	1538.1, 1538.9, 1539.7, and 1540.5
4MD-42.1	1542.1, 1542.9, 1543.7, and 1544.5
4MD-46.1	1546.1, 1546.9, 1547.7, and 1548.5
4MD-50.1	1550.1, 1550.9, 1551.7, and 1552.5
4MD-54.1	1554.1, 1554.9, 1555.7, and 1556.5
4MD-58.1	1558.1, 1558.9, 1559.7, and 1560.6

Four-Band OADM Filters

The four-band OADM filter (AD-4B-xx.x) provides add/drop access to four bands of ITU-T wavelength channels (four each), covering channels on the 100 GHz or the 50 GHz grid. Each channel band consists of four channels. Thus, two AD-4B-xx.x interface units are available to cover the MSTP 32-channel C-band spectrum. The unit operates bidirectionally, with channels being multiplexed/demultiplexed using a passive cascade of interferential filters. Wavelength add/drop operates along a single East/West path, allowing for asymmetrical channel use.

Table 9-4 summarizes the product numbers and channel assignments for the two AD-4B-xx.x product codes.

Table 9-4   *AD-4B-xx.x Wavelength Assignments*

OADM Filter	Channel Band	ITU-T Grid	ITU-T Wavelength Assignments
AD-4B-30.3	B30.3	100 GHz	1530.3, 1531.2, 1531.9, and 1532.6
		50 GHz	1530.7, 1531.5, 1532.2, and 1533.3
	B34.2	100 GHz	1534.2, 1535.0, 1535.8, and 1536.6
		50 GHz	1534.6, 1535.4, 1536.2, and 1537.0
	B38.1	100 GHz	1538.1, 1538.9, 1539.7, and 1540.5
		50 GHz	1538.5, 1539.3, 1540.1, and 1540.9
	B42.1	100 GHz	1542.1, 1542.9, 1543.7, and 1544.5
		50 GHz	1542.5, 1543.3, 1544.1, and 1544.9
AD-4B-46.1	B46.1	100 GHz	1546.1, 1546.9, 1547.7, and 1548.5
		50 GHz	1546.5, 1547.3, 1548.1, and 1548.9
	B50.1	100 GHz	1550.1, 1550.9, 1551.7, and 1552.5
		50 GHz	1550.5, 1551.3, 1552.1, and 1552. 9
	B54.1	100 GHz	1554.1, 1554.9, 1555.7, and 1556.5
		50 GHz	1554.5, 1555.3, 1556.1, and 1556.9
	B58.1	100 GHz	1558.1, 1558.9, 1559.7, and 1560.6
		50 GHz	1558.5, 1559.3, 1560.2, and 1561.0

Each of the four optical band input ports on the AD-4B-xx.x is equipped with an automatic VOA. Express, pass-through ports also use automatic VOAs. The maximum insertion loss for each input channel band, assuming the minimum VOA attenuation, is 3.5 dB on input and 4.5 dB on output. Pass-through insertion loss is 4.8 dB at minimum VOA attenuation. Wavelength channels are accessed on the AD-4B-xx.x card using eight LC-PC-II connectors on the front faceplate. The AD-4B-xx.xx card is a single-slotted interface card. It can be inserted in the ONS 15454 shelf assembly Slots 1–6 or 12–17. Figure 9-16 shows the faceplate layout of the AD-4B-xx.x card.

Figure 9-16 *AD-4B-xx.x Card Faceplate Diagram*

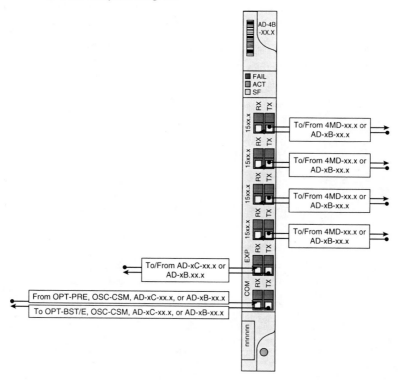

One-Band OADM Filters

The one-band OADM filter (AD-1B-xx.x) provides add/drop access to one band of ITU-T wavelength channels (four each, covering channels on the 100 GHz or the 50 GHz grid. Eight AD-1B-xx.x interface units are available to cover the MSTP 32-channel C-band spectrum. The unit operates bidirectionally, with channels being multiplexed and demultiplexed using a passive cascade of interferential filters. Wavelength add/drop operates along a single East/West path, allowing for asymmetrical channel use.

Table 9-5 summarizes the product numbers and channel assignments for the two AD-1B-xx.x product codes.

Table 9-5   *AD-1B-xx.x Wavelength Assignments*

OADM Filter	Channel Band	ITU-T Grid	ITU-T Wavelength Assignments
AD-1B-30.3	B30.3	100 GHz	1530.3, 1531.2, 1531.9, and 1532.6
		50 GHz	1530.7, 1531.5, 1532.2, and 1533.3
AD-1B-34.2	B34.2	100 GHz	1534.2, 1535.0, 1535.8, and 1536.6
		50 GHz	1534.6, 1535.4, 1536.2, and 1537.0
AD-1B-38.1	B38.1	100 GHz	1538.1, 1538.9, 1539.7, and 1540.5
		50 GHz	1538.5, 1539.3, 1540.1, and 1540.9
AD-1B-42.1	B42.1	100 GHz	1542.1, 1542.9, 1543.7, and 1544.5
		50 GHz	1542.5, 1543.3, 1544.1, and 1544.9
AD-1B-46.1	B46.1	100 GHz	1546.1, 1546.9, 1547.7, and 1548.5
		50 GHz	1546.5, 1547.3, 1548.1, and 1548.9
AD-1B-50.1	B50.1	100 GHz	1550.1, 1550.9, 1551.7, and 1552.5
		50 GHz	1550.5, 1551.3, 1552.1, and 1552.9
AD-1B-54.1	B54.1	100 GHz	1554.1, 1554.9, 1555.7, and 1556.5
		50 GHz	1554.5, 1555.3, 1556.1, and 1556.9
AD-1B-58.1	B58.1	100 GHz	1558.1, 1558.9, 1559.7, and 1560.6
		50 GHz	1558.5, 1559.3, 1560.2, and 1561.0

The optical band input port on the AD-1B-xx.x is equipped with an automatic VOA. Express, pass-through ports also use automatic VOAs. The maximum insertion loss for each input channel band, assuming the minimum VOA attenuation, is 2.2 dB on input and 3.0 dB on output. Pass-through insertion loss is 2.8 dB at minimum VOA attenuation.

Wavelength channels are accessed on the AD-1B-xx.x card using two LC-PC-II connectors on the front faceplate. The AD-1B-xx.xx card is a single-slotted interface card. It can be inserted in the ONS 15454 shelf assembly Slots 1–6 or 12–17. Figure 9-17 shows the faceplate layout of the AD-1B-xx.x card.

Figure 9-17 *AD-1B-xx.x Card Faceplate Diagram*

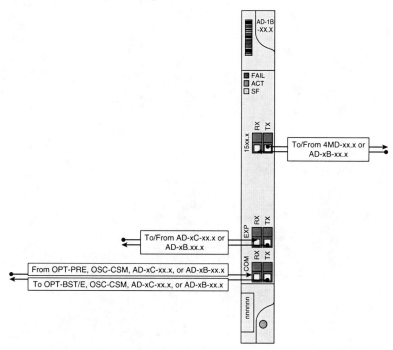

Channel OADM Cards

Three different types of channel OADM interface cards exist for the 15454 MSTP: the four-channel OADM filter, the two-channel OADM filter (AD-2C-xx.x), and the one-channel OADM filter (AD-1C.xx.x). Each provides add/drop access to adjacent ITU-T wavelength channels, covering channels on the 100 GHz or the 50 GHz grid. The MSTP 32-channel C-band spectrum can be add/dropped using 8 different four-channel OADM filters, 16 separate two-channel OADM filters, or 32 discreet one-channel OADM filters. The units operate bidirectionally, with channels being multiplexed and demultiplexed using a passive cascade of interferential filters.

Each of the optical input ports on the channel OADM units is equipped with an automatic VOA. Express, pass-through ports also use automatic VOAs. Wavelength channels are accessed on the channel OADM units using LC-PC-II connectors on the front faceplate. The channel OADM units are single-slotted interface cards. They can be inserted in the ONS 15454 shelf assembly Slots 1–6 or 12–17. Figure 9-18 shows the faceplate layout of the AD-1C-xx.x, AD-2C-xx.x, and AD-4C-xx.x cards.

Figure 9-18 *AD-1C-xx.x, AD-2C-xx.x, and AD-4C-xx.x Cards Faceplate Diagrams*

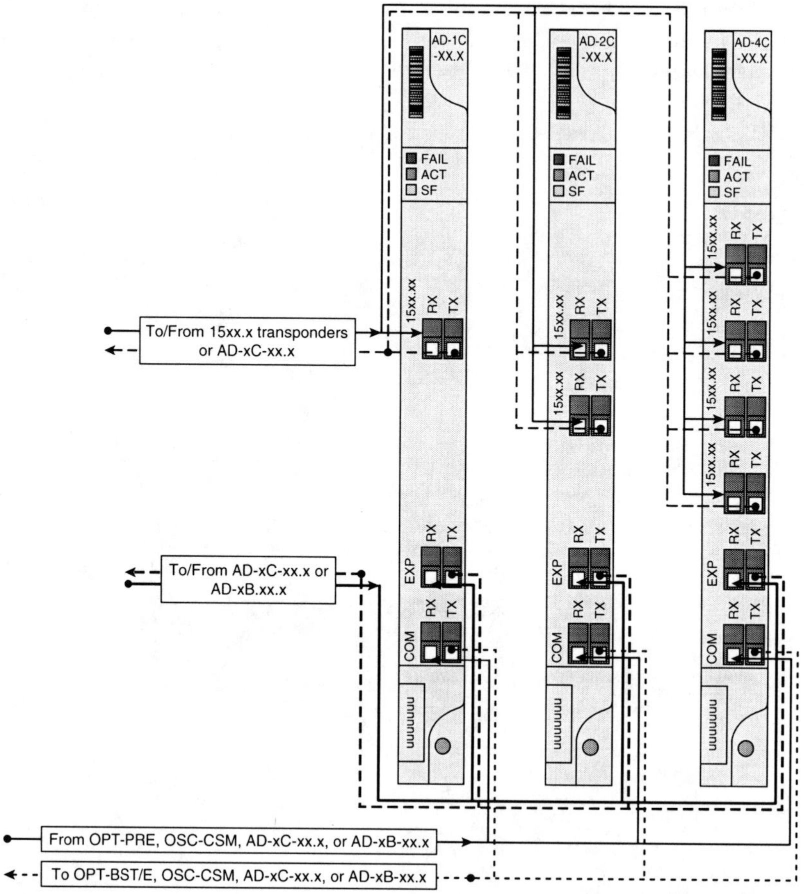

Optical insertion loss characteristics vary among OADM units. Table 9-6 summarizes the insertion loss characteristics for each of the OADM units. Each of the loss figures assumes minimum VOA attenuation.

Table 9-6 *Insertion Loss Characteristics for OADM Units*

OADM Type	Add Channel Insertion Loss	Drop Channel Insertion Loss	Pass-Through Channel Insertion Loss
AD-1C-xx.x	2.6 dB	2 dB	2.2 dB
AD-2C-xx.x	3.1 dB	2.4 dB	2.7 dB
AD-4C-xx.x	4.9 dB	5.4 dB	2.5 dB

ONS 15454 MSTP ROADM

The OADMs discussed in the earlier two sections are "fixed-channel" OADMs. The implementation of fixed-channel OADMs requires careful planning consideration to ensure that the appropriate wavelength channels are efficiently used at each add/drop site. To correctly design a system, a DWDM engineer must be aware of present and future wavelength channel requirements at each node within the system. Careful thought and consideration must be given to how and where banded OADM filters should be used so that future wavelength additions at a node do not require service interruptions. In addition, the DWDM engineer is required to manage an inventory of diverse fixed-channel OADM band filters and fixed-wavelength OADMs. For example, if the DWDM system is designed for full wavelength add/drop flexibility, up to 32 one-channel OADM filters would be present in the network, requiring 32 separate spare parts. Finally, as the DWDM network grows (that is, channel requirements increase), the addition of wavelengths to the system requires the manual fine-tuning of amplifiers at intermediate nodes, driving up the operations cost of ownership for DWDM networks.

To mitigate the operational complexities, network design constraints, and traffic engineering limitations posed by traditional DWDM deployment, carriers are increasingly implementing ROADM in their DWDM deployments. For the ONS 15454 MSTP solution, the ROADM consists of the 32 WSS and the 32 DMX. With the ONS 15454 MSTP ROADM solution, wavelength channel capacity can be remotely distributed to add/drop locations as service demands dictate, without the expense and time associated with re-engineering the network and manually dispatching operations personnel to fine-tune the DWDM parameters at each wavelength add/drop or amplifier site. With the ROADM architecture, the DWDM transport network is designed once, although a wide array of services across the network can be added, rearranged, or deleted infinitely, as capacity exists. This gives carriers a true "next-generation" network, whereby the services provided on the network do not depend upon the design and redesign of the transport technology. Moreover, the ROADM approach provides the flexibility to distribute wavelength channels in a ring, hub-and-spoke, or mesh pattern—all within the framework of a single physical design.

Figure 9-19 details a sample network using traditional DWDM versus DWDM with ROADM technology. By nature, traditional DWDM is a point-to-point technology. To interconnect DWDM systems, carriers typically deploy back-to-back terminals, whereby all the DWDM channels are dropped and ring interconnection is accomplished using transponders. This scenario requires double the normal capital expenditure outlay, in that twice the number of transponders is required for each wavelength channel. In addition, ring-to-ring interconnection requires manual intervention for transponder placement and fiber jumper routing. In contrast, the ROADM approach mechanizes wavelength channel hand-off between DWDM systems, negating the capital and operational expense requirements of the traditional DWDM approach.

Figure 9-19 *Traditional DWDM versus DWDM/ROADM Ring Interconnect*

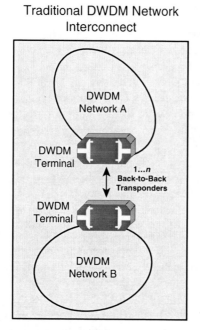

Traditional DWDM Network
Interconnect

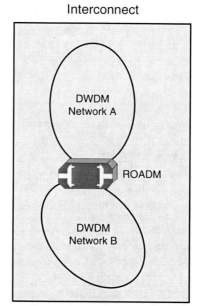

DWDM/ROADM
Interconnect

ONS 15454 MSTP Transponder/Muxponder Interfaces

Client access to ONS 15454 MSTP wavelength channels is accomplished using transponders and muxponders. These interface units function as wavelength converters, in that they convert client interface signals—operating typically at 1310 nm—to an ITU-T DWDM wavelength channel within the ONS 15454 MSTP C-band spectrum. The muxponder interfaces perform the additional function of aggregating multiple client signals into a single DWDM channel.

2.5G Multirate Transponder

The 2.5-Gbps multirate transponder (MR-L1-xx.x, MRP-L1-xx.x) provides a flexible, multirate interface for customer signals. The customer input port can operate transparently in 2R (regenerate, reamplify) or 3R (regenerate, retime, reamplify) mode, depending upon the type of signal it processes. Additionally, the 2.5-Gbps multirate transponder provides a 3R+ mode of operation. In this operating mode, the incoming client signal is mapped into an ITU-T G.709 digital wrapper for DWDM transport. The G.709 digital wrapper provides SONET/Synchronous Digital Hierarchy (SDH)–like overhead functions for improved DWDM signal performance monitoring. It also facilitates the employment of forward error correction (FEC) to enhance the optical line reach of the DWDM trunk signal.

The client port can be provisioned to accept an incoming signal ranging from 8 Mbps to a full 2.5 Gbps SONET/SDH rate. Client port access is accomplished using a wide variety of small form factor pluggable (SFP) interfaces. These SFP interfaces vary depending upon whether customer signals are carrying native SONET/SDH signals, Ethernet, or storage protocols, such as Enterprise Systems Connection (ESCON) or Fibre Channel (FC), and so on.

The DWDM trunk-facing port is software tunable within 400 GHz, covering four ITU-T channels on the 100 GHz grid or eight ITU-T channels on the 50 GHz grid. The DWDM trunk is formatted as a SONET/SDH, long-reach OC-48/STM-16 signal. Eight 2.5-Gbps multirate transponders are available to cover the ONS 15454 MSTP C-band spectrum.

The 2.5-Gbps multirate transponder can operate in the ONS 15454 MSTP shelf assembly as a single, unprotected unit, or the DWDM trunk can be protected using an on-board passive optical splitter. Thus, two product codes are available for the 2.5-Gbps multirate transponder: one for unprotected DWDM trunk operation and another for protected DWDM trunk operation. Optionally, a Y-cable can be used to split the output of a customer's signal and distribute each output to two discreet, unprotected 2.5-Gbps multirate transponders installed in the ONS 15454 MSTP chassis to accomplish DWDM trunk protection.

The 2.5-Gbps multirate transponder can operate in Slots 1–6 or 12–17 of the ONS 15454 MSTP shelf assembly.

10G Multirate Transponder

The 10-Gbps multirate transponder (10E-L1-xx.x) provides a flexible, multirate, multiprotocol 10-Gbps interface for customer signals. The customer input port can operate transparently or can be provisioned to terminate SONET/SDH line data communications channels (LDCC). The interface card supports ITU-T G.709 digital wrapper for 10-Gbps Ethernet and OC-192/STM-64. Enhanced forward error correction (E-FEC) for DWDM transport is also available.

The client port provides a 1310-nm short-reach interface that accepts a 10-Gbps Ethernet WAN/LAN PHY signal, 10-Gbps FC input, or an OC-192/STM-64 SONET/SDH signal. Client port access is accomplished using a 10-Gbps small form factor pluggable (XFP) interface. Presently, the XFP receptacle supports only short-reach, 1310-nm interfaces for all supported protocols.

The DWDM trunk-facing port is software tunable within 400 GHz, covering four ITU-T channels on the 100 GHz grid or eight ITU-T channels on the 50 GHz grid. The DWDM trunk is formatted as a SONET/SDH, long-reach OC-192/STM-64 signal, 10-Gbps WAN/ LAN PHY, or 10-Gbps FC. Eight 10-Gbps multirate transponders are available to cover the ONS 15454 MSTP C-band spectrum.

The 10-Gbps multirate transponder can operate in the ONS 15454 MSTP shelf assembly as a single, unprotected unit, or can be deployed in pairs, where client port protection is provided using the customer-attached 10-Gbps equipment. Optionally, a Y-cable can be used to split the output of a customer's signal and distribute each output to two discreet, unprotected 10-Gbps multirate transponders installed in the ONS 15454 MSTP chassis to accomplish DWDM trunk protection. The 10-Gbps multirate transponder can operate in Slots 1–6 or 12–17 of the ONS 15454 MSTP shelf assembly.

4x2.5G Enhanced Muxponder

The 4x2.5-Gbps enhanced muxponder (10ME-xx.x) accepts four OC-48/STM-16 client signals and multiplexes them into a single 10-Gbps, ITU-T 100 GHz–spaced DWDM channel. The interface card can be provisioned to operate transparently or can terminate client SONET/SDH line and section overhead bytes. In transparent mode, client endpoints connected across the DWDM network can communicate using the SONET/SDH section/multiplex data communications channel (SDCC/MSDCC). This facilitates the support of end-to-end client 1+1 automatic protection switching (APS) or bidirectional line switch ring (BLSR)/Multiplex Section Shared Protection (MS-SPR). Additionally, end-to-end client SONET/SDH unidirectional path-switched ring (UPSR)/Subnetwork Connection Protection (SNCP) is supported in both transparent and terminating port modes. The interface card supports ITU-T G.709 digital wrapper and enhanced forward error correction (E-FEC) for DWDM transport.

Client port access is accomplished using a wide array of SFP interfaces. SFP interfaces are available to support short-reach and intermediate-reach OC-48/STM-16 client inputs. In addition, DWDM SFP modules can be supported on the client interfaces.

The DWDM trunk-facing port is software tunable within 400 GHz, covering four ITU-T channels on the 100 GHz grid or eight ITU-T channels on the 50 GHz grid. The DWDM trunk is formatted as a 10-Gbps, G.709 digital wrapper, long-haul DWDM wavelength channel. Eight 4x2.5-Gbps enhanced muxponders are available to cover the ONS 15454 MSTP C-band spectrum.

The 4x2.5-Gbps enhanced muxponder can operate in the ONS 15454 MSTP shelf assembly as a single, unprotected unit, or can be deployed in pairs, where client port protection is provided using SONET/SDH standard point-to-point or ring protocols. Optionally, client ports can be protected using a Y-cable that splits a customer's input signal and distributes each output to two separate muxponder interfaces. The 4x2.5-Gbps enhanced muxponder can operate in Slots 1–6 or 12–17 of the ONS 15454 MSTP shelf assembly.

2.5G Multiservice Aggregation Card

The 2.5-Gbps multiservice aggregation card (DM-L1-xx.x, DMP-L1-xx.x) multiplexes up to eight client input signals into a single 2.5-Gbps ITU-T DWDM wavelength channel.

Client port access is accomplished using a wide variety of SFP interfaces. These SFP interfaces vary depending upon whether customer signals are carrying Ethernet or storage protocols, such as fiber connection (FICON) or FC, and so on. Presently, 850-nm/1310-nm multimode/single-mode SFP interfaces are supported for Gigabit Ethernet (GigE), 1- or 2-Gbps FC, and 1- or 2-Gbps FICON.

The DWDM trunk-facing port is software tunable within 400 GHz, covering four ITU-T channels on the 100 GHz grid or eight ITU-T channels on the 50 GHz grid. The DWDM trunk is formatted as a SONET/SDH, long-reach OC-48/STM-16 signal. Therefore, two GigE, 1-Gbps FC/FICON multiplexed ports are supported. Correspondingly, one 2-Gbps FC/FICON multiplexed signal is also supported. Eight 2.5-Gbps multiservice aggregation cards are available to cover the ONS 15454 MSTP C-band spectrum.

The 2.5-Gbps multiservice aggregation card can operate in the ONS 15454 MSTP shelf assembly as a single, unprotected unit, or the DWDM trunk can be protected using an on-board passive optical splitter. Thus, two product codes are available for the 2.5-Gbps aggregation card: one for unprotected DWDM trunk operation and the other for protected DWDM trunk operation. Optionally, a Y-cable can be used to split the output of a customer's signal and distribute each output to two discreet, unprotected 2.5-Gbps multiservice aggregation cards installed in the ONS 15454 MSTP chassis to accomplish DWDM trunk protection.

The 2.5-Gbps multiservice aggregation card can operate in Slots 1–6 or 12–17 of the ONS 15454 MSTP shelf assembly.

Table 9-7 details the service mix supported using the transponder and muxponder interfaces for the ONS 15454 MSTP shelf assembly.

Table 9-7   *ONS 15454 MSTP Transponder/Muxponder Service Capabilities*

Service Type	Transponder/Muxponder Interface
GigE	MR/P-L1-xx.x and DM/P-L1-xx.x
ISC1	MR-L1-xx.x
ISC3	MR/P-L1-xx.x
Optical T3	MR-L1-xx.x
OC-3/STM-1	MR/P-L1-xx.x
OC-12/STM-4	MR/P-L1-xx.x
OC-48/STM-16	MR/P-L1-xx.x and 10ME-xx.x
OC-192/STM-64	10E-L1-xx.x
10 GigE (LAN/WAN PHY)	10E-L1-xx.x
D1 Video	MR/P-L1-xx.x
DV6000	MR/P-L1-xx.x
High-Definition Television (HDTV)	MR-L1-xx.x

continues

Table 9-7 *ONS 15454 MSTP Transponder/Muxponder Service Capabilities (Continued)*

Service Type	Transponder/Muxponder Interface
External Time Reference (ETR) for IBM Sysplex Timer	MR-L1-xx.x
ESCON	MR/P-L1-xx.x
Fast Ethernet	MR-L1-xx.x
FC	MR/P-L1-xx.x and DM/P-L1-xx.x
FICON	MR/P-L1-xx.x and DM/P-L1-xx.x

ONS 15454 MSTP Optical Amplifiers

Optical amplifiers are used in the ONS 15454 MSTP system for two primary purposes:

- Preconditioning of the incoming composite optical signal for insertion into the MSTP optical filters. This is accomplished using the optical preamplifier (OPT-PRE).

- Increasing the power level of the outgoing composite signal before insertion into the transmission fiber. This function is attributed to the optical booster amplifier (OPT-BST) in the ONS 15454 MSTP system.

Each function demands specific characteristics from the amplifier. Typically, a preamplifier is optimized for low power and low noise. Conversely, a booster amplifier is usually characterized by a high-output, high-noise gain profile. Additionally, for high bit-rate optical signals (for example, OC-192 and above), chromatic dispersion compensation is typically needed to recover severely impaired optical pulses because of fiber nonlinear effects.

Although dispersion compensation can be accomplished anywhere within the DWDM network, it is typically done at amplifier sites to mitigate the optical attenuation loss associated with dispersion compensation fiber. The OPT-PRE amplifier is designed to allow for dispersion compensation insertion between amplifier gain stages.

Finally, amplifiers must be capable of dynamically adjusting to the addition and insertion of DWDM wavelength channels so that amplifier power gain is always uniformly distributed among the ITU-T channels. This section examines the pre/booster amplifier cards for the ONS 15454 MSTP system as they relate to these functional characteristcs.

OPT-PRE

The OPT-PRE is an EDFA operating in the ITU-T DWDM C-band at 50 GHz stability. The ONS 15454 currently operates at only the 32-channel 100 GHz ITU-T grid; however, the amplifiers used in the system are scalable to accommodate an upgrade to 64 channels. The OPT-PRE employs a two-stage gain operation that allows for mid-stage access for dispersion compensation insertion. Because the DWDM wavelength channels within the composite optical signal might have uneven power levels, the OPT-PRE employs an automatic VOA on the incoming signal to ensure even amplified optical gain among each wavelength channel, to control gain tilt. The amplifier also employs fast transient suppression to adjust for DWDM channel count changes and power fluctuations because of network disruptions.

The OPT-PRE can operate in two user-provisionable modes:

- **Constant power mode**—The power output of the amplifier is kept at a fixed rate, regardless of the input power range. Note that when in constant power mode, gain tilt cannot be controlled.

- **Constant gain mode**—The amplifier output power is linearly related to the input power.

Table 9-8 details the optical characteristics of the OPT-PRE. Figure 9-20 illustrates a block diagram of preamplifier operation in the ONS 15454 MSTP system.

Table 9-8   *OPT-PRE Transmissions Characteristics*

Optical Characteristic	Value			
Input power Single channel Fully loaded channels	−39.5 dBm minimum, −6 dBm maximum −21.5 dBm minimum, 12 dBm maximum			
Output power	17.5 dBm			
Gain range (constant power mode)	5 dB to 38.5 dB			
Noise figure	6.5 dB			
Insertion loss range for midstage DCU	3 dB to 9 dB			
Transient suppression	Input power excursion	Undershoot/ overshoot (max)	Setting time (max)	Gain error
	15 dB	3.7 dB	500 us	2.5 dB
	6 dB	0.5 dB	150 us	0.5 dB
	3 dB	0.5 dB	150 us	0.5 dB

Figure 9-20 *OPT-PRE Signal Flow*

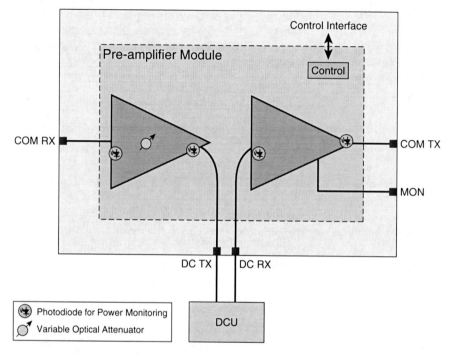

The OPT-PRE is a single-slotted interface card. It can be inserted in the ONS 15454 shelf assembly Slots 1–6 or 12–17.

OPT-BST

As with the OPT-PRE, the OPT-BST is an EDFA operating in the ITU-T DWDM C-band at 50 GHz stability. The OPT-BST is a single-stage amplifier that provides the additional function of extracting and inserting the optical service channel (OSC) from the composite fiber signal and distributing it to the OSCM card. Because the DWDM wavelength channels within the composite optical signal might have uneven power levels, the OPT-BST employs a VOA on its import port to ensure even amplified optical gain among each wavelength channel, to control gain tilt. The amplifier also employs fast transient suppression to adjust for DWDM channel count changes and power fluctuations because of network disruptions.

Two versions of the OPT-BST interface card exist: OPT-BST and OPT-BST-E. The OPT-BST-E provides an extra 3 dB of gain, in comparison to the OPT-BST. The additional power output can be used to double the ONS MSTP DWDM channel capacity or to increase the span length between amplifier sites. No other differences exist between the BST and BST-E amplifier cards.

The OPT-BST can operate in two user-provisionable modes:

- **Constant power mode**—The power output of the amplifier is kept at a fixed rate, regardless of the input power range. Note that when in constant power mode, gain tilt cannot be controlled.

- **Constant gain mode**—The amplifier output power is linearly related to the input power.

Table 9-9 details the optical characteristics of the OPT-BST. Figure 9-21 illustrates a block diagram of the booster amplifier operation in the ONS 15454 MSTP system.

Table 9-9   *OPT-BST/E Transmissions Characteristics*

Optical Characteristic	Value for OPT-BST			Value for OPT-BST-E
Input power Single channel Fully loaded channels	–21 dBm minimum, –6 dBm maximum –3 dBm minimum, 12 dBm maximum			–26 dBm minimum, –8 dBm maximum –6 dBm minimum, 12 dBm maximum
Output power	17.5 dBm			20.5 dBm
Gain range (constant power mode)	5 dB to 20 dB			8 dB to 23 dB
Noise figure	6 dB			6 dB
Transient suppression	Input power excursion	Undershoot /overshoot (max)	Setting time (max)	Gain error
	15 dB	3.5 dB	500 us	0.5 dB
	6 dB	0.5 dB	150 us	0.5 dB
	3 dB	0.5 dB	150 us	0.5 dB

The OPT-BST/E is a single-slotted interface card. It can be inserted in the ONS 15454 shelf assembly Slots 1–6 or 12–17.

Figure 9-21 *OPT-BST/E Signal Flow*

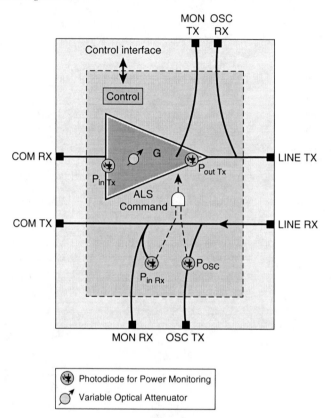

ONS 15454 MSTP Dispersion Compensation Unit

Chromatic dispersion is a fiber nonlinear condition (anomaly) that causes wavelength channels to spread. Because chromatic dispersion varies by wavelength frequency, some optical frequencies spread faster than others, resulting in optical pulses that have spread over several bit periods between transmitter and receiver. The result can be transmission degradation attributed to bit errors, as optical receivers interpret 0 bit pulses and 1s, and vice versa.

To mitigate this effect, DCUs are intermittently placed in the optical fiber path. These DCU modules typically exhibit the exact opposite polarity of the fiber, resulting in a compression effect on the optical pulse as the chromatic dispersion effects become reversed.

The DCU used in the ONS 15454 MSTP system mounts in a separate 1RU chassis, which is typically mounted atop the ONS 15454 MSTP shelf assembly. The DCU chassis contains two slots for DCUs: one for the East path and one for the West path. The DCU consists of a spool of dispersion-compensating fiber, with optical connectors. It is an entirely passive system, requiring no power connections or active management.

Table 9-10 lists the DCU units available for the ONS 15454 MSTP system.

Table 9-10   *ONS 15454 MSTP DCU Products*

Module Description	Compensation Level	Attenuation
DCU-100	100 ps/nm	2.1 dB
DCU-350	350 ps/nm	3.0 dB
DCU-450	450 ps/nm	3.5 dB
DCU-550	550 ps/nm	3.9 dB
DCU-750	750 ps/nm	5.0 dB
DCU-950	950 ps/nm	5.5 dB
DCU-1150	1150 ps/nm	6.2 dB

ONS 15454 MSTP Supported Network Configurations

The ONS 15454 MSTP supports a wide array of network configurations, ranging from point-to-point (linear) to various ring topologies. The design rules and plug-in interface requirements also vary per topology type. Chapter 10, "Designing ONS 15454 MSTP Networks," covers design rule considerations. This section details some of the interface card requirements, where appropriate, but is not intended to provide an exhaustive list of card/node requirements for each network topology implementation.

Linear Topologies

The ONS 15454 MSTP can be deployed in linear, point-to-point applications, in which the A and Z ends of the topology spans use 32-channel terminal nodes: ROADM or fixed-channel add/drop. The linear topology can employ OADM sites midspan or can be characterized by a single amplified/unamplified link from terminal to terminal. For single-span implementations, the terminal-to-terminal distance is usually most limited by the optical reach of the OSC card because it operates outside the EDFA amplifier operation pass-band. Both the OSCM/OSC-CSCM interface cards support a 37-dB end-to-end link budget. Additionally, the OSCM interface card can support a 37-dB end-to-end link budget.

Figure 9-22 illustrates the two variations of linear topologies supported on the ONS 15454 MSTP system. Figure 9-23 represents a typical ONS 15454 MSTP shelf layout for the linear topology.

Figure 9-22 *ONS 15454 MSTP Linear Topologies*

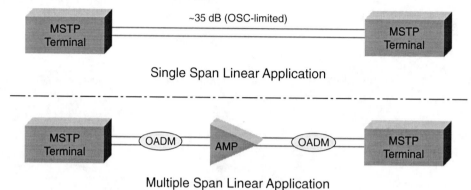

Single Span Linear Application

Multiple Span Linear Application

Figure 9-23 *ONS 15454 MSTP Terminal Shelf Layout*

Ring Topologies

The ONS 15454 MSTP provides support for a number of ring-based topologies. Ring systems are characterized by single or multiple 32-channel terminal nodes, networked with intermediate OADM sites. The ring configurations provide the additional functionality of wavelength channel protection, depending on the type of client transponder/muxponder interface card used for wavelength channel insertion. The ring topologies can be segregated into the following categories:

- Hubbed/multihubbed

- Meshed

- Reconfigurable

Hubbed/Multihubbed Rings

Hubbed rings are typified by a single 32-channel terminal node, in which all ring wavelength channels are terminated. Intermediate ring sites are OADM or pass-through nodes. Wavelength channels cam be protected or unprotected. Unprotected DWDM channels can be reused on any ring East/West span. Multihubbed rings operate in the same manner; however, this topology uses two or more hub nodes on the ring. Protected channels must terminate at a minimum of two hub nodes on the multihubbed ring.

Figure 9-24 illustrates the hubbed ring topologies supported on the ONS 15454 MSTP system. Figure 9-25 represents a typical ONS 15454 MSTP shelf layout for the hubbed ring topology.

Figure 9-24 *ONS 15454 MSTP Hubbed Ring Topologies*

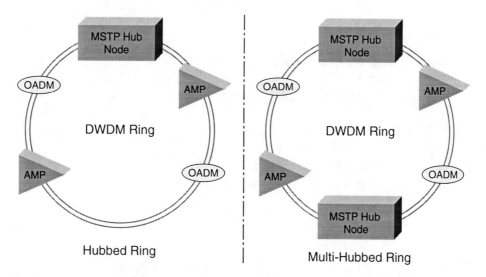

Figure 9-25 *ONS 15454 MSTP Hub and OADM Shelf Layout*

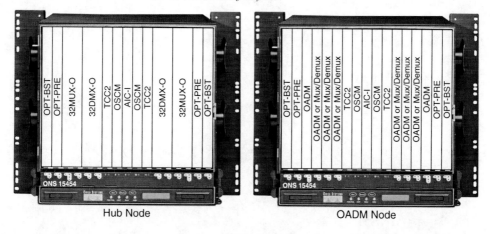

Meshed Rings

Unlike hubbed rings, meshed rings do not use terminal/hub nodes in the ring topology. All the nodes in a meshed ring topology are OADM sites. DWDM wavelength channel protection and unprotected channel reuse is available, as with the hubbed ring scenarios. The meshed ring introduces a new network node type: anti-ASE (amplified spontaneous emissions) node. The anti-ASE node is used to mitigate the potential effect of noise accumulation on unused, amplified channels, to keep from destroying the quality of DWDM channel transmissions. The anti-ASE node can be deployed using one of two methods:

- Regenerate the DWDM channels by equipping an OADM site on the ring with 32-channel multiplexer/demultiplexer units

- For 10 or fewer channels, optically patch the channels through from East to West OADM units

Reconfigurable Rings

The reconfigurable optical add/drop (ROADM) concept was covered earlier in this chapter. With this topology, any-to-any DWDM wavelength channel assignment is possible. DWDM wavelength channel protection and unprotected channel reuse is supported with this topology also. The ROADM shelves use the 32 WSS/32 DMX multiplexer/demultiplexer combination for OADM and terminal shelves, or a mixture of ROADM and fixed-channel add/drop nodes.

Figure 9-26 illustrates the meshed ring and ROADM topologies supported on the ONS 15454 MSTP system. Figure 9-27 represents a typical ONS 15454 MSTP shelf layout for the ROADM topology.

Figure 9-26   *ONS 15454 MSTP Meshed Ring and ROADM Topologies*

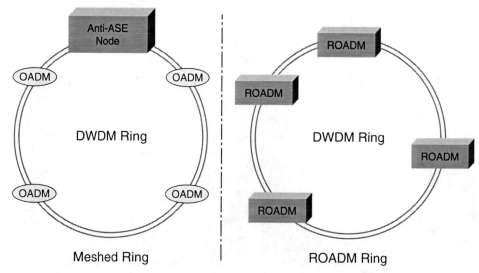

Figure 9-27 *ONS 15454 MSTP ROADM Shelf Layout*

Summary

This chapter introduced the Cisco ONS 15454 Multiservice Transport Platform (MSTP). It detailed the common and interface cards that comprise the ONS 15454 MSTP system, including the purpose for which each card is used in the MSTP system. The ONS 15454 allows for flexible deployment of either TDM/Ethernet or DWDM services. For the DWDM application, a wide variety of fixed-channel optical add/drop units are available for services delivery. However, the most flexible aspect of the ONS 15454 MSTP resides in its ROADM capability. ROADM deployments provide many benefits to DWDM networks, among them reduced equipment-sparing requirements, "SONET-like" services delivery, and streamlined DWDM system engineering. Moreover, the ONS 15454 MSTP allows for choice and flexibility in DWDM topology layouts. From the simplest point-to-point topology to more advanced ring or mesh topologies, the MSTP chassis layout can easily be tailored to fit within the DWDM services constraints.

This chapter and subsequent chapters cover the 32-channel, C-band operations of the ONS 15454 MSTP. At press time, a future software/hardware release of the ONS 15454 MSTP would increase the system's capability by making use of the L-band DWDM wavelength channels. The L-band implementation is primarily intended to operationalize DWDM with dispersion-shifted fiber (DS-fiber). Because of excess fiber nonlinearities with DS-fiber, the C-band DWDM channels are severely performance-limited, so most carriers do not use DS-fiber for C-band DWDM implementations.

Additionally, the ONS 15454 MSTP service capacity is being enhanced with new 10-Gbps transponder/muxponder interface cards. To alleviate equipment-sparing requirements, each of the 10-Gbps transponder/muxponder interface cards allows for entire C-band and L-band tunability, versus the current four-channel, C-band tunability. In addition, the ONS 15454 MSTP service mix will be enhanced using the availability of a new 10-Gbps transponder for Ethernet and storage interconnect protocols.

This chapter covers the following topics:

- ONS 15454 MSTP DWDM Design Considerations
- ONS 15454 MSTP DWDM Design Rules Examples
- ONS 15454 MSTP Manual DWDM Design Example
- ONS 15454 MSTP MetroPlanner Design Tool

CHAPTER 10

Designing ONS 15454 MSTP Networks

The design of dense wavelength-division multiplexing (DWDM) networks can encompass a wide variety of tasks and design factors. Depending upon network type and scale, the DWDM design can be very simple—as is the case with point-to-point, unamplified systems—or can be very complex. Many factors should be considered when planning a DWDM network. This chapter examines the general design considerations for DWDM networks and explores their importance for ONS 15454 Multiservice Transport Platform (MSTP) DWDM system deployment.

This chapter is not intended to provide a complex, thorough analysis of DWDM fundamentals. It is assumed that the reader has a basic knowledge of DWDM transmitters, receivers, amplifiers, fiber nonlinearities, and so on. However, where appropriate, some detail is presented on these topics. For a general review of DWDM fundamentals, refer to the technical documentation referenced at Cisco.com

ONS 15454 MSTP DWDM Design Considerations

In general, DWDM designers and planners must be cognizant of several factors that influence system topology and transmissions capability. Among them are physical limitations, such as structure locations, and fiber plant type. Additionally, the system bandwidth requirements and scale should be taken into account, along with the DWDM equipment limitations such as amplifier noise limitations (OSNR) and receiver minimum power levels.

For existing DWDM deployments, physical structure locations are usually already predetermined for amplifier/regenerator locations. For example, existing DWDM systems may use 60 km amplifier spacing rules; thus, because outside plant enclosures and power are already present at these locations, it is prudent for the DWDM designer to engineer the new system within the present amplifier spacing rules. The drawback associated with using existing structures and spacing is that it might be economically inadvisable or systematically impossible to design the new DWDM system within old physical constraints.

In addition to physical structure requirements, when providing DWDM over existing fiber infrastructure, the design is usually constrained by the transmission characteristics of the existing optical fiber. Essentially, three types of fiber are used today: standard, single-mode fiber;

nonzero, dispersion-shifted fiber (NZDF); and dispersion-shifted fiber (DS-fiber). Figure 10-1 details the typical attenuation and chromatic dispersion profiles for each fiber type.

Figure 10-1 *Optical Fiber Performance Profile*

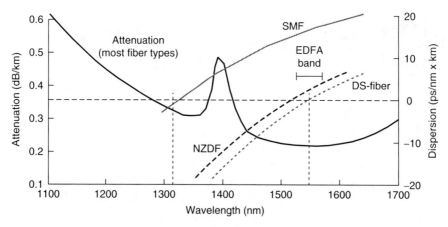

From an attenuation standpoint, DWDM C-band and L-band operation occurs in the lowest optical loss region at 0.21 to 0.25 dB/km. The major transmissions difference among the fiber types revolves around their chromatic dispersion characteristics. As mentioned in Chapter 9, "Using the ONS 15454 Platform to Support DWDM Transport: MSTP," chromatic dispersion causes an optical signal to spread into adjacent bit periods, which can cause bit errors at optical receivers. Low-cost transmitter sources, which typically have wide signal spectral widths, exhibit more sensitivity to this factor over distance and time than higher-cost transmitters, which typically are constructed with tighter signal spectral widths. Figure 10-2 illustrates this phenomenon.

Figure 10-2 *Wavelength Signal Broadening Caused by Chromatic Dispersion*

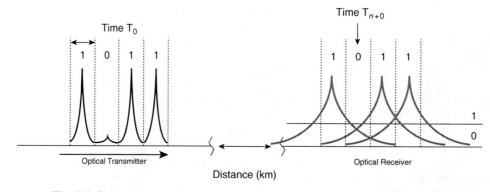

The DS-fiber mitigates this issue by shifting the zero-dispersion parameter from the 1310 nm region to the 1550 nm region. This facilitates long-distance, high-bit-rate transmission for a single wavelength at 1550 nm because only optical attenuation is a signal-impairment factor. However, the zero-dispersion property of DS-fiber causes many transmission abnormalities for high-powered, multiplexed wavelengths in the 1550 nm region, such as four-wave mixing (FWM), which makes it unsuitable for DWDM.

Certain fiber anomalies, such as FWM and channel cross-talk, are offset by the chromatic dispersion effect. However, if the signal spreads too fast, dispersion compensation must be used to compress the wavelength spectrum to near its original form. Therefore, the right balance of optical power, signal spectral width, and chromatic dispersion is necessary for optimal DWDM distance extension. For single-mode fiber (SMF), wavelength spreading occurs at an average rate of 17 picoseconds (ps)/nm-km; for NZDF, the typical chromatic dispersion rate is around 4.4 ps/nm-km. It should be noted that the tolerance of chromatic dispersion decreases as the transmission's bit rate increases. As detailed in the following formula, each time-division multiplexing (TDM) increase in bandwidth results in a [1/16] reduction in transmissions distance.

$$(BR)^2 \times |D| \times L > 104{,}000 \ (Gbps)^2 ps/nm$$

Where:

- (BR) is the transmission bit rate (such as gigabits per second)
- $|D|$ is the average chromatic dispersion value for the transmission wavelength— measured in picoseconds per nanometer-kilometer (ps/nm-km)
- L is the length of the fiber—measured in kilometers

Thus, to increase channel capacity from 2.5 Gbps to 10 Gbps (a fourfold increase), chromatic dispersion tolerance decreases by a factor of 16.

Many other fiber-transmissions factors must be considered as well. The amount of polarization mode dispersion (PMD) in fiber can be a relevant factor in the maximum bit rate that can be transmitted. The PMD effect causes light signals to travel in two orthogonal modes; as a result, a portion of the signal pulse travels on a fast axis, and a portion travels on a slow axis. The downstream optical receiver is tasked with reconstructing both pulse axes into a single bit stream. Optical-fiber PMD can be caused by a number of elements, including glitches in the fiber-manufacturing process or mishandling of fiber during cable manufacturing or installation.

The effect of other fiber nonlinearities, such as stimulated brillion scattering, stimulated raman scattering, cross-phase modulation, and self-phase modulation also must be taken into account when designing DWDM systems; however, these effects can be controlled with proper equipment design.

Amplified systems present their own complexities in terms of optical signal-to-noise ratio (OSNR) and channel gain tilt. For example, when engineering each amplified span, the DWDM engineer must compute the OSNR between each span to ensure that the signal level is within receiver specifications for each amplifier in use. The following formula is used for OSNR measurement for a multistage amplifier:

$$OSNRout_i = \frac{1}{(1/OSNRin_i) + (h^* \ f^* \ B^* NF_i / Pin_i)}$$

Where:

- $OSNR_{OUT}$ is the OSNR, in decibels, at the output power
- $OSNR_{IN}$ is the OSNR, in decibels, of the previous amplifier
- f is the noise figure in decibels
- h is Planck's constant (6.626×10^{-34} joule-seconds)
- v is the frequency of light in fiber at 1550 nm (1.935×10^{14} THz)
- B is the BW measuring the noise figure (0.1 nm)
- P_{IN} is the input power, in decibels referenced to 1 milliwatt (dBm), to the amplifier

As you can see, manually computing the effects of each fiber transmission effect for each span/wavelength set can be a cumbersome task. For this reason, the ONS 15454 MSTP is broken into a set of design rules for each topology type. The design rules take into account the limitations of each optical component used within the system so that the DWDM engineer need only compute the optical fiber limitations, which vary from span to span.

In the next sections, you examine a subset of the design rules associated with each of the topologies detailed in Chapter 9.

ONS 15454 MSTP DWDM Design Rules Examples

The various transponder/muxponder interface cards used in the ONS 15454 MSTP exhibit unique transmitter and receiver characteristics. A typical DWDM design uses a wide range of these transponder/muxponder interfaces at varying bit rates. For reference, Table 10-1 details the ONS 15454 MSTP transponder/muxponder transmissions characteristics.

Depending on which transponder/muxponder type is being used, the optical span design rules slightly differ. Tables 10-2 through 10-4 enumerate the ONS 15454 MSTP design rules for the linear and ring topologies specified in Chapter 9. Note the following assumptions/details associated with the design rules:

- The given design rules cover SMF fiber only.
- Full 32-channel loading is assumed.
- Where dispersion compensation units (DCUs) are placed, the DCU fiber attenuation is assumed to be 9 dB (worst case for the OPT-PRE).
- Both PRE/booster amplifiers are used on amplified spans.
- Where no optical add/drop multiplexers (OADMs) are used, only the OPT-PRE is installed.
- Assume equal span losses, with fixed OADM nodes having a loss of 16 dB.
- The optical preamplifier (OPT-PRE) is engineered for constant gain mode but switches to constant power mode when span losses exceed 27 dB. (This is not applicable to reconfigurable add/drop multiplexers [ROADM].)

Table 10-1 *ONS 15454 MSTP Transponder/Muxponder Trunk Optics Specifications*

Characteristic	10ME-xx.x (standard FEC) 10E-L1-xx.x (standard FEC)	10ME-xx.x (no FEC)	MR-L1-xx.x (standard FEC) MRP-L1-xx.x (standard FEC)	MR-L1-xx.x (no FEC) MRP-L1-xx.x (no FEC) DM-L1-xx.x DMP-L1-xx.x	MR-L1-xx.x (2R, no FEC) MRP-L1-xx.x (2R, no FEC)	10ME-xx.x (enhanced FEC) 10E-L1-xx.x (enhanced FEC)
BER	10^{-15}	10^{-12}	10^{-15}	10^{-12}	10^{-12}	10^{-15}
OSNR Sensitivity (dB) at 0.1 nm bandwidth resolution	23 (maximum, power limited) 8 (minimum, OSNR limited)	19	14 (maximum, power limited) 5 (minimum, OSNR limited)	14 (maximum, power limited) 10 (minimum, OSNR limited)	15	20 (maximum, power limited) 6 (minimum, OSNR limited)
Power Sensitivity (dBm)	−24	−22	−31	−30	−24	−26
Power Overload (dBm)	−8	−9	−9	−9	−9	−8
Transmit Power (dBm)	3.5	3.0 to 6.0	−1.0 to 1.0 (MR) −4.5 to 3.5 (MRP)	−1.0 to 1.0 (MR) −4.5 to 3.5 (MRP) 2.0 to 4.0 (DM) −1.5 to 0.5 (DMP)	−1.0 to 1.0 (MR) −4.5 to 3.5 (MRP)	3.0
Chromatic Dispersion Tolerance (ps/nm)	±1000	±1000	−1200 +5400	−1200 +5400	−1200 to +3300	±800

Table 10-2 details the ONS 15454 MSTP design rules for linear networks, without OADM placement.

Table 10-2 *ONS 15454 MSTP Linear (No OADM) Span Loss Design Rules*

Number of Spans	10ME-xx.x (standard FEC) 10E-L1-xx.x (standard FEC)	10ME-xx.x (no FEC)	MR-L1-xx.x (standard FEC) MRP-L1-xx.x (standard FEC)	MR-L1-xx.x (no FEC) MRP-L1-xx.x (no FEC DM-L1-xx.x DMP-L1-xx.x	MR-L1-xx.x (2R, no FEC) MRP-L1-xx.x (2R, no FEC)	10ME-xx.x (enhanced FEC) 10E-L1-xx.x (enhanced FEC)
1x	35 dB	25 dB	37 dB	33 dB	30 dB	37 dB
2x	27 dB	19 dB	30 dB	26 dB	23 dB	29 dB
3x	24 dB	17 dB	26 dB	23 dB	20 dB	25 dB
4x	22 dB	14 dB	24 dB	21 dB	19 dB	23 dB
5x	21 dB	—	22 dB	20 dB	18 dB	22 dB
6x	20 dB	—	21 dB	19 dB	17 dB	21 dB
7x	19 dB	—	20 dB	18 dB	16 dB	20 dB

With this configuration, there is no DWDM wavelength channel insertion/drop capacity at midspan nodes. For multispan sites, the midspan shelf is configured as a line amplifier. Table 10-3 details the configuration that allows for intermediate DWDM wavelength channel add/drop.

Table 10-3 *ONS 15454 MSTP Linear with OADM Span Loss Design Rules*

Number of Spans	10ME-xx.x (standard FEC) 10E-L1-xx.x (standard FEC)	10ME-xx.x (no FEC)	MR-L1-xx.x (standard FEC) MRP-L1-xx.x (standard FEC)	MR-L1-xx.x (no FEC) MRP-L1-xx.x (no FEC DM-L1-xx.x DMP-L1-xx.x	MR-L1-xx.x (2R, no FEC) MRP-L1-xx.x (2R, no FEC)	10ME-xx.x (enhanced FEC) 10E-L1-xx.x (enhanced FEC)
1x	35 dB	25 dB	37 dB	33 dB	30 dB	37 dB
2x	29 dB	20 dB	31 dB	27 dB	25 dB	30 dB
3x	26 dB	15 dB	29 dB	25 dB	23 dB	28 dB
4x	24 dB	—	26 dB	23 dB	20 dB	25 dB
5x	23 dB	—	25 dB	22 dB	16 dB	24 dB
6x	21 dB	—	24 dB	19 dB	—	23 dB
7x	20 dB	—	23 dB	16 dB	—	22 dB

Networks that use fixed-channel OADMs allow for planned DWDM channel insertion/deleting. However, as noted in Chapter 9, the most flexible ONS 15454 MSTP DWDM topology consists of ROADMs. The ROADM ring and linear topology configuration rules are enumerated in Table 10-4. Although fixed-channel and ROADM nodes can coexist in a ring/linear topology, Table 10-4 addresses only the case in which all add/drop nodes are configured as ROADM.

Table 10-4   *ONS 15454 MSTP ROADM Span Loss Design Rules*

Number of Spans	10ME-xx.x (standard FEC) 10E-L1-xx.x (standard FEC)	10ME-xx.x (no FEC)	MR-L1-xx.x (standard FEC) MRP-L1-xx.x (standard FEC)	MR-L1-xx.x (no FEC) MRP-L1-xx.x (no FEC DM-L1-xx.x DMP-L1-xx.x	MR-L1-xx.x (2R, no FEC) MRP-L1-xx.x (2R, no FEC)	10ME-xx.x (enhanced FEC) 10E-L1-xx.x (enhanced FEC)
1x	35 dB	25 dB	37 dB	33 dB	30 dB	37 dB
2x	30 dB	20 dB	34 dB	28 dB	25 dB	32 dB
3x	28 dB	17 dB	32 dB	26 dB	23 dB	30 dB
4x	26 dB	—	30 dB	24 dB	21 dB	28 dB
5x	25 dB	—	29 dB	23 dB	20 dB	27 dB
6x	24 dB	—	28 dB	22 dB	18 dB	26 dB
7x	23 dB	—	27 dB	21 dB	14 dB	25 dB
8x	22 dB	—	26 dB	20 dB	—	25 dB
9x	21 dB	—	25 dB	19 dB	—	24 dB
10x	21 dB	—	25 dB	18 dB	—	23 dB
11x	18 dB	—	24 dB	17 dB	—	23 dB
12x	17 dB	—	24 dB	15 dB	—	22 dB
13x	15 dB	—	23 dB	—	—	22 dB
14x	—	—	23 dB	—	—	21 dB
15x	—	—	22 dB	—	—	21 dB

The design rule summaries given in Tables 10-2, 10-3, and 10-4 do not represent the full capabilities of the ONS 15454 MSTP. The MSTP DWDM system is fully capable of operating on NZDF fiber, with various channel count and OADM combinations. The DWDM design engineer should consult the latest installation and operations guide for the ONS 15454 MSTP for an exhaustive list of design rules.

ONS 15454 MSTP Manual DWDM Design Example

When planning and designing DWDM networks, the span losses often are unequal because of such factors as varied optical fiber lengths, differing optical fiber splices/losses, condition of the fiber plant, and so on. In those cases, Tables 10-2, 10-3, and 10-4 serve as a good reference but cannot be used to ensure the operability of the DWDM design. Therefore, the DWDM engineer must manually compute the optical characteristics of each span to measure ONS 15454 MSTP design rule conformity.

This section examines a two-span, linear ONS 15454 MSTP design, using fixed-channel OADMs. Table 10-5 provides the DWDM channel requirements and pertinent design details. Figure 10-3 illustrates the DWDM network requirements.

Table 10-5 *ONS 15454 MSTP Network Design Example Requirements*

Site Name	DWDM Channel Requirements		Protected (Y/N)	Transponder Type	SFP
	Channel Number	**Z Location**			
Node A	1–2	Node B	N	DM-L1-xx.x	1310 nm
Node A	3–4	Node B	N	DM-L1-xx.x	1310 nm
Node A	5–6	Node B	N	DM-L1-xx.x	1310 nm
Node A	7–8	Node B	N	DM-L1-xx.x	1310 nm
Node A	9–10	Node C	N	DM-L1-xx.x	1310 nm
Node A	11–12	Node C	N	DM-L1-xx.x	1310 nm
Node A	13–14	Node C	N	DM-L1-xx.x	1310 nm
Node A	15–16	Node C	N	DM-L1-xx.x	1310 nm

Figure 10-3 *ONS 15454 MSTP Network Design Example*

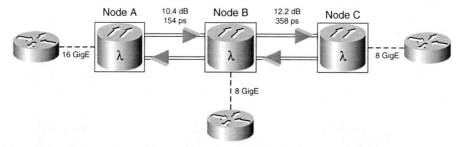

In the next sections, you examine three critical performance aspects of the proposed network:

- Attenuation
- Chromatic dispersion
- OSNR

The sections that follow detail the attenuation from site to site.

Attenuation

Attenuation is the reduction in optical signal power as the fiber pulse travels from one point to another. Consequently, attenuation worsens in direct proportion to the length of optical fiber required for point-to-point transmission. Using optical amplifiers to intermittently energize the optical photons during transmission can mitigate the effects of attenuation. Attenuation can be caused by a combination of natural optical scattering and light-absorption events, such as fiber splices, fiber connectors and terminators, and so on.

Table 10-6 provides the optical attenuation calculations, per span, for the sample network design. This information is helpful in determining whether optical amplifiers are needed and where they should optimally be inserted into the network.

Table 10-6   *Optical Attenuation Calculations, Per Span, for Sample Network Design*

Node A to Node B (West to East)	Measurement	Comments
Channel input power	+2 dBm	Per-channel power
32-channel multiplexer insertion loss	−8.5 dB	—
Composite signal power	+14 dBm	Power to multiplex 16 channels at +2 dBm per channel
Input power into fiber	7.5 dBm	—
Fiber plant/splice loss	−10.4 dB	—
Power to Node B	-7.9 dBm	—
Node B Drop Channels	—	—
Input power to OPT-PRE	−7.9 dBm	—
Gain added by OPT-PRE	17.5 dB	—
Drop loss of 4-channel add/drop filter	−5.4 dB	—
4-channel demultiplex power	−11 dBm	Power to demultiplex eight channels at +2 dBm per channel
Channel power to Gigabit Ethernet (GigE) device	−1.8 dBm	—
Node B to Node C (West to East)	—	—
Input power to OPT-PRE	−2.9 dBm [−5 dBm]	Use input VOA to attenuate to −5 dBm
Gain added by OPT-PRE	17.5 dB	—
Express power loss from 4-channel add/drop filter	−2.5 dB	—
Input power into fiber	10 dBm	—

continues

Table 10-6 *Optical Attenuation Calculations, Per Span, for Sample Network Design (Continued)*

Node A to Node B (West to East)	Measurement	Comments
Fiber plant/splice loss	−12.2 dB	—
Input power to OPT-PRE at Node C	−2.2d Bm	—
Gain added by OPT-PRE at Node C	17.5 dB	—
32-channel demultiplex power	−11 dBm	Power to demultiplex 8 channels at +2 dBm per channel
32-channel demultiplexer drop loss	−8.5 dB	—
Channel power to GigE device	−4.2 dBm	—
Node C to Node B (East to West)	—	—
Channel input power	+2 dBm	Per-channel power
32-channel multiplexer insertion loss	−8.5 dB	—
Composite signal power	+11 dBm	Power to multiplex 8 channels at +2 dBm per channel
Input power into fiber	4.5 dBm	—
Fiber plant/splice loss	−12.2 dB	—
Power to Node B	−7.7 dBm	—
Node B to Node A (East to West)	—	—
Input power to OPT-PRE	−7.7 dBm	—
Gain added by OPT-PRE	17.5 dB	—
Express power loss from 4-channel add/drop filter	−1.2 dB	—
Add channels	6.1 dBm	11dBm to multiplex 8 channels—4.9 dB channel-insertion loss)
Input power into fiber	14.7 dBm	—
Fiber plant/splice loss	−10.4 dB	—
Input power to OPT-PRE at Node A	4.3 dBm	—
Gain added by OPT-PRE at Node C	17.5 dB	—
32-channel demultiplex power	−14 dBm	Power to demultiplex 16 channels at +2 dBm per channel
32-channel demultiplexer drop loss	−8.5 dB	—
Channel power to GigE device	−0.7 dBm	—

Chromatic Dispersion

For DWDM wavelength channels that operate at 2.5 Gbps or below, chromatic dispersion typically does not present a transmission problem. However, in some design cases DWDM wavelength channels operate on aged or deteriorated fiber that could create chromatic dispersion and polarization mode dispersion issues. For the example, the DM-L1-xx.x transponder cards operate with a ±5400 ps/nm chromatic dispersion tolerance. To measure accumulated end-to-end chromatic dispersion, the following formula should be used:

$$D_{link} = D_f * L_{link}$$

Where:

- D_f is the nominal dispersion of the fiber.
- L_{link} is the length of the fiber.

In the example, actual chromatic dispersion measurements are given, so compliance measurement is a simple addition operation (that is, 154 ps + 358 ps). With a total chromatic dispersion end-to-end measurement of 512 ps, the network easily fits within the transponder specifications.

OSNR

As an optical signal is cascaded through multiple amplifiers, the signal accumulates noise. If the OSNR at the receiver is less than the interface card specifications, transmissions errors occur. For the example, the DM-L1-xx.x transponder, designed to +5400 ps/nm chromatic dispersion specifications, has an OSNR tolerance of 10 dB. You must calculate the accumulated OSNR to ensure that it is within 10 dB at the terminating optical receiver. Using the OSNR calculation formula given earlier, the following is true:

$$OSNRout_i = \frac{1}{(1/OSNRin_i) + (h* f* B*NF_i/Pin_i)}$$

- The OSNR measurement from Node A to Node B is:

  $1 / ([1/630957.3445] + [7.15884 \times 10^{-9} / 0.000512861]) = 48.1$ dB

  Where initial OSNR input is 58 dB (630957.34445 mW)

  Input to OPT-PRE is −2.9 dBm (0.000512861 W)

- The OSNR measurement from Node B to Node C is:

  $1 / ([1/64565.4229] + [7.15884 \times 10^{-9} / 0.00060256]) = 45.6$ dB

  Where OSNR input is 48.1 dB (64565.4229 mW)

  Input to OPT-PRE is −2.2 dBm (0.00060256 W)

- The OSNR measurement from Node C to Node B is:

  $1 / ([1/630957.3445] + [7.15884 \times 10^{-9}/0.000169824]) = 43.6$ dB

  Where initial OSNR input is 58 dB (630957.34445 mW)

  Input to OPT-PRE is −7.7 dBm (0.000169824 W)

- The OSNR measurement from Node B to Node A is:

  $1 / ([1/22908.67653] + [7.15884 \times 10^{-9}/0.002691535]) = 43.3$ dB

  Where OSNR input is 43.6 dB (22908.67653 mW)

  Input to OPT-PRE is 4.3 dBm (0.002691535 W)

NOTE   The conversion from decibels to milliwatts is 10E (dBm/10).

As you can see from the example in this section, even the simplest of DWDM designs can involve some degree of complexity. The MSTP DWDM design can become even more complicated when the system transports wavelengths operating at varied bit rates. Each transponder/muxponder type exhibits varying degrees of optical transmit power, OSNR sensitivity, dispersion tolerance, and so on. Manual design of complex systems lends itself to human computational error and can be extremely time-consuming.

To alleviate the design headaches associated with ONS 15454 MSTP networks, Cisco Systems has created a Java-based design tool: MetroPlanner, discussed in the next section.

ONS 15454 MSTP MetroPlanner Design Tool

MetroPlanner is a Java-based design tool for the ONS 15454 MSTP system. It provides the DWDM designer with a computer-generated, visual DWDM design that can be used to verify DWDM spans, plan future DWDM system upgrades, and model various ONS 15454 MSTP topologies. The MetroPlanner system design rules thoroughly examine the planned DWDM topology design, from the basics of attenuation, chromatic dispersion, and OSNR, to other considerations, such as channel cross-talk, amplifier gain tilt, and so on. These rules are factored into the overall design layout. MetroPlanner offers the DWDM designer the following advantages:

- Simple, flexible controls
- Comprehensive design analysis and computations
- Node installation and turn-up assistance

To better understand the advantages of using MetroPlanner for ONS 15454 MSTP designs, consider an ONS 15454 MSTP ROADM ring-based DWDM design, in comparison to the simple linear example in the previous section. Figure 10-4 illustrates the DWDM ring design layout. Notice that the design requires a mixture of DWDM wavelength channel bit rates, in that GigE and Synchronous Optical Network (SONET) OC-192 wavelengths are required. Table 10-7 lists the transponder/muxponder requirements and associated channel plan.

Table 10-7 *ONS 15454 MSTP ROADM Ring Network Design Example Requirements*

| Site Name | DWDM Channel Requirements | | Protection (Y/N) | Transponder Type |
	Channel Number	Z Location		
Node A	1	Node B	N	10E-L1-xx.x
Node A	2	Node B	N	MR-L1-xx.x
Node B	1	Node C	N	10E-L1-xx.x
Node C	1	Node D	N	10E-L1-xx.x
Node C	2	Node D	N	MR-L1-xx.x
Node D	1	Node A	N	10E-L1-xx.x

Figure 10-4 *ONS 15454 MSTP ROADM Ring Network Design Example*

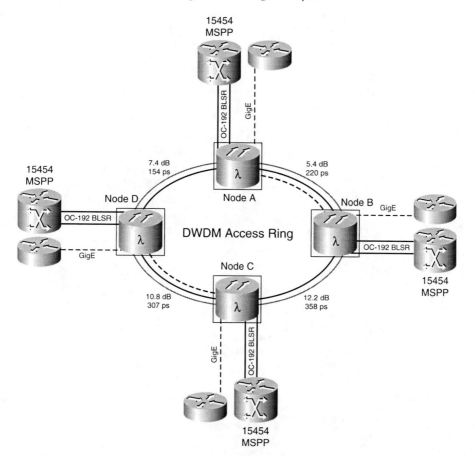

Although only two wavelength channels will be in use, the ROADM configuration allows up to 32 channels at any node to be added or dropped; therefore, the optical design constraints for a full 32-channel system are taken into account. For this reason, the ROADM system need be designed only once, while channel capacity can be added, changed, or rearranged, as necessary.

Simple/Flexible

The MetroPlanner design tool enables the user to input DWDM topologies in one of two ways:

- Using a network design wizard that automatically lays out each node
- Allowing the user to place nodes on the design screen with point-and-click motions

Regardless of which topology-input methodology is chosen, the ONS 15454 MSTP DWDM designer can input the node, fiber, and channel requirements for both simple and complex topologies in a matter of minutes. After the topology information has been input, the user has the full flexibility to insert fiber spans between nodes; modify fiber parameters such as distance, fiber type, chromatic dispersion values, and the like; modify node types (for example, ROADM, fixed-channel, pass-through, and so on); and change present and future DWDM channel requirements between nodes. Where uncertainty exists in network topology requirements, such as the type of node to be deployed at a location, the user can allow MetroPlanner to choose the nodes type(s) based upon the design input.

Figure 10-5 shows the MetroPlanner design screen summary for the ROADM ring example.

Figure 10-5 *MetroPlanner ROADM Ring Design Input Screen*

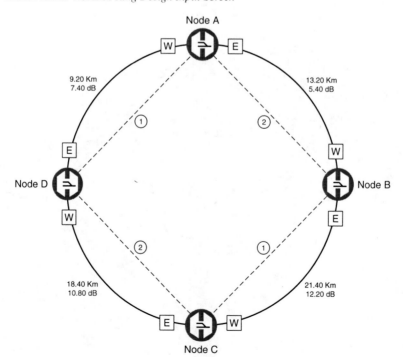

Comprehensive Analysis

For the simple linear example, you examined the DWDM topology by computing accumulated fiber attenuation, chromatic dispersion, and OSNR. These computations will suffice for most designs, but by no means should they be considered an exhaustive analysis of DWDM transmissions. By using the MetroPlanner design tool, the user is ensured of a more complete and comprehensive design analysis that includes such factors as optimized variable optical attenuator (VOA) settings, start-of-life (SOL) and end-of-life (EOL) transmission characteristics, and calculation of DCU placement and recommended DCU type/value. In addition, if the desired node/amplifier spacing impairs overall DWDM signal transmissions, the MetroPlanner design tool generates warnings to the user and then suggests alternative methods for design using port attenuation, amplifier placing/spacing, node type/placement, and so on.

Figure 10-6 shows a partial example of the MetroPlanner span design analysis for the ROADM ring example.

Figure 10-6 *MetroPlanner ROADM Ring Design Analysis Sample Screen*

TX Type	BER target	SOL OSNR (dB)	EOL OSNR (dB)	SOL OSNR margin(dB)	EOL OSNR margin(dB)	SOL RX (dBm)	EOL RX (dBm)	SOL Power margin (dB)	EOL Power margin (dB)	SOL Overload margin (dB)	EOL Overload margin (dB)
2.5G MR TXP w/FEC	1.00E-15	27.08	27.08	18.91	18.91	-16.31	-16.31	10.79	10.79	6.52	6.52
	1.00E-15	27.08	27.08	18.91	18.91	-16.31	-16.31	10.79	10.79	6.52	6.52
10G Enh MR TXP w/EFEC	1.00E-15	27.08	27.08	17.89	17.89	-16.31	-16.31	6.52	6.52	7.52	7.52
	1.00E-15	27.08	27.08	17.89	17.89	-16.31	-16.31	6.52	6.52	7.52	7.52
10G Enh MR TXP w/EFEC	1.00E-15	24.59	24.59	15.65	15.65	-15.80	-15.80	7.02	7.02	7.02	7.02
	1.00E-15	24.59	24.59	15.65	15.65	-15.80	-15.80	7.02	7.02	7.02	7.02
2.5G MR TXP w/FEC	1.00E-15	21.99	21.99	13.83	13.83	-16.34	-16.34	10.75	10.75	6.55	6.55
	1.00E-15	21.99	21.99	13.83	13.83	-16.34	-16.34	10.75	10.75	6.55	6.55
10G Enh MR TXP w/EFEC	1.00E-15	21.99	21.99	12.88	12.88	-16.27	-16.27	6.55	6.55	7.48	7.48
	1.00E-15	21.99	21.99	12.88	12.88	-16.27	-16.27	6.55	6.55	7.48	7.48
10G Enh MR TXP w/EFEC	1.00E-15	31.64	31.64	14.06	14.06	-20.90	-20.90	1.92	1.92	12.11	12.11
	1.00E-15	31.64	31.64	14.06	14.06	-20.90	-20.90	1.92	1.92	12.11	12.11

Installation/Turn-Up Assistance

To further assist in the planning and implementation of ONS 15454 MSTP networks, the MetroPlanner tool provides equipment site layouts and network material requirements as output reports. By using the MetroPlanner point-and-click navigational screens, you can view a detailed list of equipment required at each site, with an included visual diagram of the equipment layout. These screens work as visual confirmation for the planned network, or can create an equipment work order to purchase the required network element parts.

In addition to the site and network material lists, the MetroPlanner tool provides detailed channel power analysis to facilitate the initial setting of amplifier and VOA thresholds. This information can be exported to a file and uploaded to the individual ONS 15454 MSTP node for expedited site turn-up. After the initial settings have been uploaded for each node, the automatic power-monitoring capability of the ONS 15454 MSTP system takes control for future DWDM wavelength channel add/drop or rearrange activity.

Figure 10-7 displays a truncated view of the power calculations for the sample ROADM ring network, taken from the MetroPlanner design tool.

In some cases, the interface-to-interface optical fiber connectivity at an MSTP node can be confusing. To alleviate this confusion and ensure proper network element connections, the MetroPlanner tool provides you with detailed connectivity layout that describes the port-to-port connections for each interface card, along with any port attenuation requirements. This report can be used to aid or verify field installation of each MSTP node.

Figure 10-8 provides a truncated view of the field connections and installation parameters for the example network.

Figure 10-7 *MetroPlanner ROADM Ring Power Calculations*

Side	Position	Unit	Port #	Port ID	Port Label	Parameter	Value	Measure U
SideEast	Rack #1.Main Shelf.16	15454-OPT-PRE	1	LINE-16-1-RX	COM-RX	dwdm::Rx::SideEast::Amplifier::ChPower	2.0	dB
SideEast	Rack #1.Main Shelf.16	15454-OPT-PRE	1	LINE-16-1-RX	COM-RX	dwdm::Rx::SideEast::Amplifier::Tilt	0.0	d
SideEast	Rack #1.Main Shelf.16	15454-OPT-PRE	1	LINE-16-1-RX	COM-RX	dwdm::Rx::SideEast::Amplifier::WorkingMode	Control Gain	str
SideEast						dwdm::Rx::SideEast::MaxExpectedSpanLoss	5.4	d
SideEast						dwdm::Rx::SideEast::MinExpectedSpanLoss	5.4	d
SideEast						dwdm::Rx::SideEast::Power::Add-and-DropInputPower	2.0	dB
SideEast						dwdm::Rx::SideEast::Power::DropPower	-17.1	dB
SideEast	Rack #1.Main Shelf.16	15454-OPT-PRE	1	LINE-16-1-RX	COM-RX	dwdm::Rx::SideEast::Threshold::AmplifierInPowerFail	-23.1	dB
SideEast						dwdm::Rx::SideEast::Threshold::ChannelLOS	-22.1	dB
SideEast						dwdm::Rx::SideEast::Threshold::OSC-LOS	-19.4	dB
SideWest						dwdm::Rx::SideWest::MaxExpectedSpanLoss	7.4	d
SideWest						dwdm::Rx::SideWest::MinExpectedSpanLoss	7.4	d
SideWest						dwdm::Rx::SideWest::Power::Add-and-DropInputPower	-6.6	dB
SideWest						dwdm::Rx::SideWest::Power::DropPower	-21.7	dB
SideWest						dwdm::Rx::SideWest::Threshold::ChannelLOS	-11.2	dB
SideWest						dwdm::Rx::SideWest::Threshold::OSC-LOS	-18.7	dB
SideEast						dwdm::Tx::SideEast::Power::Add-and-DropOutputPower	-9.0	dB
SideEast						dwdm::Tx::SideEast::Threshold::FiberStageInput	-14.0	dB
SideWest	Rack #1.Main Shelf.01	15454-OPT-BST	6	LINE-1-3-TX	LINE-TX	dwdm::Tx::SideWest::Amplifier::ChPower	2.0	dB
SideWest	Rack #1.Main Shelf.01	15454-OPT-BST	6	LINE-1-3-TX	LINE-TX	dwdm::Tx::SideWest::Amplifier::Tilt	0.0	d
SideWest	Rack #1.Main Shelf.01	15454-OPT-BST	6	LINE-1-3-TX	LINE-TX	dwdm::Tx::SideWest::Amplifier::WorkingMode	Control Gain	str
SideWest						dwdm::Tx::SideWest::Power::Add-and-DropOutputPower	-8.0	dB
SideWest						dwdm::Tx::SideWest::Threshold::FiberStageInput	-13.0	dB
SideEast	Rack #1.Main Shelf.16	15454-OPT-PRE	1	LINE-16-1-RX	COM-RX	dwdm::Rx::SideEast::Amplifier::ChPower	2.0	dB
SideEast	Rack #1.Main Shelf.16	15454-OPT-PRE	1	LINE-16-1-RX	COM-RX	dwdm::Rx::SideEast::Amplifier::Tilt	-1.1	d
SideEast	Rack #1.Main Shelf.16	15454-OPT-PRE	1	LINE-16-1-RX	COM-RX	dwdm::Rx::SideEast::Amplifier::WorkingMode	Control Gain	str
SideEast						dwdm::Rx::SideEast::MaxExpectedSpanLoss	12.2	d
SideEast						dwdm::Rx::SideEast::MinExpectedSpanLoss	12.2	d
SideEast						dwdm::Rx::SideEast::Power::Add-and-DropInputPower	2.0	dB
SideEast						dwdm::Rx::SideEast::Power::DropPower	-16.6	dB

Figure 10-8 *MetroPlanner ROADM Ring Design Optical Connections*

Position-1	Unit-1	Port #-1	Port ID-1	Port Label-1	Attenuator	Position-2	Unit-2	Port #-2	Port ID-2
Rack #1.Main Shelf.17	15454-OSC-CSM	1	LINE-17-1-RX	COM-RX		Rack #1.Main Shelf.14	15454-32-WSS	67	LINE-14-2-T
Rack #1.Main Shelf.17	15454-OSC-CSM	2	LINE-17-1-TX	COM-TX		Rack #1.Main Shelf.16	15454-OPT-PRE	1	LINE-16-1-F
Rack #1.Main Shelf.16	15454-OPT-PRE	2	LINE-16-1-TX	COM-TX		Rack #1.Main Shelf.14	15454-32-WSS	68	LINE-14-2-F
Rack #1.Main Shelf.16	15454-OPT-PRE	4	LINE-16-2-TX	DC-TX	Att-Lpbk-4dB	Rack #1.Main Shelf.16	15454-OPT-PRE	3	LINE-16-2-F
Rack #1.Main Shelf.01	15454-OPT-BST	1	LINE-1-1-RX	COM-RX		Rack #1.Main Shelf.02	15454-32-WSS	67	LINE-2-2-T
Rack #1.Main Shelf.01	15454-OPT-BST	2	LINE-1-1-TX	COM-TX		Rack #1.Main Shelf.02	15454-32-WSS	68	LINE-2-2-F
Rack #1.Main Shelf.01	15454-OPT-BST	3	LINE-1-2-RX	OSC-RX		Rack #1.Main Shelf.08	15454-OSCM	1	LINE-8-1-T
Rack #1.Main Shelf.01	15454-OPT-BST	4	LINE-1-2-TX	OSC-TX		Rack #1.Main Shelf.08	15454-OSCM	1	LINE-8-1-F
Rack #1.Main Shelf.14	15454-32-WSS	66	LINE-14-1-RX	EXP-RX		Rack #1.Main Shelf.02	15454-32-WSS	65	LINE-2-1-T
Rack #1.Main Shelf.14	15454-32-WSS	65	LINE-14-1-TX	EXP-TX		Rack #1.Main Shelf.02	15454-32-WSS	66	LINE-2-1-F
Rack #1.Main Shelf.14	15454-32-WSS	69	LINE-14-3-TX	DROP-TX		Rack #1.Main Shelf.13	15454-32-DMX	33	LINE-13-1-F
Rack #1.Main Shelf.14	15454-32-WSS	1	CHAN-14-1-RX	RX-30.3 - 36.6 [1]		Rack #1.Main Shelf.12	15454-10E-L1-30.3	2	CHAN-21
Rack #1.Main Shelf.14	15454-32-WSS	2	CHAN-14-2-RX	RX-30.3 - 36.6 [2]		Rack #1.Main Shelf.05	15454-MR-L1-30.3	2	CHAN-25
Rack #1.Main Shelf.02	15454-32-WSS	69	LINE-2-3-TX	DROP-TX		Rack #1.Main Shelf.04	15454-32-DMX	33	LINE-4-1-F
Rack #1.Main Shelf.02	15454-32-WSS	1	CHAN-2-1-RX	RX-30.3 - 36.6 [1]		Rack #1.Main Shelf.06	15454-10E-L1-30.3	2	CHAN-26
Rack #1.Main Shelf.13	15454-32-DMX	1	CHAN-13-1-TX	TX-30.3 - 36.6 [1]		Rack #1.Main Shelf.12	15454-10E-L1-30.3	2	CHAN-21
Rack #1.Main Shelf.13	15454-32-DMX	2	CHAN-13-2-TX	TX-30.3 - 36.6 [2]		Rack #1.Main Shelf.05	15454-MR-L1-30.3	2	CHAN-25
Rack #1.Main Shelf.04	15454-32-DMX	1	CHAN-4-1-TX	TX-30.3 - 36.6 [1]		Rack #1.Main Shelf.06	15454-10E-L1-30.3	2	CHAN-26
Rack #1.Main Shelf.02	15454-OPT-PRE	1	LINE-2-1-RX	COM-RX		Rack #1.Main Shelf.01	15454-OSC-CSM	2	LINE-1-1-T
Rack #1.Main Shelf.02	15454-OPT-PRE	2	LINE-2-1-TX	COM-TX		Rack #1.Main Shelf.03	15454-32-WSS	68	LINE-3-2-F
Rack #1.Main Shelf.02	15454-OPT-PRE	4	LINE-2-2-TX	DC-TX	Att-Lpbk-4dB	Rack #1.Main Shelf.02	15454-OPT-PRE	3	LINE-2-2-F
Rack #1.Main Shelf.17	15454-OSC-CSM	1	LINE-17-1-RX	COM-RX		Rack #1.Main Shelf.14	15454-32-WSS	67	LINE-14-2-T
Rack #1.Main Shelf.17	15454-OSC-CSM	2	LINE-17-1-TX	COM-TX		Rack #1.Main Shelf.16	15454-OPT-PRE	1	LINE-16-1-F
Rack #1.Main Shelf.16	15454-OPT-PRE	2	LINE-16-1-TX	COM-TX		Rack #1.Main Shelf.14	15454-32-WSS	68	LINE-14-2-F
Rack #1.Main Shelf.16	15454-OPT-PRE	4	LINE-16-2-TX	DC-TX	Att-Lpbk-4dB	Rack #1.Main Shelf.16	15454-OPT-PRE	3	LINE-16-2-F
Rack #1.Main Shelf.01	15454-OSC-CSM	1	LINE-1-1-RX	COM-TX		Rack #1.Main Shelf.03	15454-32-WSS	67	LINE-3-2-T
Rack #1.Main Shelf.14	15454-32-WSS	66	LINE-14-1-RX	EXP-RX		Rack #1.Main Shelf.03	15454-32-WSS	65	LINE-3-1-T
Rack #1.Main Shelf.14	15454-32-WSS	65	LINE-14-1-TX	EXP-TX		Rack #1.Main Shelf.03	15454-32-WSS	66	LINE-3-1-F
Rack #1.Main Shelf.14	15454-32-WSS	69	LINE-14-3-TX	DROP-TX		Rack #1.Main Shelf.13	15454-32-DMX	33	LINE-13-1-F
Rack #1.Main Shelf.14	15454-32-WSS	1	CHAN-14-1-RX	RX-30.3 - 36.6 [1]		Rack #1.Main Shelf.12	15454-10E-L1-30.3	2	CHAN-21

Summary

This chapter examined the nuances associated with ONS 15454 MSTP system design. Many factors can linearly and nonlinearly affect the system's performance. For simple point-to-point designs using fixed-channel OADMs, it is possible to quickly and effectively design a system manually. However, when the design becomes more complicated to include ring protection, ROADMs, mixed bit rates, and so on, it is more prudent and less time-consuming to employ the MetroPlanner design tool for system design.

This chapter presented only the basics for ONS 15454 MSTP system design. For complex scenarios, technical documentation for the ONS 15454 MSTP is available at Cisco.com for download and use.

This chapter covers the following topics:

- Types of Wavelength Services
- Wavelength Services Protection Options
- Implementing Wavelength Services on the ONS 15454 MSTP
- Managing Wavelength Services on the ONS 15454 MSTP

CHAPTER 11

Using the ONS 15454 MSTP to Provide Wavelength Services

Traditionally, dense wavelength-division multiplexing (DWDM) has been used to relieve congested fiber networks when the option of deploying additional fiber cable strands is cost-prohibitive. An example of the "fiber relief" strategy is the deployment of point-to-point DWDM within a fiber transmission section to create "virtual" fiber strands in the network. These virtual fiber strands are then deployed as fiber sections for Synchronous Optical Network (SONET) rings or for point-to-point optical fiber systems.

As DWDM systems become more intelligent, DWDM system deployment is becoming more expansive to include optical service drops, such as storage protocol extension and high-bandwidth Ethernet. These types of services are bandwidth-intensive and tend to strain the limits of typical SONET OC-48/192 system deployments. Therefore, the most optimal method for offering these and other services is with dedicated wavelengths.

This chapter examines the tenets of offering such "wavelength services" over the ONS 15454 Multiservice Transport Platform (MSTP) and explores the different categories and characteristics of wavelength services as they relate to ONS 15454 MSTP features/functions. The chapter concludes with a discussion on the management capabilities of ONS 15454 MSTP wavelength services.

Types of Wavelength Services

The ONS 15454 MSTP offers transponder/muxponder interfaces that enable a wide array of services. These interfaces provide DWDM wavelength services that operate from 8 Mbps to 10 Gbps. In general, you can categorize the types of DWDM wavelength drops as one of the following:

- SONET/Synchronous Digital Hierarchy (SDH)
- Storage-area networking (SAN)
- Ethernet
- Variable bit-rate interface

Table 11-1 summarizes the DWDM wavelength service capabilities for the ONS 15454 MSTP system.

Table 11-1 *ONS 15454 MSTP DWDM Wavelength Services Support*

	2.5-Gbps Multirate Transponder (MR/P-L1-xx.x)	2.5-Gbps Multiservice Aggregation Interface (DM/P-L1-xx.x)	10-Gbps Multirate Transponder (10E-L1-xx.x)	4x2.5-Gbps Enhanced Muxponder (10ME-xx.x)
SONET/SDH	Yes	No	Yes	Yes
SAN	Yes	Yes	Yes	No
Ethernet	Yes	Yes	Yes	No
Variable Bit Rate	Yes	No	No	No

SONET/SDH Services

One of the key aspects of providing this type of service over DWDM is transparency. In effect, the DWDM system must operate as a virtual optical fiber extension, in that the SONET/SDH communications overhead bytes and performance-monitoring data should be preserved during transmissions. Both the 2.5-Gbps and 10-Gbps multirate transponder interfaces for the ONS 15454 MSTP allow for SONET/SDH services to operate transparently. The client SONET/SDH communications overhead bytes are not terminated on the transponder interface card in this mode of operation. This feature is also present on the 4x2.5-Gbps enhanced muxponder interface. To preserve end-to-end client SONET/SDH performance monitoring and fault isolation, key overhead bytes, such as the B1 overhead and the J0 overhead bytes, are also passed across the ONS 15454 MSTP system from end to end. In this manner, client-to-client unidirectional path-switched ring (UPSR)/subnetwork connection protection (SNCP), or bidirectional line-switched ring (BLSR)/multiplex section shared protection rings (MS-SPR) can be deployed over ONS 15454 MSTP DWDM wavelength channels.

Storage-Area Networking Services

Characteristics for storage interconnect interfaces vary, depending on the type of storage protocols being distributed across the network. Where real-time data replication is required, full line-rate throughput for protocols such as Fibre Channel (FC) are essential. Consequently, distance-extension techniques such as buffer-to-buffer credits and flow control must be incorporated into the DWDM client interface. Conversely, asynchronous, non-real-time storage requirements are less demanding on bandwidth use and distance extension.

A wide range of storage interconnect protocols can be supported using the ONS 15454 MSTP transponder interfaces. The 2.5-Gbps multirate transponder can support industry-standard 1- or 2-Gb FC or fiber connection (FICON), in addition to lower-bit-rate protocols such as Enterprise Systems Connection (ESCON) or geographically dispersed parallel sysplex (GDPS). The interface card is Small Form Factor Pluggable (SFP)–based, so each of these channel varieties can exist simultaneously on a single interface, with each port operating independently of the others. Optionally, the 2.5-Gbps multiservice aggregation interface can be used to provide the FC/FICON extension services. Where higher-bit-rate storage services are required, the 10-Gbps multirate transponder provides the option for 10-Gbps FC extension over the ONS 15454 MSTP network.

Ethernet Services

Ethernet interfaces are quickly becoming the predominant growth engine for transport networks. Because of its ubiquitous interface parameters and low-cost equipment, Ethernet is replacing services that time-division multiplexing (TDM) rates such as T1/T3 and OC-n typically served. As Ethernet services become more reliable, tremendous growth in Gigabit, fractional Gigabit, and 10-Gbps Ethernet services will dominate the market place.

The ONS 15454 MSTP provides a full suite of transponder interfaces to effectively extend Ethernet networks across a DWDM transport network. Both the 2.5-Gbps multiservice aggregation card and the 2.5-Gbps multirate transponder provide SFP interfaces for native Gigabit Ethernet (GigE) transport. Both cards allow for low latency and packet loss, in that Ethernet Layer 2 source/destination switching is not required for transport. Additionally, Ethernet performance monitoring is available through frame/packet counters on the client interface ports.

For high-bit-rate Ethernet transmissions, the 10-Gbps multirate transponder can be utilized to support 10-GigE service extensions. The Ethernet services marketplace is experiencing a high rate of growth for 10-Gbps Ethernet capability to effectively trunk data between high-traffic Ethernet switching devices. These data rates cannot be supported with SONET/SDH because they exceed the maximum SONET OC-192/STM-64 line-rate capacity of 9.9532 Gbps. Thus, aside from direct fiber transport, wavelength-division multiplexing (WDM) transmission provides the only viable alternative for 10-GigE port extensions.

Variable Bit-Rate Services

A critical aspect of DWDM wavelength services is flexibility and transparency. Some customer transport requirements don't easily fall into a well-known category or bit-rate scheme, such as SONET/SDH or Ethernet. For those cases, the DWDM wavelength system must exhibit the flexibility to interface with a variety of nonstandard or seldom-used transport technologies.

For the ONS 15454 MSTP DWDM system, the 2.5-Gbps multirate transponder allows for 2R provisioning. In this mode, the incoming client signal is transparently passed through the DWDM system without specific bit-rate or formatting requirements. As such, the ONS 15454 DWDM system does not provide performance-monitoring capability for 2R-transported signals. However, the 2R mode enables the carrier to offer DWDM wavelength transport of asynchronous signals, such as video and asynchronous transfer mode (ATM).

Wavelength Services Protection Options

Typical DWDM systems are deployed in a point-to-point topology with no means of protecting the DWDM trunks themselves, aside from duplicating the channel transmission in a diverse physical path. The ONS 15454 MSTP allows for this type of traditional DWDM channel protection, but also provides more intelligent approaches to DWDM failover that transform the DWDM system into a more SONET-like deployment. For the ONS 15454 MSTP, client interface transponders for wavelength services can be protected using one of three methods:

- A Y-cable
- Dual transponders
- DWDM trunk line split-routing

Y-Cable Protection

With this protection scheme, a single-client DWDM wavelength input channel is passively split into two discreet input signals through the Y-cable. The cable is constructed with one client-input port, which is connected to the client device, and two output ports, which are connected to the ports of the ONS 15454 MSTP transponders. Consequently, two transponder/muxponder interfaces are required on the ONS 15454 MSTP shelf assembly to protect against interface card failure. Only one of the two transponder cards can transmit back to the client device at any given time; thus, the transponder/muxponder interface cards are designated as either "working" or "protect."

Optical transmissions from the protect interface card back to the client device are disabled, to avoid transmissions error from combining the two transponder/muxponder signals into the Y-cable. Each of the DWDM trunk wavelengths associated with each transponder/muxponder interface is diversely routed in an ONS 15454 MSTP DWDM ring topology to provide network-level protection. Where linear systems are deployed, a single fiber-span failure results in network outages. Figure 11-1 illustrates the Y-cable protection scheme for a ring-based ONS 15454 MSTP topology. Y-cable protection is available only when the unprotected transponder/muxponder interface cards are used in the ONS 15454 shelf assembly.

Figure 11-1 *Y-Cable Protection Example*

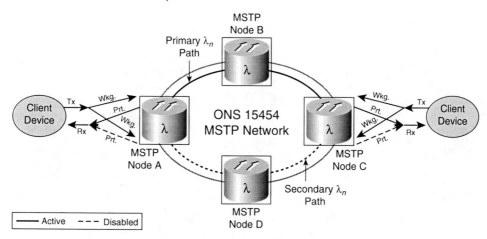

Dual-Transponder Protection

Similar to the Y-cable protection scheme, this protection methodology employs a set of two transponder/muxponder interface cards in the ONS 15454 MSTP shelf assembly. However, the transponder/muxponders operate independently of one another instead of as a protection group, as in the Y-cable example. Because two active signals are present between the client device and ONS 15454 transponder/muxponder at all times, the client device must manage the reception of two active wavelength signals. This can be accomplished using 1+1 automatic protection switching (APS), SONET/SDH ring protocols, or channel load balancing for data devices.

Dual-transponder interface protection can be combined with linear or ring network-level protection to provide additional DWDM system resiliency. Where linear systems are deployed, a single fiber-span failure can result in network outages; however, because the dual interfaces operate independently of one another, each transponder/muxponder can be installed in diverse linear systems to mitigate span-failure limitations.

The DWDM trunk wavelengths used for protection might or might not use the same channel frequency. Figure 11-2 illustrates the dual-transponder protection scheme for a linear-based ONS 15454 MSTP topology. Combined dual transponder/DWDM ring protection is illustrated in Figure 11-3. Dual-transponder protection is available only when the unprotected transponder/muxponder interface cards are used in the ONS 15454 shelf assembly.

Figure 11-2 *Dual-Transponder/Linear Topology Protection Options*

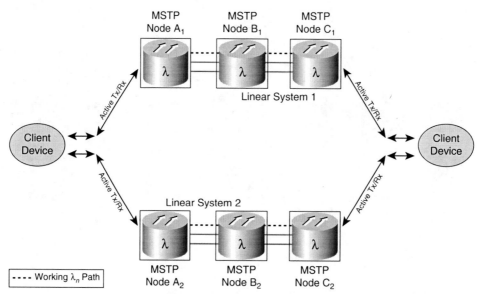

Figure 11-3 *Dual-Transponder/DWDM Ring Topology Protection Options*

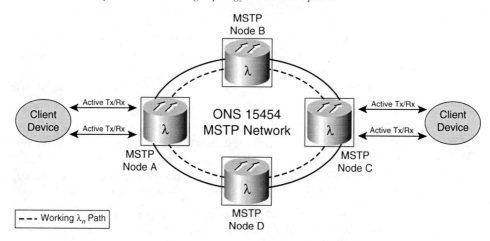

DWDM Trunk Split-Routing

The 2.5-Gbps transponder interfaces for the ONS 15454 MSTP allow for both unprotected and protected DWDM trunk operation. For protected trunk operation, the transponder uses an on-board optical splitter to separate the incoming client signal into two identical DWDM wavelength channels. Both channels traverse the ONS 15454 DWDM network in opposite directions. At the receiving DWDM trunk interface, the transponder uses an optical switch to choose the better-quality signal from the two input paths, based upon performance-monitoring measurements (loss of signal [LOS], loss of frame [LOF], single failure [SF], and signal degrade [SD]). The better of the two signal paths is then transmitted out the transponder client port to interface with an external device. This operation facilitates rapid failure recovery in case of a fiber-span failure. However, because the optical splitting/selection is performed in a single transponder card, this protection mechanism does not allow for interface card protection. Figure 11-4 illustrates the DWDM trunk split-routing protection scheme for a ring-based ONS 15454 MSTP topology.

Figure 11-4 *DWDM Trunk Split-Routing Topology*

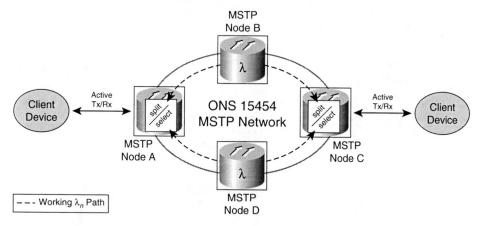

Implementing Wavelength Services on the ONS 15454 MSTP

From an industry perspective, wavelength services are implemented either as a managed service offering or by resale of DWDM facilities. The managed service approach supports customer ownership of the DWDM network facilities; however, the service provider handles day-to-day maintenance and operations. This implementation is usually more attractive to customers who need the bandwidth flexibility of DWDM wavelengths but have little or no operational experience with DWDM design and implementation. Conversely, customers experienced with DWDM network installation, maintenance, and operations typically choose a resale option, in which the DWDM service provider installs a dedicated DWDM network and resells it to the customer.

Regardless of the deployment methodology, wavelength services are typically differentiated by three key characteristics:

- Channel protection
- Variety of services offered
- Speed of implementation

The previous sections covered the first two characteristics. It is clear from these sections that the ONS 15454 MSTP provides the operator with a variety of bit rates and services capability for DWDM wavelength deployment. Additionally, flexible transponder/muxponder and network-protection options can be implemented to provide tiered levels of service resiliency.

Many factors affect the time required to implement DWDM wavelength services. First, the physical network must be engineered and constructed to provide the DWDM services infrastructure. For the ONS 15454 MSTP, the DWDM network can be engineered manually, for simple network topologies, or it can be engineered with the MetroPlanner design tool. The manual approach provides several challenges, in that each interdependent design constraint creates the need for network redesign or reoptimization. For example, if a DWDM node is planned as a pass-through initially, but engineering reveals that the site requires optical amplification, that node and others must be re-engineered to accommodate the high optical power levels for amplified channels, versus low-powered, nonamplified signals. Additionally, changes among any of the factors discussed earlier regarding DWDM design—attenuation, optical signal-to-noise ratio (OSNR), fiber nonlinear effects, and so on—require manual redesign/reoptimization. Thus, the time required to complete and finalize a design for ONS 15454 MSTP DWDM wavelength infrastructure can vary from as little as a few hours to as much as a few weeks.

Conversely, the MetroPlanner design tool can be used to mitigate the time gap associated with DWDM design so that both simple and complex designs can be finalized within, essentially, the same span of time. Using the MetroPlanner tool for wavelength services infrastructure design enables the carrier to achieve economies of scale for DWDM deployments, thereby improving the service provider's profit margins and/or resulting in lower-priced services to the end user.

When the physical DWDM network is in place, DWDM channel-provisioning time becomes a critical factor in the total amount of time required to implement wavelength services. Optimization of the channel-provisioning time greatly depends upon the type of DWDM optical add/drop multiplexing (OADM) channel technology selected during the physical design phase. Essentially, two choices exist for optical add/drop design: fixed-channel add/drop and reconfigurable optical add/drop multiplexing (ROADM).

Fixed-Channel Optical Add/Drop

The fixed-channel optical add/drop scenario allows DWDM wavelength channels to be dropped either singularly or as groups (bands) of channels. With this approach, DWDM channel assignments are preplanned for each node on the DWDM network. Therefore, where bands of channels are allocated to a DWDM node, the possibility of "stranded" bandwidth exists if the site channel requirements never reach the planned capacity.

For example, traditional DWDM networks used four- or eight-channel band fixed OADMs. Deployment of these band filters implies that either four or eight DWDM channels are ultimately required at the DWDM node. If the actual channel capacity required is less than the number of channels dropped on the band filter, DWDM wavelengths are dropped at the services node but are not used. Because these channels are dedicated to the dropped node, they cannot be reused elsewhere in the network without re-engineering the DWDM infrastructure. Thus, the possibility exists that a 32-channel-capacity DWDM wavelength system cannot be fully used because of channel bandwidth stranding. Moreover, to implement the re-engineered network, the DWDM network most often requires partial disassembly, resulting in temporary, protracted network outages.

The fixed-channel approach also lends itself to operational inefficiencies for DWDM channel growth or modifications. To successfully deliver an end-to-end wavelength service, OADM and transponder/muxponder equipment must be installed and provisioned at nearly every node on the DWDM network to accommodate pass-through and add/drop channels. For intermediate sites, DWDM channel pass-through equipment/fiber must be installed to provide wavelength continuity across the network. Optical amplifiers and variable optical attenuators (VOAs) within the network also require manual adjustments to accommodate the change in DWDM wavelength channel assignments. From an operations perspective, these equipment additions/adjustments result in time, capital, and personnel expenditures associated with the dispatch of qualified DWDM technicians to administer the network.

ROADM

Alternatively, the DWDM wavelength services network can be engineered with ROADMs at each services drop site. The ROADM approach allows for the drop/insertion of individual DWDM wavelength channels within the 32-channel system spectrum. Each of the channels operates independently of one another, allowing for separate protection schemes and channel-routing characteristics for each DWDM wavelength service. Because the system is initially designed to accommodate 32-channels add/drop at each node site, changes in wavelength assignments do not necessitate a corresponding change in design. Additionally, the ONS 15454 DWDM/ROADM system is self-optimizing. When the initial equipment settings are in place (for example, amplifier power and VOA settings), each affected equipment component self-adjusts for the addition/subtraction of wavelength channels. Thus, to provision an end-to-end wavelength service, only the DWDM channel endpoints require a site visit.

By providing both operational and capital expenditure efficiencies, the ROADM-based network tends to be the most effective deployment for wavelength services delivery. Cisco Systems, Inc., recently conducted a study to determine the economic savings of using ROADM-based DWDM networks versus the traditional point-to-point DWDM approach. The results indicate that networks requiring redundant DWDM wavelength channel assignments and carrier-class resiliency are more economically suitable for ROADM deployment. This is illustrated in Figure 11-5. The graphs represent the normalized capital cost to deploy five typical DWDM networks using ROADM versus traditional fixed-channel add/drop.

Figure 11-5 *OADM Technology Deployment Comparison*

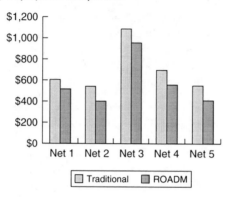

Managing Wavelength Services on the ONS 15454 MSTP

A key element in delivering wavelength services in a timely manner is the capability to quickly and easily provision the DWDM channels. Additionally, when the DWDM wavelength channels are in place, the associated management system must be capable of providing full accounting and performance monitoring of each wavelength circuit for compliance with service-level agreements (SLAs). The ONS 15454 MSTP can be managed using multiple operations support structure architectures. Depending on the deployment environment, ONS 15454 MSTP provisioning, fault reporting, and performance monitoring can be accomplished using TL1, by way of the Cisco Transport Manager (CTM) element manager system, or by third-party network/element managers capable of communicating through the Common Object Request Broker Architecture (CORBA) or the Simple Network Management Protocol (SNMP). Figure 11-6 illustrates the management solution possibilities for the ONS 15454 MSTP.

The level of management support that the ONS 15454 MSTP DWDM system provides varies, depending on the type of wavelength service provided. The highest level of management support is provided for SONET/SDH wavelengths or for wavelength trunks that use the G.709 digital wrapper overheads. For these types of wavelengths, fault management, configuration, accounting, performance, and security (FCAPS) utilities are provided. Other services, such as Ethernet, are monitored via client packet/frame counters and Ethernet remote-monitoring (RMON) statistics.

Figure 11-6 *ONS 15454 MSTP Management Architectures*

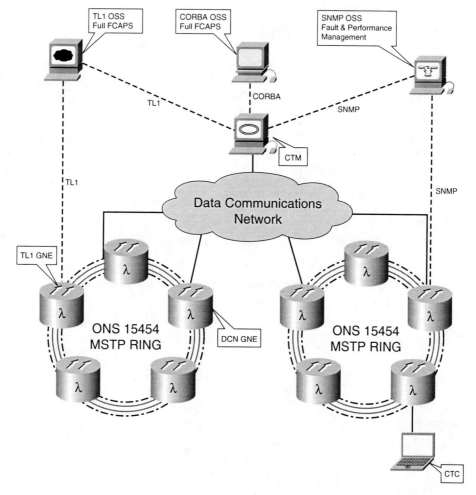

Fault Management

Fault isolation within the ONS 15454 MSTP system is achievable at both the client and DWDM trunk wavelength levels. Each client input is constantly monitored for signal integrity. The MSTP system provides continuous surveillance of laser bias current, with user-configurable alarming thresholds. This feature helps isolate transmission problems caused by client laser degradation or malfunction. Additionally, the optical power level for each client signal is monitored at ingress to the ONS 15454 MSTP system, along with aggregate channel power measurements. This data is used not only for system power optimization, but also to determine the specific network failure or trouble points for individual wavelength channels.

The digital wrapper functionality allows for DWDM wavelength visibility to transparent wavelength services. With the digital wrapper overhead monitoring, signal degradations in the ONS 15454 MSTP transport network can be discovered and corrected before network channel failure. Used in concert with the automatic power control feature, the ONS 15454 MSTP can self-optimize itself as network fiber conditions deteriorate because of fiber aging or equipment malfunctions.

Moreover, the ONS 15454 MSTP supports SONET-like alarm-correlation functions. Instead of flooding the network with numerous interrelated alarms and conditions, the ONS 15454 MSTP system aggregates and correlates alarms to DWDM network gateway network elements (GNEs). This feature allows alarm-monitoring and surveillance centers to quickly and accurately act upon degrading network conditions with pinpoint precision.

Figure 11-7 illustrates a sample of the fault-management capability of the ONS 15454 Cisco Transport Controller (CTC) graphical user interface (GUI).

Figure 11-7 *ONS 15454 MSTP Fault Management Sample*

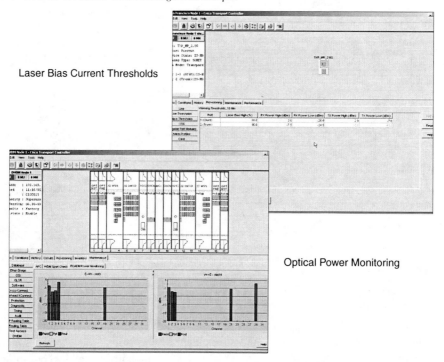

Laser Bias Current Thresholds

Optical Power Monitoring

Configuration

The ONS 15454 MSTP is designed to maximize the skill set of those already familiar with ONS 15454 Multiservice Provisioning Platform (MSPP) operations. As such, it employs the same CTC GUI that is used to configure MSPP SONET/SDH networks. The DWDM channels for the ONS 15454 MSTP system are provisioned in a similar manner to SONET/SDH circuits. Consequently, MSPP circuit functions, such as detailed A–Z circuit views and point-and-click provisioning, are characteristic of ONS 15454 MSTP circuits. Figure 11-8 displays the detailed view of DWDM wavelength circuit.

Figure 11-8 *ONS 15454 MSTP Wavelength Circuit Detail*

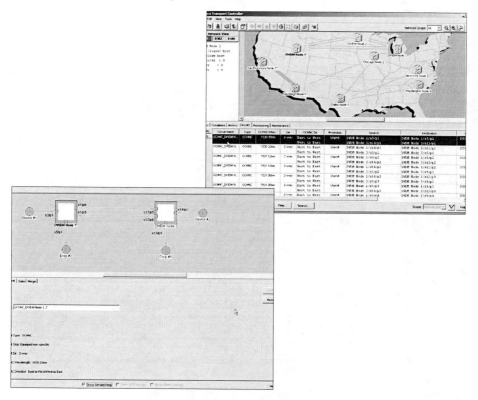

The MetroPlanner design tool also interfaces with the CTC GUI. The initial DWDM interface card settings from a MetroPlanner design can be uploaded into a DWDM node using CTC, thereby significantly speeding up system turn-up time. Additionally, the MetroPlanner design parameters, such as expected fiber loss and VOA settings, are compared against actual measured values in CTC to ensure overall design integrity.

Performance

When transparent SONET/SDH client signals are provisioned on the ONS 15454 MSTP, the MSTP DWDM system uses the SONET/SDH B1/B2 bytes for end-to-end client monitoring. Standard performance-monitoring parameters, such as severely errored seconds and code violations, are available and can be retrieved and viewed using the CTC GUI or the CTM element manager. The B1/B2 monitoring does not provide health and state information for the associated DWDM wavelength circuit. The DWDM wavelength circuit uses its own communications overhead bytes when digital wrapper encapsulation is turned on. The digital wrapper provides an additional set of wavelength-specific performance-monitoring parameters that can be used to troubleshoot the DWDM network and determine compliance with SLAs.

Table 11-2 details the types of performance-monitoring parameters available for client interfaces and DWDM wavelengths. In addition to the values listed in Table 11-2, optical power statistics are available for all the multiplexer/demultiplexer interface cards and the optical service channel interfaces.

Table 11-2 *ONS 15454 MSTP DWDM Performance-Monitoring Parameters*

Parameters	Client Interface SONET	G.709 Encapsulated DWDM Wavelength SDH
Background Block Errors (BBE)	—	√
Background Block Error Ratio (BBER)	—	√
Code Violations–Section (CV-S)	√	—
Errored Seconds (ES)	—	√
Errored Second Ratio (ESR)	—	√
Errored Seconds–Section (ES-S)	√	—
Errored Seconds–Line (ES-L)	√	—
Failure Count (FC)	—	√
Failure Count–Line (FC-L)	√	—
Severely Errored Seconds (SES)	—	—
Severely Errored Seconds–Section (SES-S)	√	—
Severely Errored Seconds–Line (SES-L)	√	—
Severely Errored Seconds Ratio (SESR)	—	√
Severely Errored Framing Seconds–Section (SEFS-S)	√	—
Unavailable Seconds (UAS)	—	√
Unavailable Seconds–Line (UAS-L)	√	—

Security

The ONS 15454 MSPP security features are also applicable to the ONS 15454 MSTP. The CTC GUI allows for user-provisionable security levels ranging from read-only access to superuser authority. The CTM element manager can also be used to further augment the CTC security features by restricting user access to predefined domains, which can be determined by geographic location, ring identification, and so on. Additionally, the Timing Communications Control (TCC) card stores an audit trail that keeps track of each login attempt/failure, and records the activity of each user. A total of 640 audit trail log entries is kept in each node. These entries can be retrieved and archived to any available database source.

Summary

This chapter discussed the capability to provide wavelength services using the ONS 15454 MSTP. It also detailed the types of wavelength services available and their associated characteristics.

Additionally, each wavelength service provided on the ONS 15454 MSTP can employ various protection schemes ranging from unprotected circuits to both client- and DWDM trunk-protected circuits. The most economically viable means of offering wavelength services on DWDM physical infrastructure is to use ROADM nodes at each wavelength insertion/drop point.

The ROADM provides the service provider with the most flexibility in determining DWDM channel utilization and also provides speedy installation, maintenance, and operations, in comparison to traditional fixed-channel OADM solutions.

Finally, the ONS 15454 MSTP provides a wide array of industry-standard management interfaces that allow the DWDM wavelength system to be managed locally, using the craft interface or using Element/Network Manager solutions.

PART V

Provisioning and Troubleshooting ONS 15454 Networks

This chapter covers the following topics:

- Turning Up the ONS 15454
- Operating and Supporting an ONS 15454 MSPP Network Element

CHAPTER 12

Provisioning and Operating an ONS 15454 SONET/SDH Network

The Cisco ONS 15454 is described as a Multiservice Provisioning Platform (MSPP). The Cisco ONS 15454 is designed to offer many different types of services for the telecommunications industry. Today's market requires service providers and enterprise customers to have communication networks designed to support voice, data, video, and storage services. The ONS 15454 can meet the needs of today's requirements and also operate with legacy telecom equipment for those older provider services, such as DS-1s and DS-3s. The ONS 15454 supports services from DS-1 to OC-192, as well as 10-/100-Mbps and Gigabit Ethernet (GigE), Fibre Channel (FC), and Enterprise System Connection (ESCON) for storage applications.

This chapter was written at a high level; it is not intended to be a step-by-step manual to turn up and operate a Cisco ONS 15454 MSPP. It provides enough information to give anyone interested in this MSPP product working knowledge of the installation and operation of the Cisco ONS 15454 MSPP.

The ONS 15454 provides a graphical user interface (GUI), called Cisco Transport Controller (CTC), that makes it easy to provision, maintain, and operate the system. CTC is an element-management system designed to control a single network element, but it also has the capability to "see" and control multiple network elements on a Synchronous Optical Network (SONET) network.

NOTE   The Cisco Transport Manager (CTM) is designed to control, manage, and monitor thousands of network elements in a very large network.

CTC also enables the user to provision a circuit automatically from one node to another on a network. This is an important feature because the autoprovisioning functionality automatically creates the circuits that pass through the intermediate nodes on a network without manual intervention. This is a desirable feature for many organizations; however, some organizations might not own all the nodes on a network, so this could prohibit an end-to-end circuit assignment. The ONS 15454 can just as easily provision a circuit within a single node for those applications.

TL1 (Transaction Language 1) is also an option for provisioning of the ONS 15454.

TL1 is the most common man-machine language used in the telecommunications industry. It is common among all major manufacturers of telecommunications equipment, including digital cross-connects, SONET transport, digital loop carrier (DLC), and legacy pre-SONET optical transport equipment. TL1 can be used via a terminal session through a serial connection on the TCC2 interface module on the 15454 or through a TL1 window in the CTC GUI.

The ONS 15454 also provides the capability to automatically recognize and provision the various common modules and the various interface modules as they are installed in the slots of the shelf.

Turning Up the ONS 15454

This section explains how to provision and turn up a Cisco ONS 15454 for service, including assigning the node name, date and time, timing references, and network information, such as the Internet Protocol (IP) address, subnet mask, default router, users and security levels, and interface module protection groups. This section discusses how to install the shelf and external cables to the backplane, install interface modules, and connect a PC and log into the ONS 15454 using the GUI called CTC.

Installing and Powering the Shelf

The Cisco ONS 15454 comes shipped from the factory in a sturdy, well-protected cardboard box. In the top of the larger box is a smaller box that contains miscellaneous installation material, including various tie-wraps, screws, instruction manuals, mounting studs, plastic fiber/cable-routing guides, and an access tool.

To install the ONS 15454 shelf, you need to install the two mounting studs from the miscellaneous installation material into holes in the rack or bay. These studs can be installed in two ways: You can hang the ONS shelf from them, or they can be used to support the weight of the shelf while you are installing the shelf-mounting screws.

To power the ONS 15454 shelf, you first need to ensure that the office ground has been installed at the top of the bay or frame, according to local electrical specifications. Then, to install the frame ground from the ONS shelf to the frame or bay, make up the grounding cable using a length of #6 AWG stranded copper wire and two double-hole compression lugs. After this, you attach one end of the frame ground wire to the back of the frame or bay, and attach the other end to the studs marked "Frame Ground," under the plastic cover on the lower-right side of the ONS shelf. Figure 12-1 illustrates ONS 15454 frame-grounding example.

To connect the DC power to the ONS shelf, you use #10 AWG stranded copper wire (red and black) and determine the source of the DC power. The source could be a fuse and alarm panel, a battery distribution fuse bay (BDFB), or, in some cases, the DC output of a small rectifier. The ONS shelf has a DC terminal strip in the lower-left corner of the back of the shelf. This strip supports two DC power connections (redundant); each set is labeled as Bat 1, Ret 1 and Bat 2, Ret 2. Figure 12-2 shows ONS 15454 DC power connections.

Figure 12-1 *ONS 15454 Frame-Grounding Example*

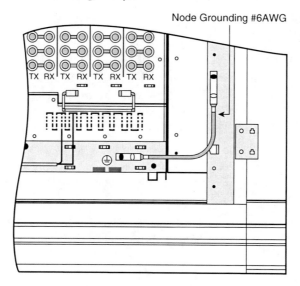

Figure 12-2 *ONS 15454 DC Power Connections*

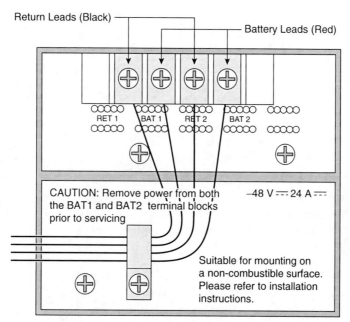

Using a red wire for Bat 1 and a black wire for Ret 1, connect the Bat 1 and Ret 1 of the ONS shelf to the same terminal designations on the source. If a redundant power feed is desired, repeat this for Bat 2 and Ret 2, using a second set of red and black wires. Use a voltmeter to verify that the voltages for both A and B battery are between –42 VDC (volts, direct current) and –57 VDC.

Initial Configuration

The ONS 15454 turns up in an initial configuration as a unidirectional path-switched ring (UPSR) node. It has a default IP address and subnet mask, a target identifier, date, time, and time zone. It also has a default timing configuration. All the service interface modules (when installed) initially are working or active, without any modules assigned as protect or standby.

The individual assigned to the provisioning of the system must make the proper changes to all the appropriate parameters. This process is usually controlled by an engineering or information technology (IT) group that assigns the appropriate parameters and slot assignments for all interface modules.

Detailed shelf turn-up procedures can be found in the ONS 15454 documentation located at http://www.cisco.com/en/US/products/hw/optical/ps2006/tsd_products_support_series_home.html.

Installing Common Equipment Cards

The common equipment cards (interface modules) consist of the cards in Slots 7, 8, 9, 10, and 11. Slots 7 and 11 contain the timing communication and control cards; one is active and the other is standby.

Slots 8 and 10 contain the cross-connect cards, either the XCVT, XC10G, or XC10GVX, depending on the application and assignment by engineering or IT documentation. One module is active, and the other is standby.

Slot 9 contains the Alarm Interface Card (AIC). This card is used for providing housekeeping alarm contact interfaces to an external alarm-polling system located in an unmanned location. An example of a housekeeping alarm might be a commercial power failure or an open door.

The procedure for installing the common equipment cards is described in the installation and turn-up practices of the ONS 15454, which is located at www.cisco.com/en/US/products/hw/optical/ps2006/tsd_products_support_series_home.html.

Timing Communications and Control Card

The Advanced Timing Communications and Control (TCC2) card performs system initialization, provisioning, alarm reporting, maintenance, diagnostics, Internet Protocol (IP) address detection/resolution, SONET section overhead (SOH) data communication channel/generic communication channel (DCC/GCC) termination, and system fault detection for the ONS 15454. The TCC2 also ensures that the system maintains Stratum 3 (Telcordia GR-253-CORE) timing requirements. It monitors the supply voltage of the system. The TCC2 card can be either active or standby, which means that there are two TCC2 cards in the ONS shelf in Slots 7 and 11.

The TCC2 card can be installed in either Slot 7 or Slot 11 by unlocking the extractors on the top and bottom edges of the card, sliding the card into Slot 7 or Slot 11, fully seating the card in the slot, and then closing the extractors to lock the card into the slot. When the TCC2 card is in the slot, the light emitting diodes (LEDs) on the face of the card begin to blink and flash in the following sequence of events:

1 All LEDs turn on briefly.

2 The red FAIL LED, the yellow ACT/STBY LED, the red REM LED, the green SYNC LED, and the green ACO LED turn on for about 10 seconds.

3 The red FAIL LED and the green ACT/STBY LED turn on for about 40 seconds.

4 The red FAIL LED blinks for about 10 seconds.

5 The red FAIL LED turns on for about 5 seconds.

6 All LEDs (including the CRIT, MAJ, MIN, REM, SYNC, and ACO LEDs) blink once and turn off for about 10 seconds.

7 The ACT/STBY LED turns on.

The second TCC2 card can be installed at this time using the same method as with the first card. The LEDs on the second card also perform the same sequence as the first TCC2 card.

Cross-Connect Card (XCVT/XC10G/XC-VXC-10G)

The XCVT card establishes cross-connections at the STS-1 and VT levels. The XCVT provides nonblocking STS-48 capacity to Slots 5, 6, 12, and 13, and nonbidirectional blocking STS-12 capacity to Slots 1 to 5 and 14 to 17. Any STS-1 on any port can be connected to any other port, meaning that the STS cross-connections are nonblocking.

The XC10G card cross-connects STS-1, STS-12, STS-48, and STS-192 signal rates, as well as concatenated signal rates of STS-3c, 6c, 9c, 12c, 24c, 48c, and 192c. The XC10G allows up to four times the bandwidth of current XCVT cards (used for OC-3, OC-12, and OC-48 systems). The XC10G provides a maximum of 1152 STS-1 cross-connections. Any STS-1 on any port can be connected to any other port, meaning that the STS cross-connections are nonblocking.

The XC-VXC-10G card establishes connections at the STS and VT levels. The XC-VXC-10G provides STS-192 capacity to Slots 5, 6, 12, and 13, and STS-48 capacity to Slots 1 to 4 and 14 to 17. Any STS-1 on any port can be connected to any other port, meaning that the STS cross-connections are nonblocking.

The XCVT/XC10G/XC-VXC-10G can be installed using the extractors, as described in the TCC2 section, in Slots 8 and 10, which also provides an active standby arrangement. When the cross-connect card is in the slot, the LEDs on the face of the card begin to blink and flash in the following sequence of events:

1 The red LED turns on and remains lit for 20 to 30 seconds.

2 The red LED blinks for 35 to 45 seconds.

3 The red LED remains lit for 5 to 10 seconds.

4 All LEDs blink once and turn on.

5 The ACT/STBY LED turns on.

Alarm Interface Controller-International (Optional)

The optional Alarm Interface Controller–International (AIC-I) card provides customer-defined alarm input/output (I/O) and supports user data and local and express orderwire. It provides 12 customer-defined input and 4 customer-defined input/output contacts. The physical connections are through the backplane wire-wrap pin terminals. A power-monitoring function monitors the supply voltage (–48 VDC) for the ONS shelf.

This card is optional and typically is installed only when "housekeeping" alarms are required to be transported across the network to an alarm center or network operations center (NOC).

Housekeeping alarms are usually associated with alarms or events that are external to the MSPP system. Examples of these types of alarms or events are open doors, high temperatures, a commercial power failure, and so on.

Optical and Electrical Interface Modules

The various types of optical or electrical interface modules can be installed in the system as determined by documentation from engineering or IT. These interface modules can be assigned to service Slots 1 to 6 or 12 to 17 of the ONS 15454 shelf. Installing these modules at this time facilitates the assignment of system turn-up parameters associated with synchronization or DCCs, mentioned later in this chapter.

General Network Element Information

When all the common cards are installed and the appropriate LEDs are lit, you can begin to provision the ONS 15454 shelf.

To begin the provisioning process or turn-up of the ONS 15454, you must have some general information for each network element (NE) or shelf in a network. This information usually can be provided by a engineering or IT group and consists of the following:

- Node name, or target identifier (TID)
- Location of the node
- Contact information
- Current date and time
- Usernames and passwords (used for user security purposes)

You also need to know the network requirements:

- IP address
- Subnet mask
- IP address of the default router
- IP address of the DHCP server, if required

You also need to know the types of services in the system, such as the DS3, DS1, OCN, and so on. You need to know how the protection groups will be assigned as well.

ONS 15454 provides protection to interface modules and ports of interface modules with protection groups. This gives added value to a module or port that might be providing service (such as DS-3 or OC-*N*). Providing protection is often a billable item to a customer who needs a reliable circuit or service. This means that a customer who subscribes to a protected service will have a "working" circuit and a "protect" circuit, in case the working circuit fails for some reason. All interfaces of the ONS 15454 can be either protected or unprotected.

Protection groups are assigned as 1:1 or 1:*N* for electrical interfaces, and 1+1 for optical interfaces. When you assign protection groups, you assign a protect interface module to a single working interface module, as in a 1:1 relationship, or to multiple working interface modules, as in a 1:*N* relationship with *N* being 1 to 5.

In the case of assigning protection groups to optical interface modules, the option choice is 1+1. Optical interface modules can also be provisioned as UPSR or BLSR, but this process isn't a part of assigning protection groups. Optical interfaces are defaulted as UPSR. If the optical interfaces need to be configured as bidirectional line switch ring (BLSR), a user-friendly feature of CTC called the BLSR Wizard is used to create a BLSR from a UPSR.

Creating a UPSR or BLSR is a separate task that uses the individual network elements (NE) that have been provisioned and connected with other NEs using fiber-optic cable to form a ring or network.

The Cisco ONS 15454 is provisioned using a GUI-based tool that is typically loaded onto a PC. CTC is a preloaded software package GUI that is factory installed on the TCC2. A

Java Runtime Environment (JRE) file and an online help file are on a CD and must be loaded onto a PC; then an Internet browser can be used to access CTC on the TCC2 card.

Each TCC2 card comes from the factory with a default IP address (192.1.0.2) and a default subnet mask of 255.255.255.0. Each network element must be provisioned with a unique IP address, to avoid IP address conflicts in the network. Set the IP address of the network interface connection (NIC) in the PC to be associated with the same network as the TCC2 card address. The NIC of the PC is connected to the local-area network (LAN) port on the TCC2 card using a Category 5 (CAT 5) straight LAN cable with RJ-45 connectors.

You can then launch an Internet browser, which uses the JRE file to connect to the CTC GUI on the TCC2 card. Then a CTC login window appears. The TCC2 card comes from the factory with a default username and password. These are provided in the "Detailed Procedures" section in the Cisco documentation, located at www.cisco.com/en/US/ products/hw/optical/ps2006/tsd_products_support_series_home.html. At the CTC login window, the user will fill in the spaces for username and password and select the Login button, as shown in Figure 12-3.

Figure 12-3 *Cisco Transport Controller Login Window*

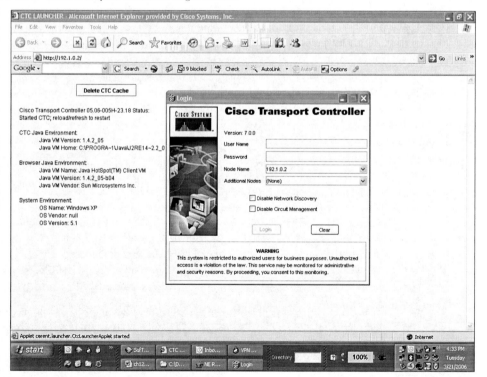

This connects the user to the ONS 15454 shelf and the TCC2 card in the shelf, revealing a graphical representation of the ONS 15454 MSPP shelf and all cards installed in the shelf, as shown in Figure 12-4.

Figure 12-4 *Cisco Transport Controller Shelf View*

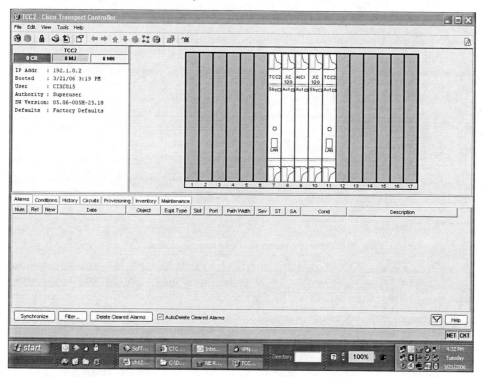

IP Addressing

When you have logged into the ONS 15454 and you have a GUI representation of the shelf, notice that you have various tabs across the top of this section and down the left side of the lower half of the window. If you use the computer mouse to select one of the tabs across the top of this section, you will notice that the tabs on the left side also change. By selecting the various tabs, you change the lower half of the window, which enables you to set and change the various parameters of the ONS shelf.

To change the network settings from the default settings to those assigned by your network administrator, select the Provisioning tab and the Network tab. These selections change the lower half of the window to show various fields that can be changed. The basic network information, provided by the network administrator, consists of the IP address, the subnet mask, and possibly the default router. After you change these options, click the Apply button, and answer Yes to the confirmation response, both of the TCC2 cards reboot individually, and you are logged off the shelf.

You must reset the NIC on your PC to match the new network information that you set for the shelf—that is, the IP address, subnet mask, and default router. After these are reset, launch the browser and log in to the shelf for additional provisioning options, such as general information, security, synchronization, and DCC.

ONS 15454s can be connected in many different ways within an IP environment, including these:

- Using LANs through direct connections or a router.

- Using IP subnetting to create ONS 15454 node groups that allow you to provision non-DCC-connected nodes in a network.

- Using different IP functions and protocols to achieve specific network goals. For example, Proxy Address Resolution Protocol (ARP) enables one LAN-connected ONS 15454 to serve as a gateway for ONS 15454s that are not connected to the LAN.

- Creating static routes to enable connections among multiple CTC sessions with ONS 15454s that reside on the same subnet with multiple CTC sessions.

- Connecting ONS 15454s to Open Shortest Path First (OSPF) networks so that ONS 15454 network information is automatically communicated across multiple LANs and wide-area networks (WANs).

- Using the ONS 15454 proxy server to control the visibility and accessibility between CTC computers and ONS 15454 element nodes. Proxy servers are covered in Chapter 15, "Large-Scale Management."

General node information can be provisioned by selecting the Provisioning tab and the General tab on the window. This changes the window to display various fields to be changed. This is the window used to provision the node name or target identifier (TID), description of the node, latitude and longitude, date, time, time zone, daylight saving time, contact information, and NTP/SNTP choice, as shown in Figure 12-5.

Security and Users

The CISCO15 ID (Superuser) is provided with the ONS 15454 system, but this user ID is not prompted when you sign into CTC. This ID can be used to set up other ONS 15454 users.

Figure 12-5 *CTC Shelf View General Tab*

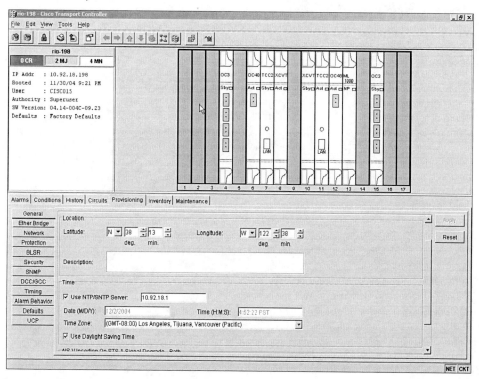

You can have up to 500 user IDs on one ONS 15454. Each CTC or TL1 user can be assigned one of the following security levels:

- **Retrieve**—Users can retrieve and view CTC information but cannot set or modify parameters.

- **Maintenance**—Users can access only the ONS 15454 maintenance options.

- **Provisioning**—Users can access provisioning and maintenance options.

- **Superusers**—Users can perform all the functions of the other security levels, as well as set names, passwords, and security levels for other users.

By default, multiple users can log into a node using the same user ID. However, you can provision the node to allow only a single login per user and prevent concurrent logins for all users.

To provision users and security, select the Provisioning tab and the Security tab in the window. This opens a window for Security that enables you to provision new users and their security levels, as shown in Figure 12-6.

Figure 12-6 *CTC Shelf View Security/Users Tab*

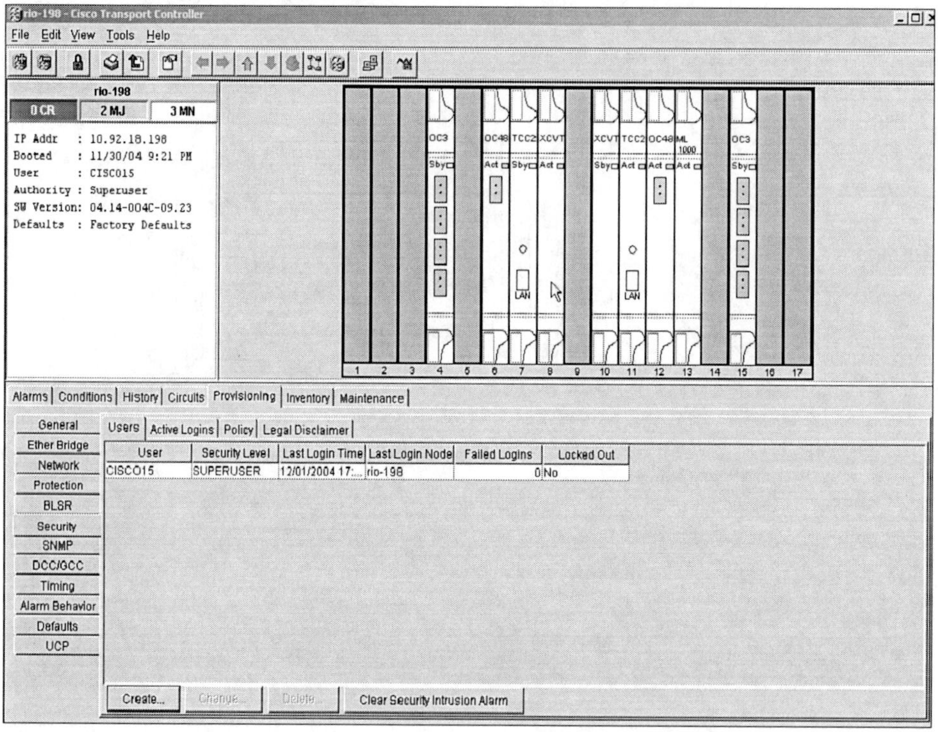

Selecting the Users tab at the top opens a window that shows the current users. Clicking the Create button in the lower-left corner of the window opens a Create User window, as shown in Figure 12-7. Input the information for the User Name, Password, Confirm Password, and Security Level fields, and then click the OK button to confirm the selections.

DCC

SONET provides four DCCs for network element operation, administration, maintenance, and provisioning: one on the SONET Section layer (D1 to D3 bytes) and three on the SONET Line layer (D4 to D12 bytes). The ONS 15454 uses the Section Data Communications Channel (SDCC) for ONS 15454 management and provisioning.

Figure 12-7 *CTC Create New User Window*

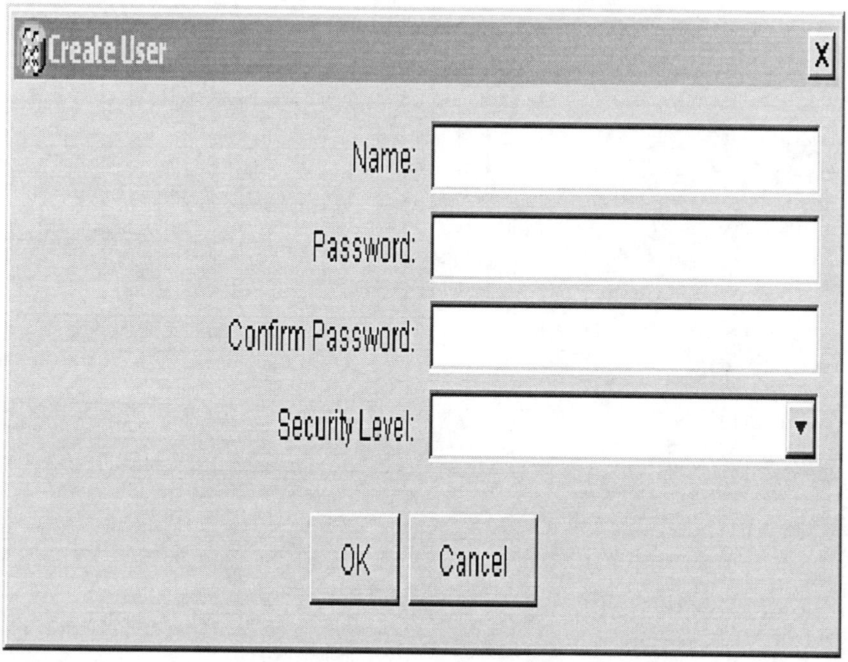

The ONS 15454 uses the DCC to transport information about operation, administration, maintenance, and provisioning (OAM&P) over a SONET interface. You can locate DCCs in the SDCC or Line DCC (LDCC). Unused SDCCs and LDCCs can be used to tunnel DCCs from third-party equipment across ONS 15454 networks.

A SONET link that carries payload from an ONS 15454 node to a third-party SONET node also has an SDCC defined in the Section overhead. However, the third-party node does not recognize OAM&P messages, and the SDCC should not be enabled. Disabling the SDCC has no effect on the DS3, DS1, and other payload signals carried between nodes.

To provision DCC on the ONS 15454, select the Provisioning tab and the DCC/GCC in the window. This opens a window to enable you to provision the DCCs for as many SONET interfaces as needed, as shown in Figure 12-8.

Figure 12-8 *CTC Shelf View—DCC Terminations*

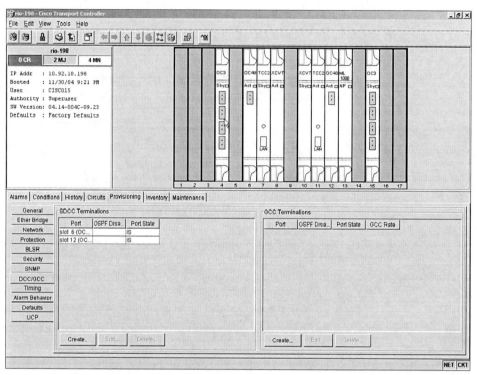

Clicking the Create button opens a new window, which shows a list of all the SONET interfaces that have the capability to have the DCC provisioned and placed in service, as shown in Figure 12-9.

Synchronization and Timing

Network synchronization and timing are critical elements within a SONET network. The goal is to create a fully synchronous optical hierarchy by ensuring that all ONS 15454 nodes derive timing traceable to a primary reference source (PRS). An ONS 15454 network can use more than one PRS. A PRS is equipment that provides a timing signal whose long-term accuracy is maintained at 10^{-11} or better with verification to Universal Time Coordinated (UTC), and whose timing signal is used as the basis of reference for the control of other clocks within a network. Standards define a primary reference source as one with a frequency offset of less than 1×10^{-11} (ITU-T G.811 and ANSI T1.101).

A Stratum 1 timing source is typically the PRS within a network. Other types of clocks used in the synchronized network include stratum 2, 3, and 4. In most cases, these clocks are components of a digital synchronization network and are synchronized to other clocks within that network using a hierarchical master-slave arrangement.

Figure 12-9 *CTC Create DCC Window*

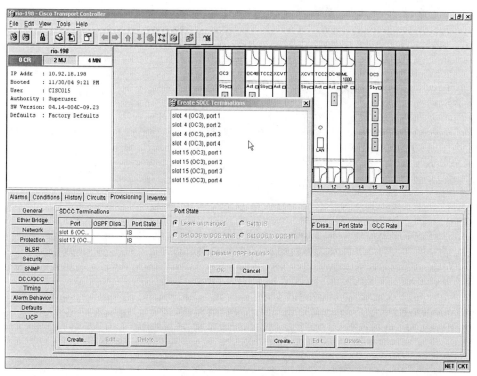

The clocks used to synchronize ONS 15454 nodes must be Stratum 3 (or better quality) to meet ANSI T1.101 synchronization requirements.

NOTE   What is ANSI? The American National Standards Institute (ANSI) is a voluntary membership organization, run with private funding, that develops national consensus standards for a wide variety of devices and procedures.

In the United States, a Building Integrated Timing Supply (BITS) clock is commonly used to distribute timing signals from Stratum clocks to an ONS 15454 node. BITS timing references run over the working and protect SONET paths.

Local cesium clocks (often referred to as atomic clocks) can also be used to provide Stratum 1–quality synchronization. The advantage of a cesium clock is that it never needs recalibration. However, the cost for each unit is very high.

Compared to cesium clocks, the Navstart Global Positioning Satellites provide a lower-cost alternative source for network synchronization. These satellites are accessible throughout North America and have internal cesium clocks that can be used as a Stratum 1 source. A Global Positioning System (GPS) receiver costs much less than a cesium clock.

The Cisco ONS 15454 is designed to operate in networks compliant with the Telcordia GR-253-CORE and GR-436-CORE.

Timing guidelines for the ONS 15454 evolve around the following conditions:

- Where BITS timing is available, configure the ONS 15454 node to be externally timed from the BITS clock.

- Where no BITS timing is available, configure the ONS 15454 node to be lined-timed from an OC-*N* signal.

- Where both external and line-time references are to be used, configure the ONS 15454 for mixed timing.

Timing and synchronization in the ONS 15454 are controlled by the TCC, which is Stratum 3–compliant. A redundant architecture protects against failure or removal of one TCC card. For timing reliability, the TCC card selects either a recovered clock, a BITS clock, or an internal stratum as the system timing reference. You can provision any of the clock inputs as primary or secondary timing sources. If you identify two timing references, the secondary reference provides protection. A slow-reference tracking loop allows the TCC to track the selected timing reference, synchronize to the recovered clock, and provide holdover in case the reference is lost. In a failover scenario, selection of the next timing reference is governed by the availability of the next-best (clock quality) timing reference, as defined by the stratum hierarchy (discussed in the next section). The timing modes available on the ONS 15454 include the following:

- External (BITS) timing
- Line (optical) timing
- Mixed (both external and internal) sources
- Holdover (automatically provided when all references fail)
- Free-running (a special case of holdover)

The external timing input signal on the Cisco ONS 15454 must come from a synchronization source whose timing characteristics are better than the Stratum 3 internal clock. The TCC tracks the external reference with the internal clock.

The BITS signal is a DS-1 level, 1.544-MHz signal, formatted either as Superframe (SF), which contains 12 frames per superframe, or Extended Superframe (ESF), which contains 24 frames per superframe. For the ONS 15454, the default setting for the BITS framing reference is ESF.

The ONS 15454 can accept reference timing from any optical port. For increased reliability, optical cards with multiple ports (such as four-port OC-3) can have only one of its ports provisioned as a timing reference. The optical cards divide the recovered clock to 19.44 MHz and transmit it to the active and standby TCC2 cards, where it is qualified for use as a timing reference.

Synchronization status messaging (SSM) can be optionally enabled or disabled on an optical port. A controller on the optical card monitors the received SSM and reports any changes to the TCC synchronization process. If an optical port (receiver) is selected as the active timing reference, the SSM value DUS (Don't Use for Synchronization) is transmitted (on the transmit port) to help prevent timing loops. If SSM is disabled, the controller does not monitor the received SSM value and transmits the SSM value STU (Synchronized, Traceability Unknown).

Mixed-mode timing enables you to select both external- and line-timing sources. However, Cisco does not recommend its use because it can create timing loops. Caution must be used when using mixed-mode timing because it can result in inadvertent timing loops. The most common reason for using mixed-mode timing is so that an OC-*N* timing source can be provisioned as a backup for the BITS timing source.

The ONS 15454 is considered to be in free-running state when it is operating on its own internal clock. The ONS 15454 has an internal clock in the TCC that is used to track a higher-quality reference or, in case of node isolation, provide holdover timing or a free-running clock source. The internal clock is a certified stratum clock with enhanced capabilities that match the Stratum 3E specifications.

To provision the timing of the ONS 15454, select the Provisioning tab and the Timing tab in the window. This opens a new window to allow provisioning of the timing parameters, as shown in Figure 12-10.

Figure 12-10 *CTC Shelf View-Synchronization Tab*

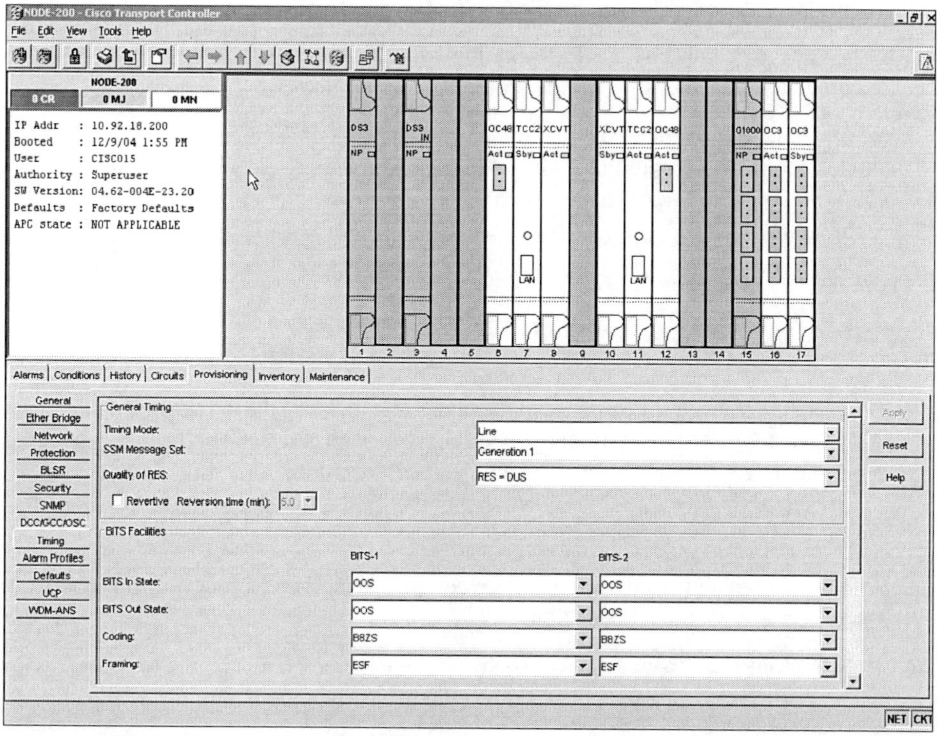

Connecting the Optics

Ensure that the optical cards have been installed in the appropriate slot in the ONS 15454 shelf.

The cards automatically are turned up, and the associated ports, Transmit and Receive, are automatically placed in the OOS-AINS (Out of Service—Automatic in Service state).

The fiber jumpers should be thoroughly cleaned, using an optical fiber–cleaning kit, and tested for the proper output and receive levels, using an optical power meter.

Connect the assigned fiber jumpers from the optical connectors on the optical card ports to the optical connectors in the fiber patch panel or the LGX (Light Guide Cross-connect) panel.

The fiber jumpers should be connected as indicated in the Cisco installation documentation for the particular configuration or topology required. These topologies are unidirectional path-switched ring (UPSR), bidirectional line switch ring (BLSR), 2-fiber BLSR and 4-fiber BLSR, terminal (which is some times referred to as PT to PT point to point, linear (add/drop multiplexer) ADM, and path-protected mesh networks (PPMN).

Final Configuration

When the ONS 15454 is initially turned up, the default topology is UPSR. After all the nodes on the network have been provisioned and connected with fiber cable and DCCs turned on, you can keep the UPSR topology or change it to either BLSR or terminal. Cisco documentation includes procedures to guide you through changing topologies.

Simply put, after the network is completely provisioned and the fibers are connected, you should be able to log into the shelf (node) that you've turned up and view the shelf configuration with CTC. Then click the Network View button of CTC, as shown in Figure 12-11.

Figure 12-11 *CTC Toolbar Network View Button*

The Network view of CTC shows all the nodes that make up the network you built, whether that is 2 nodes or 16 nodes. If all the nodes have been provisioned and connected with fiber cable and the DCCs have been turned on, you will see all of your nodes. A network with no alarms shows all green nodes as well as the fiber links connecting them, as shown in Figure 12-12.

If you don't see all the nodes in the network, you must retrace your turn-up provisioning procedures of the nodes in question.

Click the Alarm tab to make certain there are no existing alarms in any of the nodes on the network. Then click the Conditions tab and make certain there are no existing conditions.

Figure 12-12 *CTC Network View*

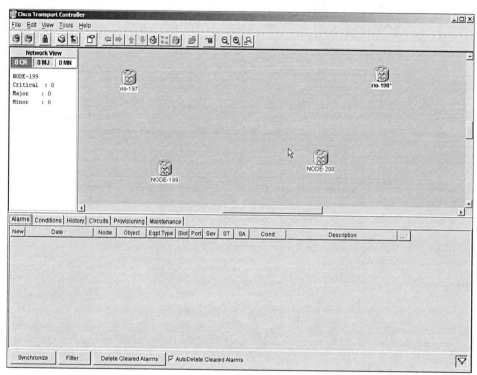

Operating and Supporting an MSPP Network

Operation is the day-to-day task to support a MSPP network. This can include the following actions:

- Monitoring alarms and conditions
- Adding or removing interface modules
- Provisioning circuits
- Troubleshooting any alarms or conditions

Monitoring Alarms and Conditions

Monitoring alarms and conditions is most easily accomplished using CTC, but you can also use a serial connection to the shelf and TL1 language. In addition, the ONS 15454 shelf has three colored LEDs (yellow, orange, and red) on the front of the shelf fan tray that illuminate for each respective alarm severity: yellow for minor, orange for major, and red for critical. Using CTC permits the user to view alarms and conditions from Shelf view as well as Network view. In each case, the alarms or conditions are related to the respective view.

Alarms

For alarms, if the user is viewing in Network view of CTC and any alarms are present, the alarm information is displayed under the Alarms tab in the bottom half of the CTC display. The affected nodes in Network view of the display also change from green to either yellow for minor alarm, orange for major alarm, or red for critical alarm.

If the user is viewing in Shelf view and any alarms are present, the alarm information is displayed under the Alarms tab in the bottom half of the CTC display. The alarm has a description and the severity is displayed in the same colors as mentioned previously. If the alarm is associated with an interface module in the shelf, the module is also colored appropriately for the severity of the alarm.

Conditions

For conditions, if the user is viewing in either Network view or Shelf view of CTC and any conditions are present, there likely would be no visible indication. To view the conditions, the user clicks the Conditions tab in the bottom half of the display to display the conditions table, and then clicks the Retrieve button to display all the conditions related to the shelf.

NOTE   Conditions often reflect a "hidden" problem; if ignored, the condition can result in an alarm. Initially, this is a minor alarm; if it is still ignored, it can escalate to a major or critical alarm.

Adding or Removing Interface Modules

After a ONS 15454 system has been in operation, the time will come when a interface module needs to be added or replaced.

Typically, interface modules are added when the system needs to expand. Adding more DS-1 or DS-3 interface modules for anticipated or immediate growth is an example of the need to add interface modules.

When an interface module fails or has some technical issue that will create an alarm, it must be replaced. Various types of interface modules exist; replacing one requires following a specific procedure.

Procedures for adding or replacing interface modules of various types, and for various reasons, can be found in the Procedure Guide, located at http://www.cisco.com/en/US/ products/hw/optical/ps2006/tsd_products_support_series_home.html.

Provisioning Service

Provisioning services is part of the operation of the MSPP. Services exist in the form of the types of circuits required to support various applications, as well as the associated interface module or port the circuits will terminate on. These types of service circuits could be as follows:

- **DS-1**—Digital Signal 1, approximately 1.5 Mbps; equivalent to 24 DS-0 voice channels.

- **DS-3 or STS-1**—DS-3, approximately 44.7 mbs; equivalent to 28 DS-1s. Synchronous Transport Signal 1 (STS-1), basic SONET rate 51.84 Mbps; equivalent to 28 DS-1s or 1 DS-3. The difference in the rates of the DS-3 and STS-1 is the SONET overhead.

- **OC-N**—For Optical Carrier N, typically N can be signal rates of OC-3 155.52 Mbps, OC-12 622.08 Mbps, OC-48 2488.32 Mbps, or OC-192 9953.28 Mbps. Notice that each N is a multiple of the basic SONET rate of 51.84 Mbps.

- **Point-to-point Ethernet**—Point-to-point Ethernet services speeds of 10 Mbps, 100 Mbps, or 1 Gbps traverse a network from one location to another, with access only at the two end locations.

- **Point-to-multipoint Ethernet**—Point-to-multipoint Ethernet services traverse a network with multiple points of access. This is sometimes referred to as a shared Ethernet service.

- **FC or ESCON**—FC is a high-performance serial link, usually used to connect mainframe computers, storage devices, or other devices that require FC connection. Typical link speeds are 133 Mbps, 266 Mbps, 530 Mbps, and 1 Gbps.

  ESCON is an IBM fiber-optic serial connection that operates at speeds up to 2 Gbps and typically connects mainframe computers and storage devices.

In addition to creating circuits, the user must provision the particular interface module or port on a module being used for the service. The user logs into the ONS 15454 shelf using CTC and is shown Shelf view of CTC. The user selects a particular interface module by double-clicking the graphical representation of the module in the shelf. This presents the module level of CTC and a graphical representation of the module and all interface ports associated with this particular module. (Interface ports can be from 1 to 56, depending on the module.) If the user clicks the Provisioning tab in the lower section of the CTC display, a table (similar to a Microsoft Excel table) displays. Each column in the table has a heading for each parameter associated with the ports of the displayed module. The most commonly used heading is "Service State" or "State." This is the column that the user selects to provision a port to a particular status. Examples of service state are Out of Service (OOS), In Service (IS), In Service Normal (IS-NR), and Out of Service—Maintenance (OOS-MT).

Creating Circuits

Provisioning service is a two-part process: provisioning the interface module or the port of an interface module, and creating a single circuit or multiple circuits. In this section, you will learn about creating circuits to provision services.

You can create circuits at the Shelf or Network view of CTC. For this chapter, we use Network view and create a DS-3 circuit from a DS-3 port on a DS-3 interface module in one location (source) to another DS-3 port on a different DS-3 interface module at a second location (destination).

CTC allows two methods of creating circuits: manually routing circuits and automatically routing circuits. (Both methods are discussed in this chapter.) Both methods use the Circuit Wizard in CTC. The Circuit Wizard creates a circuit or multiple circuits from one location (called the source) to a different location (called the destination). This is accomplished when the user selects items from drop-down menus, checks boxes, and clicks icons and Next or Finish buttons.

NOTE *Source* and *destination* are relative terms, meaning that they don't necessarily have to be two different locations or ONS 15454 shelves. For example, a DS-3 circuit is created from one port of a DS-3 interface module (source) to a different port on the same DS-3 interface module (destination), which is typically called a "hairpin circuit." *Source* and *destination* just refer to the two terminating points of the circuit.

Automatically Routed Circuits

This section describes the step-by-step method of creating a circuit by selecting the "Route Automatically" option in the Circuit Routing Preferences window of the Circuit Wizard in CTC.

Step 1 In Network view, select the Circuits tab, which changes the window. Then click the Create button at the bottom of this window.

This begins a circuit-creation wizard and opens a circuit dialog box.

Step 2 In the circuit dialog box, complete the fields labeled Circuit Type, (VT, STS, VT Tunnel, or VT Aggregation Point), Number of Circuits (the default is 1), and Auto Range. Then click the Next button, which opens the attributes box.

In the attributes box, you can assign the circuit name, set the size of the circuit, and set the state of the circuit, as shown in Figure 12-13.

Note   The Auto Range box is checked and autoranges multiple circuits with the same destination point.

Figure 12-13 *CTC Circuit Wizard—Attributes Window*

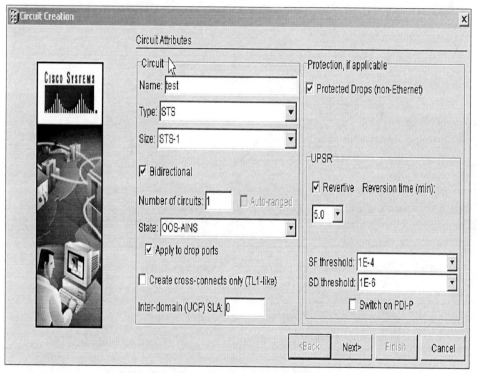

Step 3   Apply checks to the Bidirectional and Apply to Drops boxes. You can also select Protected Drops and Revertive, as needed.

Step 4   Click the Next button, which opens the Circuit Source box.

In the Circuit Source box, you can use a drop-down window to select from a list of nodes on the network to be the source node for this circuit. You can also select the card and the port on that card for this circuit, as shown in Figure 12-14.

Figure 12-14 *CTC Circuit Wizard—Circuit Source Window*

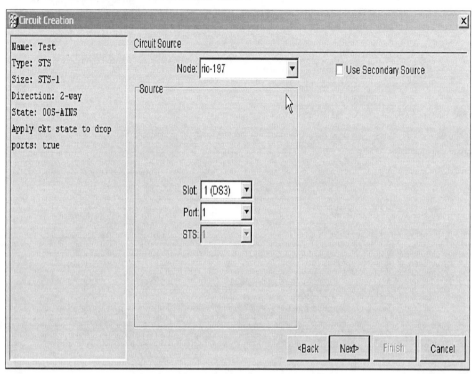

Step 5 Click the Next Button. The wizard opens a similar dialog box for the circuit destination information. Then select the destination node from the drop-down window, select the slot and port of the node, and click the Next button. The wizard opens a Circuit Routing Preferences box.

In the Circuit Routing Preferences box, you have the option to select Route Automatically or not. If you uncheck this box, you must continue to complete this circuit in manual mode. For this example, select Route Automatically, as shown in Figure 12-15.

Step 6 Click the Finish button. The wizard closes and you return to Network view of CTC. In the Circuits area in the lower half of the CTC window, you can see the circuit or circuits you created.

To create optical circuits (circuits that use optical cards and ports) and Ethernet circuits (circuits that use Ethernet cards and ports), you use the same method, with some exceptions. When provisioning optical or Ethernet circuits, in the Circuit Source and Destination boxes, you must select the STS-1 or STS-1s concatenated that you want to use on the port.

Figure 12-15 *CTC Circuit Wizard—Circuit Routing Preferences*

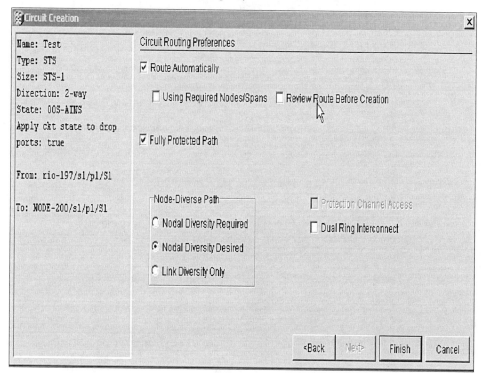

Manually Routed Circuits

Creating manually routed circuits begins the same as the process of creating automatically routed circuits:

Step 1   Use Steps 1 to 4 in the "Automatically Routed Circuits" section to create manually routed circuits.

Step 2   For a manually routed circuit, remove the check mark in the Circuit Routing Preferences box and click the next button. The wizard opens a Route Review/Edit window.

The Route Review/Edit window displays a graphical representation of the ONS 15454 network, very similar to the network view in CTC, but one of the nodes is labeled Source and another node is labeled Destination. In the lower half of the CTC display is a window labeled Included Spans, and also an area on the lower-right corner, with smaller windows, for selecting from and to, as well as a choice of an STS or VT on the span.

Step 3   Roll the mouse pointer over the green spans between nodes, and notice that the pointer changes to a hand.

You will also notice that two of the green span lines have arrowheads pointing in a specific direction but generally toward the destination node.

Step 4   Locate the source node and position the mouse hand (pointer) over the green span that points in the direction *closest* to the destination node. Click this green span line. This is referred to as the shortest route.

The span changes colors, and the From and To windows are automatically filled with information. The user can select the STS (for a DS-3) or VT (for a DS-1).

Step 5   Click the Add Span button, which enters the From and To information into the Included Spans window.

Step 6   Roll the mouse hand over the next green span line from the source in the opposite direction toward the destination node, and click the green span line. This is referred to as the long route.

The span changes colors, and the From and To windows automatically are filled with information.

Step 7   Select the STS (for a DS-3) or VT (for a DS-1).

Step 8   Click the Add Span button, which adds this information to the Included Spans window.

Notice that the arrowhead changes to another green span line, and the process repeats itself for each span that exists in the network between the source node and the destination node.

Step 9   When all the spans are in the Included Spans window, click the Finish button.

Troubleshooting Alarms or Conditions

Troubleshooting alarms and conditions begins with the user logging into the ONS 15454 with CTC and, in Shelf view, clicking the Alarms tab or the Conditions tab in the lower half of the CTC display. This displays a table similar to an Excel table with information related to the type of alarm, the date and time, the node name, the interface module and port, and the severity and description of the alarm.

To see this same information for conditions, after clicking the Conditions tab, click the Retrieve button. All conditions are displayed, with the most current at the top of the table.

Troubleshooting alarms and conditions can be a time-consuming process. It is important to recognize the description of the alarm or condition. This information can be used with the

ONS 15454 documentation to determine what the trouble or condition is and what to do to clear it. The ONS 15454 documentation is located at http://www.cisco.com/en/US/products/hw/optical/ps2006/tsd_products_support_series_home.html.

Acceptance Testing

Why do acceptance testing? MSPPs are designed to provide 99.999 percent reliability to the traffic on the system. Without this testing, you have no way of knowing that the network that you have built is performing properly.

Now that you have a circuit provisioned on your network, it's time to perform the acceptance testing of the network. Simply put, acceptance tests prove that the network is carrying traffic (such as voice and data) from one node to another and that the protection schemes that were provisioned are operating properly. To perform an acceptance test, you need two DS-1/DS-3 SONET test sets. These test sets connect to the circuits that have been provisioned between nodes on the network, with one test set at each of the nodes being tested. The DS-1/DS-3 circuits will have been terminated at some type of termination point. This is typically a DSX-1 or DSX-3 panel. The optical and Ethernet circuits can be tested either from the face of the card on the port or at some type of termination panel.

NOTE	In this example, a DS-3 circuit test application is used.

You will need to log into the node using CTC and verify that no alarms or conditions are present. Any alarms or conditions should be cleared prior to acceptance testing.

Use the test equipment setup procedures to configure the test set for DS-3 testing.

Ensure that the port on the card is in either the IS (In Service), OOS-MT (Maintenance), or, preferably, OOS-AINS (Out of Service—Automatic in Service) state.

Connect the test set at each respective node location to the appropriate circuit on the DSX-3 panel, with test leads from the test equipment labeled Transmit and Receive. Proper connection of the test set is indicated on the liquid crystal display (LCD) screen or by the LEDs that signal, pattern, and frame synch are all present. These should be in the OK state on the LCD screen or green LEDs on the test set.

With both test sets connected and transmitting valid signals, you can insert a bit error from one test set to the other. This proves that test traffic is transported over the network.

Depending on the protection scheme you have used in configuring your nodes on the network, you can simulate various "outage" conditions of optical interfaces by unplugging a fiber jumper from a port on an interface module or from a connector in the LGX panel. To simulate an outage on an electrical interface—that is, DS-1, DS-3, or STS-1—you can remove one end of a jumper, usually located at an electrical DSX-*n* panel, which creates an LOS for this particular port on an interface module.

Performing a manual or forced switch from CTC on your PC forces the traffic on a circuit, port, or interface module to move from the working path for a circuit, port, or interface module to the assigned protect circuit, port, or interface module.

You perform a manual or forced switch on a circuit when it is in Shelf view or Network view of CTC. To do this, follow these steps:

Step 1   Select the Circuits tab, which displays a table with various circuit types listed.

Step 2   Click the appropriate circuit, which highlights the entry and also changes various buttons below the circuit table.

Step 3   Click the Edit button, to reveal the Edit Circuit window.

Step 4   Click the Selectors tab, and then click the cell of the column labeled Switch State. This reveals a drop-down menu to select the type switching desired, such as forced or manual.

Step 5   Click the type of switch desired, and then click the Apply button on the right side of the display window.

Manual or forced switches for modules or ports on a module in an ONS 15454 shelf are performed in Shelf view of CTC and involve clicking the Maintenance tab.

Step 6   Click the Cross-Connect subtab to perform switching of the active and standby XCVT, XC10G, or XC-VXC-10G cross-connect modules.

Step 7   To perform a manual or forced switch to a module or port on a module, select the Maintenance tab and the Protection subtab.

This displays a window with various protection groups that were created during the turn-up process.

Step 8   Click a protection group, to highlight the group and reveal a window with options for switching the module or port on a module from working to protect.

Purposely unseating a working card from the shelf forces the traffic from a working interface module to a protect interface module, and creates an Improper Removal alarm.

NOTE   Purposely unseating an interface module is not the Cisco preferred way of conducting a switching test.

Any of these actions or simulated outage conditions causes some level of protection to take place, and your test circuit will take a few bit errors each time. However, this is a normal condition. This is what protection is all about: protecting the traffic from total failure during some type of network interruption.

Detailed protection test procedures can be found in the Cisco ONS 15454 Procedures Guide, located at http://www.cisco.com/en/US/products/hw/optical/ps2006/ tsd_products_support_series_home.html.

Maintenance

Maintenance is typically conducted on a routine basis: weekly, monthly, or quarterly. These routines include a physical inspection of the 15454 shelf and interface modules.

You can use CTC to perform a database backup and review the conditions of the system and the performance of interface modules.

Performance-monitoring and database-backup procedures are described in the next two sections of this chapter.

Physical maintenance for the ONS 15454 is limited to making sure that the system is located and operated in a clean, environmentally controlled area. Periodic cleaning of the fan filter, which is located beneath the fan tray of the 15454 shelf, is recommended.

Performance Monitoring

In Chapter 2, "Technology Foundation for MSPP Networks," you learned about SONET. Some alarms displayed in the Alarm tab window might have some reference to *section*, *line*, or *path*. These are all SONET terms and are used in the troubleshooting process of locating and clearing alarms and conditions. Understanding the terminology and having some knowledge of SONET technology will help in understanding performance monitoring and managing alarms of the Cisco ONS 15454.

Log into the ONS 15454 using your PC and CTC, and select the Alarms tab at the top of the lower window, as shown in Figure 12-16.

NOTE	For the local technician, CTC is the most important tool used for alarm and trouble management.

Figure 12-16 *CTC Shelf View Alarm Window*

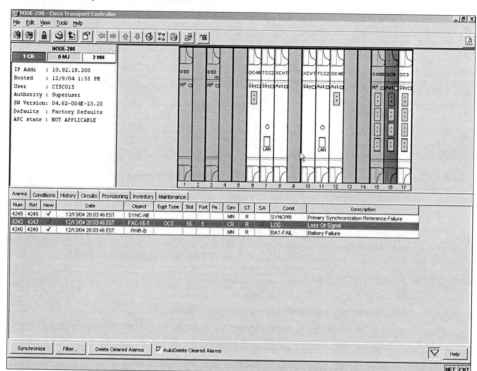

When you view the Alarms window, you will notice that there are different colors in both the graphical shelf section and the tab section of the window. Red is associated with a critical alarm. Gold is associated with a major alarm. Yellow is associated with a minor alarm. You can also see alarms in the Network view of CTC, as shown in Figure 12-17.

The Alarms tab of CTC provides information such as the date and time, the name of the node, the type of alarm, the card slot of the shelf, the port on the card, and a description of the alarm. You might notice in Figures 12-16 and 12-17 that both views of CTC (Shelf and Network) show some of the same alarms. Also in Network view, being able to see the entire network and the node that has the alarm is a great benefit for reducing the time needed to clear the alarm. Many times when an alarm occurs on a network, a large amount of time is spent trying to determine where the alarm is. Having the network "map" clearly shows by the different colors displayed which node or nodes have a problem.

When alarms are displayed in the tab area of the window with the various colors or levels of severity, usually one of the displayed alarms is the primary alarm. That means that when the primary alarm is cleared, the rest of the lower-level (minor) alarms clear as well.

For example, if a node was timed from or was receiving its clocking from the optical line instead of an external source, and there was a break in the fiber connected to that optical receiver, you would have a number of alarms in that node. Some would be more critical than others, such as loss of signal (LOS), and some would be synchronization reference alarms. After viewing all the alarms, you would have to select the most critical alarm to be cleared. In this example, you would try to determine the reason for the LOS. If the fiber cable had been cut, you would have to wait until the parties responsible for the cable repair completed the repairs. You would then see the alarms associated with this problem clear automatically because the optical receiver recognized the signal coming from the optical transmitter at the node on the other end of this particular span.

Figure 12-17 *CTC Network View Alarm Window*

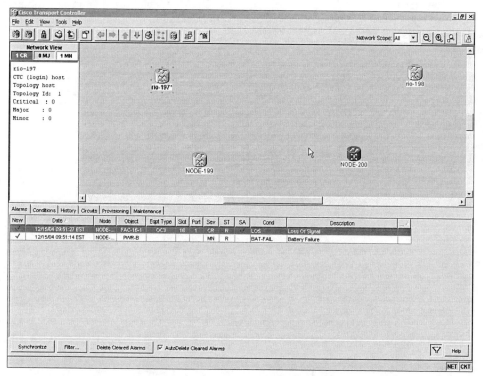

In addition to alarms, conditions are events that could become alarms or that indicate some sort of abnormality. An example of a condition that could become an alarm is a bit error condition. Bit errors can be caused by many things, including dirty fiber connectors, bad splices in the cable, too much loss in the fiber span between nodes, and a bend radius violation in a fiber jumper. Any of these could cause a "dirty" or unrecognizable signal to be received, resulting in bit errors. These bit errors are counted, and when a certain threshold is crossed, an alarm is generated. Different threshold levels generate different levels of alarms, such as minor, major, or critical.

A condition could also indicate that an abnormal condition exists in the system. An example of this type condition is an interface card, such as a DS-1, DS-3/STS-1, or optical card (OC-*N*), that might have been placed in loopback mode for testing purposes. An interface card in loopback mode is an abnormal condition and is displayed under the Conditions tab.

Database Backup

The ONS 15454 uses the two TCC2 in the shelf for redundancy. Each of these processors contains all the system software of a specific release, all the provisioning and options you input when you turn up the system, and the CTC craft interface GUI, as shown in Figure 12-18.

Figure 12-18 *Database Backup and Restore Window*

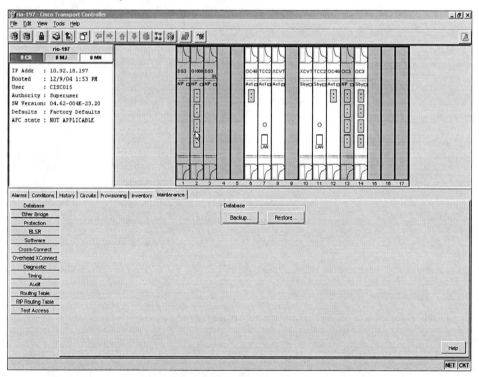

A time might come when you need to upgrade a system release, perform some type of maintenance procedure, or completely replace a node that has been destroyed by a disaster. These require "backing up" the system database in its current working configuration as a precautionary measure or to "restore" a system that has been replaced. The system database can be backed up to a PC file or a CD. Some companies set up a routine to periodically back up all node configurations in a network as part of the restoration plan. In large networks, this task is typically performed from a network operation center (NOC) or some other centrally located facility.

Database Restoration

Restoring a node typically is done after a node has been destroyed, a failure has occurred and standard troubleshooting methods do not resolve the problem, a new node has been installed, or possibly the TCC2 cards have been replaced.

Detailed procedures for database backup and restoration can be found in the Cisco 15454 ONS Procedure Guide.

Card Sparing

Card sparing means that maintenance spares of the critical cards for the ONS 15454 system are maintained in a protected location. These spares can be used in the rare case of a card failure in the system. Detailed information on replacing failed cards can be found in the Cisco 15454 ONS Procedure Guide located at http://www.cisco.com/en/US/products/hw/optical/ps2006/tsd_products_support_series_home.html.

Software Upgrades

When the ONS 15454 shelf and the common cards (TCC2, XC-VT, or XC-10G) are received from a supplier, the TCC2s come with the current system software release loaded on the cards, along with CTC. As with all systems that are software based, they might need to be upgraded to a newer release. The reason for the upgrade might be to add a new feature that the current release doesn't have, or a maintenance release might be required. In either case, the ONS 15454 can be upgraded to a newer release by using CTC and access to the new release from a CD or a file.

To perform a software upgrade, you must verify that no alarms or conditions exist in the node being upgraded. If any alarms or conditions are present, they must be cleared before software activation, as shown in Figure 12-19.

The TCC2 card has two memory partitions for performing software upgrades: There is a "working version" location and a "protect version" location. When you log into the node in Shelf view and click the Maintenance tab and the Software tab, you will see the software upgrade window. At the top of the window, you will see fields for the node name, node status, working version, protect version, and download status, along with the Download, Cancel, Activate, and Revert buttons. Two memory locations allow the TCC2 card to be loaded with a newer release in the protect version location while the system continues to operate on the working version location.

Figure 12-19 *Software Upgrade View*

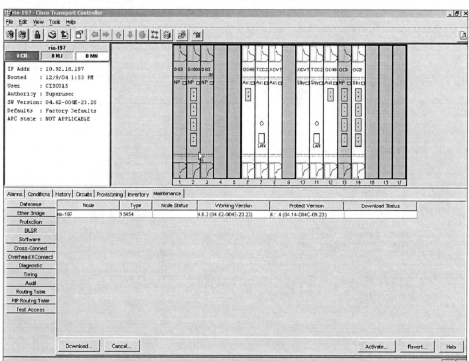

Clicking the Download button opens a window to browse for the location of the new release. The download process then begins, and the new release is loaded into the protect version location. When the download is complete, the next task is to activate the new release by clicking the Activate button.

Software Activation

Activation of the new release causes the node to reboot, which logs you out of the node. When the reboot is complete and you have logged into the node, click the Maintenance tab and the Software tab to view the upgrade window. You'll notice that the new release is in the working version location and that the prior release is now in the protect version location. The system is now operating on the new release.

Software Reversion

If problems occur with the new release, you have the option to revert to the protect version by clicking the Revert button. This causes the system to reboot onto the prior release, which is still in the TCC2 card. This is the ONS 15454 method of recovering a system that might have failed a software upgrade. Keep in mind that any new provisioning additions are lost through the reversion process.

Summary

The Cisco ONS 15454 is a carrier-class MSPP that is easy to install, turn up, and operate. It can be said that CTC makes all this possible. CTC is one of the most unique user interfaces on the market for MSPP. The most important aspect of this GUI is that it is a real-time GUI. Thus, what you see on the screen is what is going on with the network or shelf. For instance, this could be in the way alarms are displayed (red, gold, or yellow), with changes made to cards or ports (in service, out of service), or in Circuit Wizard, for the creation of circuits end to end automatically.

When turning up a system, having the general information in hand makes this process much easier. Most individual nodes can be provisioned in a matter of minutes. The physical aspects of turning up the system could take longer. External cables for DS-1 or DS-3 terminations will take some time. Ensuring that the fiber cable pairs and jumpers meet the required specifications adds to the turn-up time as well.

Pre-engineering and planning go a long way in making the installation and provisioning of the Cisco ONS 15454 a relatively easy task.

For day-to-day operations of the Cisco ONS 15454, a PC connected to a node on a network, with CTC displaying the Network view, is probably the most convenient method of monitoring the network. If anything causes an alarm, it quickly is displayed on the screen, without any human interaction. With other MSPP GUIs, a person must poll the network to provide updates to changes in a network.

Having a Circuit Wizard in the GUI is also a great tool. Circuits can be created individually or as a group simply by following the wizard and inputting the information into the proper fields. For more information on circuits, refer to documentation located at http://www.cisco.com/en/US/products/hw/optical/ps2006/tsd_products_support_series_home.html.

This chapter covers the following topics:

- Resources for Troubleshooting the ONS 15454
- Locating the Problem and Gathering Details Using CTC
- Possible Causes to Common Issues

CHAPTER 13

Troubleshooting ONS 15454 Networks

This chapter provides you with a general but high-level approach to troubleshooting ONS 15454 SONET networks, including the most common problems and issues found during turn-up of an individual node, and network issues.

NOTE   Given the great flexibility, multiple types of services, and multiple possible configurations of the ONS 15454, it is not feasible to provide an exhaustive troubleshooting guide in a single chapter of a book. Therefore, for more information, see the "Documentation" section later in this chapter for details on gaining access to the complete ONS 15454 troubleshooting guide and other related documentation.

Resources to Troubleshoot the ONS 15454

Various resources are available for troubleshooting the ONS 15454. Understanding these resources greatly facilitates and reduces the amount of time it takes to troubleshoot an issue.

Documentation

Unlike most other vendors in the industry, Cisco Systems freely provides all product documentation online (both Hypertext Markup Language [HTML] and PDF formats). The most current and complete Cisco ONS 15454 documentation is readily available on the World Wide Web. To take advantage of this rich and useful resource, go to http://www.cisco.com/univercd/cc/td/doc/product/ong/15400/index.htm.

At the website, identify the software release your ONS 15454 node/network is running, and then select the type of guide that best fits your needs. Some of the most commonly used guides include these:

- **Procedure Guide**—Provides step-by-step instructions to perform any task to configure the ONS15454 node or multiple nodes as part of a network.

- **Troubleshooting Guide**—Provides instructions on troubleshooting and fixing issues. Describes the alarm codes reported by the ONS 15454 and the corresponding possible solutions. Also addresses Simple Network Management Protocol (SNMP) parameters and performance monitoring.

- **Reference Manual**—Provides reference material to every aspect of the ONS 15454 (such as card specifications, Synchronous Optical Network [SONET] topologies, and so on).

- **Release Notes**—Addresses software release–specific new features/functionality, closed issues, and any known caveats.

If you have limited access to the Internet, or prefer to have the documentation CD or the paper versions of the documentation contact your reseller. Both CD and paper versions can also be purchased directly from Cisco Systems by going to http://www.cisco.com/univercd/cc/td/doc/es_inpck/pdi.htm.

Cisco Transport Controller Online Help

Cisco Transport Controller (CTC) is the software used by the end user. CTC is user-friendly and enables the user to graphically provision every aspect of the ONS 15454. CTC resides on the Timing, Communications, and Control card (TCC), and it automatically downloads to your PC or laptop the first time you connect to an ONS 15454. CTC is a Java-based application and requires the appropriate Java Runtime Environment (JRE) on your computer.

NOTE	Software releases prior to (and including) R4.6 are compatible with JRE 1.3. Newer releases since (and including) R4.6 are compatible with JRE 1.4.

In both the documentation and the software CDs for the ONS 15454, Cisco includes the appropriate JRE for the corresponding software version your node is running. The JRE is one of the items you might opt to install on your computer as part of the CTC Installation Wizard. Another item you might want to install is the Online Help (or Online User Manuals), which makes it easy to access help and additional information.

NOTE	CTC Online Help is always available to you because the documentation resides on your computer. It is useful to load this as part of the CTC Installation Wizard.

Installing Online Help

You need to install Online Help only once (per software release). To do so, perform the following steps:

Step 1 Insert the documentation or software CD into your CD-ROM drive.

The wizard autoruns and starts automatically.

Figure 13-1 displays the CTC Installation Wizard screen that you see when inserting the documentation or software CD. Note that options can vary slightly, depending on release.

Figure 13-1 *Cisco Transport Controller Installation Wizard*

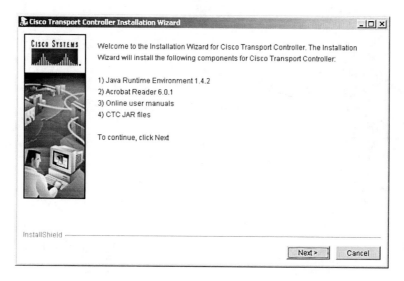

Step 2 Click the Next button. You have the option to select a Typical or Custom install, as shown in Figure 13-2.

Figure 13-2 *Typical or Custom Install*

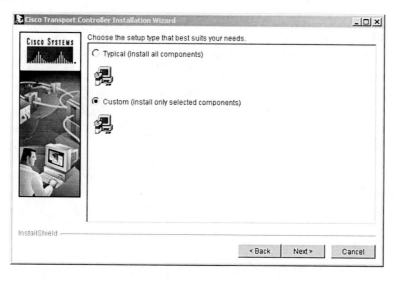

Step 3 Click the Next button. The wizard provides you with the various options to select from. At this point, select Online User Manuals (or Online Help), as shown in Figure 13-3.

Figure 13-3 *Online User Manuals*

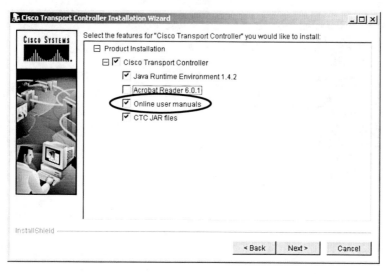

NOTE If you have not yet logged in to an ONS 15454 (that is using this computer) and you are not sure if you have the appropriate JRE, it is a good time to install JRE and CTC JAR files.

Step 4 Click Next and follow the basic instructions provided by the wizard. Online Help and the manuals are then installed on your computer (along with any other options you select).

Congratulations! You can now take advantage of Online Help! What exactly is Online Help? Well, when you are logged in to an ONS 15454, you have complete access to the ONS 15454 documentation. Figure 13-4 shows you how to gain access to this convenient resource.

Cisco Technical Assistance Center

If you require further assistance and want to engage Cisco support, Cisco Technical Assistance Center (TAC) is available 24 × 7 to support you. You can contact TAC via phone at 800-553-2447. Have your contract number ready.

You can also go to the Technical Support Website at http://www.cisco.com/techsupport.

The Technical Support URL provides you with additional troubleshooting documents and tools to help you resolve your network issues. A Cisco.com user ID and password are required to access this website. If you have a valid service contract and do not have a Cisco.com user ID, you can register at http://tools.cisco.com/RPF/register/register.do.

Figure 13-4 *CTC Menu Bar, Help*

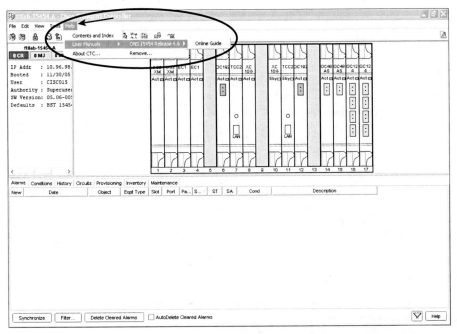

Locating the Problem and Gathering Details Using CTC

It is human nature to attempt to correct an issue immediately after it is discovered—sometimes by connecting and reconnecting cabling, and other times by making configuration change upon configuration change, in an effort to solve the issue quickly. Sometimes this approach works, but often it doesn't—or worse, configuration changes create an even bigger problem or issue.

The first step in troubleshooting is to identify the issue and gather as much detail as possible—before making any changes. Not only is this approach useful to you, but if you need to call for additional support, you also now have documented detail that might help solve your issue more quickly.

CTC is a great tool for gathering all kinds of detailed information that will help you determine the potential causes of your ONS 15454 network problems or issues. The following tools and approaches can help you gather data during the initial stages of troubleshooting:

- CTC Alarms tab
- CTC Conditions tab
- CTC History tab
- CTC Performance tab
- Other data

Alarms Tab

While on CTC, you can be looking at either one of three views:

- **Card view**—Visibility and access to every port on a single card on a single node.

- **Node view**—Visibility and access to an entire node.

- **Network view**—Visibility to the entire network (every node on the interconnected network).

Each of the three CTC views has an Alarm tab. You select the Alarm tab by clicking it. On it, you can see the various alarms for that particular view (for example, selecting Card view displays all alarms for all ports in the card you select, and clicking the Alarm tab from Network view displays all alarms for all cards on all nodes on the network).

In Figure 13-5, you can see a snapshot of the Alarm tab while in Network view. You can quickly identify which two nodes are in alarm by just looking at the network diagram. In the top-left pane, you can see how many and which types of alarms this network has. For detailed alarm information, take a look at the bottom pane. In this pane, you can see the type and description of the alarm, and port/card/node and other details.

Figure 13-5 *CTC Network View, Alarm Tab*

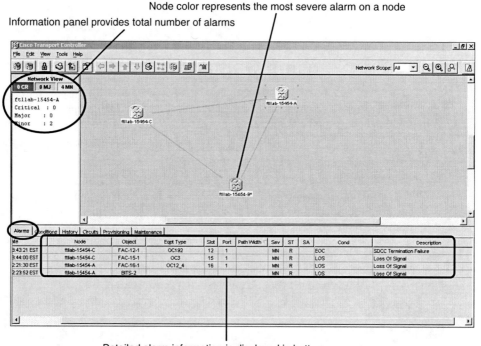

Node color represents the most severe alarm on a node

Information panel provides total number of alarms

Detailed alarm information is displayed in bottom pane
(e.g., node, card, port, type of alarm, description, etc.)

Figure 13-6 provides a snapshot of the Alarm tab while in Node view. Notice that the card in alarm is yellow (representing a minor alarm on this card). This view also provides details on each of the alarms on its bottom pane.

Figure 13-6 *CTC Node View, Alarm Tab*

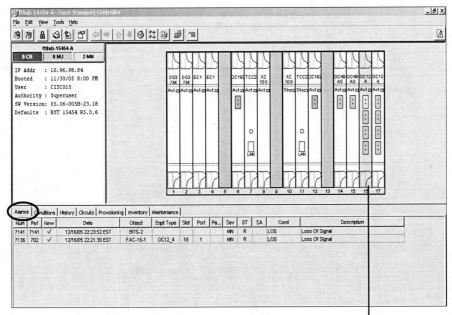

Notice the card in alarm changes
color to match the alarm severity

If you need further detail on any given alarm displayed by CTC, go to the *ONS 15454 Troubleshooting Guide* and see the "Alarm Troubleshooting" chapter. All alarms are explained in the guide.

As you begin to gather details for the problem or issue that you are troubleshooting, take a snapshot of the alarms for every node (in alarm) on the network. The Export feature can also be used to retrieve alarms. This can be found on CTC's menu bar by selecting File, Export. Be sure to name this file so that it is easily recognizable.

Conditions Tab

The Conditions tab (available in all three CTC views) provides details of retrieved fault conditions. A condition is the status or a fault detected by either hardware or software. The Conditions tab displays all current events, even related alarm/conditions that might have been superseded by a more critical related alarm.

The Conditions tab enables you take a deeper, more detailed look at activity on the node or network. In Figure 13-7, you can view the additional details provided by the Conditions tab (Network view), compared to the details provided by just looking at the Alarms tab (see Figure 13-5) on the same network.

Figure 13-7 *CTC Network View, Conditions Tab*

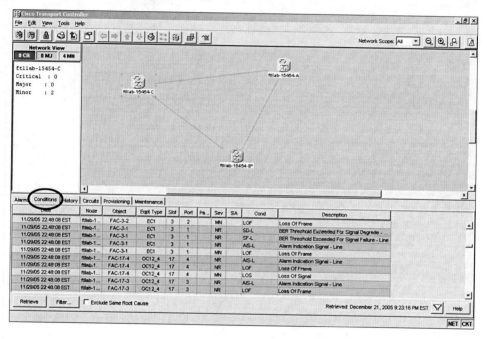

History Tab

Another useful source of information is the History tab. The historical alarm data can be retrieved from all three CTC views (Network, Node, Card) and provides you with both alarming and nonalarming activity (for example, threshold crossing alerts, ring switch, and so on).

As with most other screens in CTC, you can export this data to a file. This information might become very useful during troubleshooting.

The ONS 15454 can store up to 640 critical alarms, 640 major alarms, 640 minor alarms, and 640 condition messages. As these limits are reached, the oldest events on each category are discarded to allow new events to be recorded.

Figure 13-8 displays the History tab information from Network view. Notice that the historical data for both alarming and nonalarming events is stored.

Figure 13-8 *CTC Network View, History Tab*

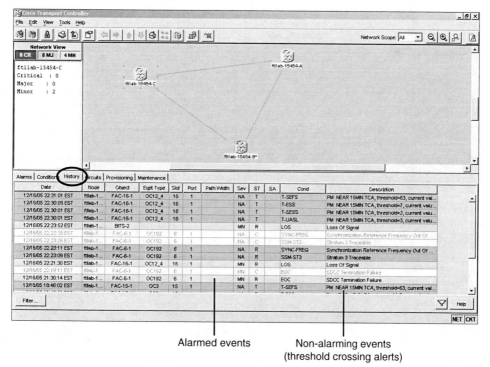

Alarmed events   Non-alarming events
(threshold crossing alerts)

Performance Tab

You can retrieve detailed performance-monitoring (PM) data from the various service cards. Because this is information that is specific to the service ports, it can be retrieved only from Card view. The performance-monitoring values vary depending upon the type of card being monitored.

While looking at this screen, you might see errors, but an alarm might not be reported. This is because the errors/values are below the predetermined threshold crossing alerts (TCA) values; therefore, they do not raise a flag.

Values are recorded for both near and far ends (depending on the type of service), and in 15-minute and 24-hour intervals. This information can be used as a proactive tool to identify potential future issues.

Figure 13-9 displays the PMs for a four-port OC3 card. Details are shown for Port 1. In the figure, notice the values for all PMs on that port—for example, the Error Seconds Section (ES-S) value for each 15-minute interval is 900. This means that during all the 15-minute intervals, this port has seen errors every second.

Figure 13-9 *CTC Card View, Performance Tab*

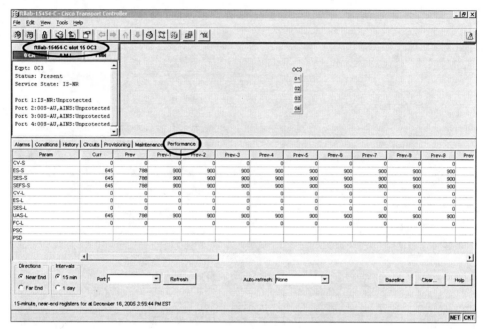

Other Data and Items to Check

Any available information related to this node or network facilitates troubleshooting.
For example, a detailed ring map displaying nodes, fibers, LXG ports, IP addresses, pass-
through locations, and so on can help determine the root cause of an issue by identifying
all potential points of failure.

Diagnostics File

If your issue appears to be complex and you will be calling Cisco TAC for support, you can
download a diagnostics file from the node. This file can be interpreted only by TAC and
provides you with no useful information.

You can download the diagnostics file from Node View > Maintenance > Diagnostic.

Database Backup

Many companies have an automated mechanism for backing up the complete database
of each node routinely. Other companies either don't automatically back up their node's

databases or don't perform this task often enough to keep a current copy. The risk involved with not having a current backup copy of every node's database can be significant. If a customer site is flooded or damaged by fire and requires complete hardware replacement, it could take days to reconfigure that node (including potentially hundreds of circuits) from the time the new hardware is installed. Having a current copy of a node's database enables you to completely restore that node in a matter of minutes.

It is also a good practice to back up the database of the nodes you are about to start making configuration changes to. This ensures that, if you need to restore a node to its original state, you have that database backup readily available. It is unlikely that you will ever need that backup, but if you did need it, it would save you a considerable amount of time and a few headaches.

Backing up a node's database is safe and simple, and it only takes a few seconds. The most common issue with backing up a node's database is the fact that many technicians forget where in their PC or laptop's hard drive they saved this file.

Backing up the node's database is performed from Node View > Maintenance > Database, Backup (at which point you select where in your hard drive you want this file.

TIP	If you are backing up multiple nodes, use a naming convention that will easily identify each node, time, and date of backup.

NOTE	Do not restore the database to a node unless this is required. Restoring the database to a node reboots the TCCs. This causes a temporary connectivity loss to the node through CTC. If you are restoring the database to multiple nodes, wait 1 minute after the TCC has completed before proceeding to the next node.

Card Light Emitting Diodes

The light emitting diodes (LEDs) on each of the cards (both common and service cards) provide useful and instant information. Take a few minutes to familiarize yourself with the various LEDs. These LEDs are marked and are self-explanatory. For specific detail on each LED on each card, see the *ONS 15454 Reference Guide*.

As an example of the LEDs' usefulness, you might be waiting for a protection card that is currently active to revert and become the standby card once again. Also, you can easily see whether you have a signal fail on a service card by quickly looking at the LEDs. Don't overlook this simple but useful feature.

Cabling

As an estimate, based on personal experience, issues related to cabling and fiber top the list of most common troubles. These issues include improperly labeled cables and fibers; dirty, crimped, or cut cable or fibers; incorrect type of fibers or cables; lack of or inadequate attenuation; air gaps; and cable length vs. Line Build-Out (LBO) settings, just to name a few.

Always check the physical medium because it is often at fault. Checking the Signal Fail (SF) LEDs on the service cards can often indicate a potential cabling/fiber issue.

A common straight CAT5 cable is required for an end user to log on directly from a PC to the RJ45 jack on the TCC card of an ONS 15454.

If it does not affect traffic, light readings should be taken to and from optic cards to ensure that the transmitting and receiving light levels are within specification.

Power

WARNING   Do not attempt to work with live power unless you are qualified to do so. When working with live power, always use proper tools and eye protection.

The ONS 15454 has an operating range of −40.5 to −56.7 VDC (−48 VDC nominal). There are two power feeds (A and B) to the ONS 15454; you can easily monitor these by visually checking the dedicated power-monitoring LEDs on each of the TCC cards.

NOTE   Only the TCC2 and later versions of the TCC cards provide power-monitoring LEDs on their faceplate. Older TCC and TCC+ cards do not provide this feature, and nodes with these older cards require the Alarm Interface Controller (AIC) to monitor A and B power.

Although maximum draw is about 22 amps at nominal voltage, 30-amp fuses are recommended (assuming a maximum draw at low voltage while still providing a buffer).

Both the TCC2 and TCC2P controller cards support alarming of extreme voltage thresholds and also provide the capability to select lesser thresholds at which to report alarms. It is not uncommon to see these lesser thresholds crossed (and therefore alarming) at customer sites where certain brands of rectifiers are used. These power-monitoring thresholds can easily be modified (of course, following your company's guidelines) by going to Node View > Provisioning > General > Power Monitor.

Confirm that the node is properly grounded. Taking a power reading of both A and B power is necessary when troubleshooting a power issue. Also note the thresholds currently configured on the node you are investigating, and compare them to the power levels coming into the 15454; a mismatch could be the cause for your power-related alarms.

Connectivity

CTC has a very user-friendly interface, which provides you with great detail. This enables you to quickly narrow the potential root cause for the issue you are troubleshooting. However, such a great tool will not do you any good if you cannot connect to the ONS 15454.

Confirm the Internet Protocol (IP) address of the node you are trying to connect to. Confirm that your PC is properly configured to communicate to the node you are trying to reach.

If your PC is physically connected to the ONS 15454, ensure that you have a working straight CAT5 cable without broken tabs.

If you are connecting through a network, make sure that you have the appropriate rights to gain access to the network, and confirm that any firewalls between you and the ONS 15454 network are not preventing you from accessing the network. Also ensure that the Gateway Network Element (GNE) is properly wired (if using wire-wrap pins in the back of the ONS 15454—also confirm that they were wired correctly).

Data Gathering Checklist

By simply gathering this data in the previous section, you probably already have identified the root cause of the problem.

If the issue is not obvious and you are still troubleshooting, you can find a checklist of items to gather before you start making any configuration changes or requesting technical support. Having this information ready by the time you call technical support will significantly reduce your time on the phone. Use this troubleshooting checklist:

- Refer to the *Cisco Trouble Shooting Guide* provided in the documentation for further detail, at http://www.cisco.com/univercd/cc/td/doc/product/ong/15400/index.htm.
- Create an onsite event log with the date, time, event history, and any previously completed troubleshooting.
- Verify that there is power to the bay.
- Perform a database backup for each of nodes possibly affected.
- Save an alarm log file to the hard drive.
- Record the working software version.
- Record the backup software version, if applicable.
- Print a copy of the Network view.
- Print a copy of the circuits.
- Record the ring configuration (unidirectional path switch ring [UPSR]/bidirectional line switch ring [BLSR]).
- Verify Sync configuration and wiring to the Building Integrated Timing Supply (BITS) clock.

- Verify and record any alarms on the node or network.
- Verify that there are no cards with Fail LED(s) on or blinking.
- Verify that all the fiber-optic cables are properly plugged into the cards.
- Verify that the fiber-optic cables are routed properly, to avoid micro bends and pinches.
- Verify that there are no identifiable loose cables.
- Verify that all the cables have been visually inspected for physical damage.
- Verify that all the optical power levels have been checked, to ensure that they are within the specified range.
- Verify that the air filter to the node is clear of debris.
- Verify that all the cards are properly seated in the chassis.
- Verify that there are no bent pins on the card and backplane.
- Verify that the correct card is plugged into the correct slot, to ensure proper seating of the card.
- Verify that there are no blown fuses in the bay.
- Verify that the voltage level to the chassis is within the proper range.
- Verify that the shelf is properly grounded.
- Verify that the fibers are good with a fault finder.
- Verify that the circuit(s) have not been deleted.
- Note all nodes and interfaces where bit errors are being recorded.

Troubleshooting Tools

In addition to a comprehensive set of detailed real-time status and monitoring information, the ONS 15454 provides the following troubleshooting mechanisms to further aid in finding the root cause of a trouble.

Loopbacks

CTC allows for software loopbacks at all ONS 15454 electrical cards, OC-*N* cards, G-Series Ethernet cards, muxponder cards, transponder cards, and Fibre Channel cards.

Loopbacks are useful not only for testing newly created circuits, but also for determining the source of a network failure.

Because CTC allows for software loopbacks, these can be created either locally or remotely.

NOTE	To create loopbacks in the ONS 15454, the port (where the software loopback will be created) must be changed to an Out-of-Service Maintenance service state.

CAUTION	Before patching a hard loopback to an optical interface (fiber jumper from Tx to Rx on same port), always check the specifications for the Tx and the Rx ports. If the transmit port puts out a higher light level than the receiver's maximum receive level, you must use an attenuator to lower the light level to the allowable Rx range. Irreparable damage to the Rx port in a given card can result if this caution is ignored.
	As an example, the OC-192 LR card puts out a significantly higher light level than the allowable Rx range in the sensitive receiver. If a hard loopback is used in the optical port of this card, an attenuator is necessary.

CTC provides three types of software loopbacks (more detail on loopbacks can be found in the Cisco *ONS 15454 Troubleshooting Guide*):

- **Facility loopback**—A facility loopback takes the signal coming in the Rx port of an I/O card and transmits it back out the Tx port of that same interface. Depending on the location of your test set and the type of interface tested, a facility loopback can eliminate (or identify) issues with cabling, fibers, Electrical Interface Assembly (EIA), and the interface's Tx and Rx ports.

- **Terminal loopback**—With a terminal loopback created at a given interface, a signal coming through the cross-connect cards is looped back toward the same cross-connect cards that it came from. A terminal loopback tests the entire path (from where the test signal is generated), including cross-connect cards, all the way to the port where the terminal loopback was performed. However, a terminal loopback does not test the physical ports (Tx and Rx) of that interface where the terminal loopback was performed

- **Cross-connect loopback**—Given an OC-*N* signal, a cross-connect (XC) loopback allows looping back through a given STS (or STS-*N*c) circuit without affecting other traffic on the optical port. As the name implies, the looping of the specified STS takes place at the cross-connect cards. As an example, using an XC loopback, you can loop the third STS of an OC-12 signal without affecting the remaining 11 STSs on that OC-12 signal.

STS Around the Ring

This feature allows the creation of a circuit from an originating node, having that circuit traverse every span/node around the ring and then go back to the same originating node. It provides a simple method to test all spans on a given ring. Stringent service providers commonly use this feature during ring turn-up to test all spans around the ring.

Monitor Circuit

You can create a monitor circuit to monitor a specific span of a given circuit carrying traffic. CTC enables you to identify a destination port (for this monitoring circuit) on a different node. This tool enables you to monitor spans on a circuit remotely.

Test Access

You can create nonintrusive test access points (TAPs) to monitor for errors on a circuit. Third-party broadband test units can monitor these test access points. Details on creating and deleting test access points can be found in the *Cisco ONS SONET TL1 Command Guide*. You can view test access information in CTC (Node view > Maintenance > Test Access).

Possible Causes to Common Issues

An important part of troubleshooting is gathering alarms, issues, and node/network details related to the problems and issues you are investigating. Collecting such information often results in identifying and correcting the issue. By now, you already have a detailed list of items that correspond to the configuration and state of the node or network you are working on. This section provides you with a list of a few of the most commonly encountered issues and their corresponding potential causes.

Poor or No Signal of an Electrical Circuit

Cabling and wiring issues are common, particularly in new installations. The following are potential causes for this common issue:

- Incorrect DSx panel labeling (for example, wrong labels used in wrong ports).

- Miswiring at patch panel or to the Electrical Interface Assembly (EIA) at the rear of the ONS 15454.

- Damaged cable. Thin coax is commonly used because it is easier to work with. On long lengths of coax, 734 and 735A are spliced so that the thinner cable is at the connector head (allowing much higher densities) while allowing lengths of up 450 feet. During installation, cable-pulling tools are sometimes used. If the maximum torque is exceeded, the splice could be damaged, which would affect the cable.

- Cable connector that is not fully engaged.

- Defective cable from the test set providing the test signal.

- Misconfigured test set providing the signal (check settings such as framing and timing).

- EIA on the back of the ONS 15454 that was not fully inserted during installation. Ensure that it is fully inserted and that all screws are fastened.

- Possible bent pins during installation (where the back of the service card meets the chassis, or during installation of the EIA in the rear of the ONS 15454). Perform a visual inspection.

- Circuit that was provisioned to a different port, card, or node than originally intended.

- Out of Service (OOS) state on that circuit-terminating port.

- Electrical service card (DS1, DS3, or EC1) that is not fully inserted. Ensure that both top and bottom levers are fully engaged.

- Potentially bad card. Switch traffic to the protect card, and monitor the signal on the protect card.

Errors or No Signal of an Optical Link

WARNING   All lasers in the 15454 are invisible; never look at the end of a fiber.

A significant amount of trouble on the ONS 15454 is traced back to a faulty fiber. The following list includes some of the most common causes related to optical link troubles:

- **Incorrect fibers**—Improper fiber was used (for example, mistakenly using a multimode fiber instead of single-mode fiber).

- **Incorrect fiber connections**—Fiber connectors were inserted in the wrong ports (for example, using Tx instead of Rx, and vice versa, or mistakenly mixing fibers for various ports).

- **Air gap**—This significantly can reduce the power of the received signal. Always ensure that the connector end of the fiber is fully inserted.

- **Scratched connectors**—When dealing with fibers, always ensure that connectors are properly capped.

- **Dirty fibers connectors**—Fibers should always be clean before they are inserted into an optical port. Higher-speed interfaces are far more sensitive to dirty fiber connectors. Fiber could remain in that port for years to come.

- **Kinked fibers**—Often fibers are pulled too hard, are stepped on, get caught in between the door and the cabinet's frame, or end up at the bottom of hundreds of pounds of other cabling—just to name a few. When a fiber is damaged (for example, a bend on a fiber has exceeded the maximum bending radius), it might appear to work fine for a while, but even if it is slightly moved, the signal on that fiber could severely degrade.

- **Low light levels into an Rx port on a card**—Always ensure that the Rx fiber connector going into a port has a high enough level to meet specs.

- **Too-high light levels**—Always ensure that the light coming into an Rx port does not exceed the maximum receive level specified for that card. Use of attenuators is required. Receivers on a card can be damaged if the light level exceeds its maximum receive levels.

Unable to Log into the ONS 15454

CTC is a great tool that greatly facilitates troubleshooting of the node or network—but first you must be able to log in. The following are common causes that can prevent you from logging in the ONS 15454:

- **Incorrect cable type is used**—For example, a technician in front of node who is attempting to log in directly from his PC to the TCC might be using a crossover cable instead of the required CAT5 straight cable.

- **CAT5 RJ-45 connector tabs often break**—Using a CAT5 cable with a broken tab often causes a loose connection, resulting in a bouncing connection.

- **CAT5 cables may have poor connection at the RJ-45 connector resulting in random or no connection**—When in doubt, use another cable.

- **The browser, operating system, and JRE are not compatible**—Check the documentation for the relevant ONS 15454 software release, to ensure that you have a valid browser version and OS. As an example, Release 6.0 CTC supports the following operating systems:

  — Microsoft Windows NT

  — Microsoft Windows 98

  — Microsoft Windows XP

  — Microsoft Windows 2000

  — Solaris 8

  — Solaris 9

  Browsers and JRE supported by R6.0:

  — Netscape 7 (PC or Solaris 8 or 9 with Java plug-in 1.4.2)

  — Microsoft Internet Explorer 6.0 (PC platforms with Java plug-in 1.4.2)

  — Mozilla 1.7 (Solaris only)

- **Wrong IP address**—Confirm that you have the correct IP address of the node you are trying to reach.

- **Mistyped IP address**—Confirm the address and retype it correctly.

- **Firewall configuration**—A firewall configuration can prevent your access if you are attempting to log in through a network. Check with your network administrator for access.

- **LAN connection wiring**—The LAN connection on the rear of the ONS 15454 might be wire-wrapped incorrectly.

- **Wrong PC/laptop IP configuration**—Look at the Procedure Guide for the various approaches available for connecting to the ONS 15454.

Follow these steps to troubleshoot basic connectivity:

Step 1 Open a DOS window on your PC (choose Start, Run; type **cmd**; and click OK—if you are using a Windows 98 system, type **command** instead of **cmd**).

Step 2 Find your laptop's IP address (use the **ipconfig** command on the same DOS window).

Step 3 Ping your own IP address from that same DOS window using the **ping** command (for example, ping **192.168.1.101.**—make sure you type the address correctly). This confirms that your IP protocol stack is properly configured.

Step 4 Ensure that you have a working straight CAT5 cable connected between your laptop and the ONS 15454.

Step 5 Confirm that the network interface card (NIC) link light is either blinking or turned on. If not, ensure proper cabling and connections, and make sure that the NIC is fully inserted (if PCMCIA NIC is used in a laptop). The TCC2 and TCC2P cards also have link and activity LEDs at the RJ-45 jacks, which you can check to verify connectivity.

Step 6 In the already open DOS window, ping the ONS 15454's IP address from your PC to establish network connectivity (four replies from the ONS 15454 should be displayed). Congratulations! Receiving the four replies from the ONS 15454 confirms that you have a good cable and that your IP configuration works.

Step 7 Open a new browser window (Netscape or Internet Explorer).

Step 8 Type the ONS 15454's IP address in the URL address window and click the Enter key.

Step 9 Log in at the CTC login screen. At this point, enter a valid login and password. If the message "Loading Java Applet" does not appear on your browser, follow the procedure in the troubleshooting guide called "Browser Login Does Not Launch Java."

Cannot Convert UPSR Ring to BLSR Ring

Upgrading to a BLSR ring using CTC's BLSR Creation Wizard fails if an error resides on one or more of the 15454 spans. The Performance tab in the Card view identifies any errors in the section, line, or path on the optical ring cards. The span should be free of errors before you attempt to upgrade to a BLSR ring.

Signal Degrade in Conditions Tab

When troubleshooting, it is important to look at the conditions in the 15454. Telcordia defines signal degradation (SD) as a soft failure condition. SD and signal fail (SF) both monitor the incoming bit error rate (BER) and are similar, but SD is triggered at a lower bit error rate than SF. An SD is identified in the Conditions tab and in the alarm table if *nonalarmed events* are not being filtered. An SD commonly is caused by one of the following two items:

- A fiber problem

    — A physical fiber problem, such as a faulty fiber connection

    — A bend in the fiber that exceeds the permitted bend radius

    — A bad fiber splice

- A degraded DWDM link

The SONET equipment commonly is seen as the problem, but in many situations, the dense wavelength-division multiplexing (DWDM) equipment or local fiber connections are at fault. A quick method of eliminating the SONET layer is to use a SONET test box to send a signal across the link. If errors still exist, the SONET equipment can be eliminated. At this point, you can investigate the DWDM equipment and look for faulty fiber connections.

Ethernet Circuit Cannot Carry Traffic

When traffic cannot be passed over an Ethernet circuit using G1Ks, first verify that the Gigabit interface converters (GBICs) are secure in the ports on the G1K card. The GBIC has two clips, one on each side of the GBIC, that secure the GBIC in the slot on the G1K-4 card. Also verify that the fibers are securely seated on the GBICs: A simple push on the fiber is always a good practice to ensure that you have a good fiber connection.

If the problem is not at the physical layer, start to troubleshoot at the Ethernet layer. First, however, you must understand an important concept called link integrity, which helps eliminate many problems when turning up Gigabit or fractional Gigabit Ethernet (GigE) circuits using G-series cards on an ONS 15454 ring.

End-to-end Ethernet link integrity essentially means that if any part of the end-to-end path fails, the entire path fails. In other words, a failure at any point of the path causes the G-Series card at each end to disable its transmit laser; this causes the devices at both ends to detect a link down. Also note that if one of the Ethernet ports is administratively disabled or set in loopback mode, the port is considered a failure point because the end-to-end Ethernet path is unavailable. In this example, both ends of the path are disabled because the port state is disabled or in loopback mode.

Because of link integrity, you need to check all parts of the entire Ethernet path to ensure that an active end-to-end link exists. This checklist should uncover the problem that is preventing you from sending traffic over an Ethernet link:

- Verify physical connectivity between the ONS 15454s and the attached device.
- Verify that the ports are enabled on the Ethernet cards.
- Verify that you are using the proper Ethernet cable and that it is wired correctly, or replace the cable with a known-good Ethernet cable.
- Ensure that the proper cross-connects or circuit is in place between the two G1K cards.
- Ensure that you are using the appropriate GBIC (short reach vs. long reach) and that the GBIC used at the ONS 15454 Ethernet port matches the GBIC type used at the connecting equipment (at the other end of the fiber jumper).
- Find and clear any path alarms that apply to the port.

Summary

This chapter provided a high-level approach to troubleshooting ONS 15454 SONET networks, including providing resources to troubleshoot the ONS 15454, locating the problem and gathering details using CTC, and also determining possible causes of common issues.

Certain problems were discussed in this chapter to give you a high-level checklist of common issues that you might encounter in optical networks on a day-to-day basis. Hands-on experience with ONS 15454 rings is the best way to become proficient at troubleshooting MSPP-related problems in the field.

For more information, refer to the ONS 15454 troubleshooting guide and other related documentation and resources, such as Cisco.com.

PART V

MSPP Network Management

This chapter covers the following topics:

- MSPP Fault Management
- Using SNMP MIBs for Fault Management
- Using TL1 for Fault Management
- Using CTC for Fault Management
- MSPP Performance Management
- Ethernet Performance Monitoring
- Using Local Craft Interface Application Versus EMS

CHAPTER 14

Monitoring Multiple Services on a Multiservice Provisioning Platform Network

This chapter first discusses two network management architectures in which a large enterprise or a service provider can use to monitor faults and performance levels of a Multiservice Provisioning Platform (MSPP) network. The Cisco ONS 15454, which is an MSPP, can support dense wavelength-division multiplexing (DWDM), Synchronous Optical Network (SONET), time-division multiplexing (TDM), Ethernet, and storage services. Reasons to implement a network management system can include the following:

- Alarm filtering and consolidation
- Event and alarm correlation

These functions ultimately help to reduce the average time it takes to troubleshoot a network problem. This is beneficial in a service provider environment when trouble tickets need to be created as quickly as possible so that a technician can be dispatched, thereby reducing the direct impact on operational costs. In addition, filtering and correlating data can help obtain meaningful performance information to track service-level agreements (SLAs) established between service providers and their customers.

This chapter introduces you to the fault-management and performance-management capabilities of the ONS 15454. It also addresses several areas that are essential in managing MSPP networks:

- Simple Network Management Protocol (SNMP) Management Information Bases (MIBs), for fault management.
- TL1, for fault management.
- Available alarms by service type.
- MSPP performance management for electrical, SONET, DWDM, Ethernet, and storage services.
- Key differences in using the Local Craft Interface (Cisco Transport Controller [CTC]) versus the Element Management System (EMS).

MSPP Fault Management

One fault-management architecture that service providers commonly use is an EMS based architecture. The MSPP vendor supplies the EMS to manage the services on the MSPP network. In this case, the EMS provides a user interface and a northbound interface (NBI), as shown in Figure 14-1.

Figure 14-1 *EMS with Architecture Example*

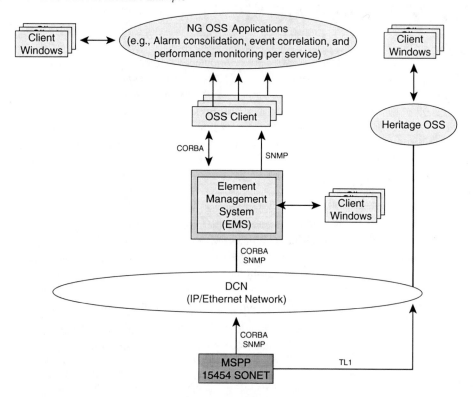

Most service providers prefer this architecture when the EMS is used to interface the MSPP network into the service provider's Operational Support System (OSS) network. In other words, the service provider's OSS sends messages, which are commonly based on an application protocol called Common Object Request Broker Architecture (CORBA), to the EMS's NBI, which instructs the EMS to carry out certain tasks. Provisioning an Ethernet circuit between two ONS 15454 nodes is an example of a task. The EMS NBI forwards alarms and events to the OSS through the NBI whenever a MSPP node on a ring generates alarms or events. This architecture is commonly used for large-scale networks; it is further explained in Chapter 15, "Large-Scale Management."

NOTE An OSS is developed either in-house (commonly referred to as a homegrown system), by a single vendor, or as an integration of commercial off-the-shelf systems. The latter is sometimes referred to as a *best-of-breed* system. The Telcordia OSS is an example of a single-vendor OSS. The Telcordia OSS consists of NMA, TIRKS, and TEMS applications, used for monitoring and provisioning large-scale networks using TL1.

In other smaller-scale networks, an enterprise or service provider might choose not to use an EMS, but instead use an OSS that contains commercial network-management software and homegrown software to monitor and provision services on MSPP rings. In the case of the ONS 15454, these existing systems would interface through an Ethernet/Internet Protocol (IP) network and retrieve alarms and events through SNMP or TL1, as shown in Figure 14-2.

Figure 14-2 *Non-EMS Interface Architecture Example*

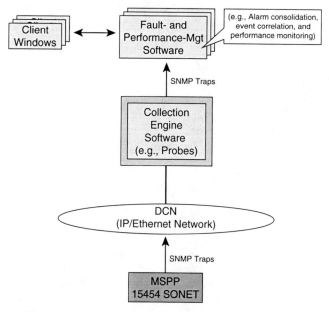

When a problem occurs in the network, the SNMP monitoring application receives messages from the MSPP network and raises the appropriate alarms. Many of the fault-management software today allow complex logic to be programmed within it, to eliminate some of the manual troubleshooting.

NOTE A person's logic and years of experience and knowledge can be useful in solving a particular problem.

This fault-management software is useful in MSPP networks when the underlying technology is SONET or DWDM and the end-to-end service is Ethernet and storage networking. In this scenario, the SONET expert might not be the same person as the Ethernet expert, so multiple organizations must work together.

A collection engine, also known as a probe, is software that can periodically poll devices for statistics and receive SNMP traps when faults occur. However, when a fault occurs, the service provider has no idea which customers are impacted. Fault-management software can help resolve this problem and also automate the fault-resolution process by doing the following:

- Detecting the problem by receiving the traps and correlating the traps from each endpoint into one alarm for the connectivity problem
- Localizing and diagnosing the problem in general
- Reconfiguring the MSPP if it is misconfigured
- Confirming that service has been re-established

Using SNMP MIBs for Fault Management

You can use different management protocols for fault management in a MSPP network. The Cisco 15454 uses two management protocols: SNMP and TL1. Certain service providers, such as incumbent local exchange carriers (ILECs), traditionally use TL1 to monitor networks; large corporate networks and Internet service providers (ISPs) traditionally use SNMP to monitor networks. SNMP has been applied to telecom interfaces such as DS1, DS3, and SONET interfaces. In other words, SNMP MIBs have been developed for these telecom interfaces to allow fault-management and performance-management systems to monitor them.

What Is an SNMP MIB?

An SNMP MIB defines the management information that is collected and stored on communications equipment, such as an MSPP. The SNMP MIB contains a list of parameters, commonly referred to as objects, that defines what information can be collected. Standard MIBs have been written by the standards Internet Engineering Task Force (IETF). Equipment vendors also write their own proprietary MIBs for their equipment to fill in the gaps of additional information that is not contained as part of standard MIBs. These proprietary MIBs are necessary to completely monitor an end-to-end service such as Ethernet traffic on a MSPP ring.

Vendors decide which MIBs to implement on their communications equipment. Reading SNMP MIBs can be overwhelming because of the large number of objects defined by the MIB. Service providers and other companies might choose to collect only the MIB objects needed to help prevent services from degrading or failing.

An SNMP-managed network has two components: SNMP agents and an SNMP manager. An SNMP agent is software that resides on the MSPP and is responsible for keeping track of information on the MSPP inventory, fault management, and performance management. The MIB provides a blueprint of this information in the form of objects. The SNMP manager can access all or a subset of these SNMP objects.

NOTE   An SNMP monitoring application contains the SNMP Manager component and is part of the fault-management software. Fault-management software is typically included in a network management system.

The SNMP manager accesses this information by either retrieving (commonly referred to as polling) this information or having the SNMP agent autonomously send notifications of events to it. These notifications, called SNMP traps, are used to notify the network-management system when certain important events occur, such as Ethernet frames being dropped. The SNMP agent sends these unsolicited traps to the network-management system. For example, traps can alert the SNMP monitoring application of a certain condition on the network, such as monitoring Ethernet link status (up or down) or other significant events. Figure 14-3 shows the SNMP agent architecture and messages.

Figure 14-3 *SNMP Agent on an MSPP*

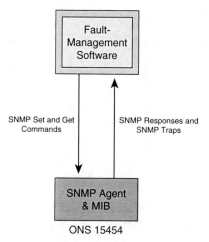

Some data and telecom equipment allows SNMP network-management systems to configure and monitor the equipment. You cannot provision services on the Cisco 15454 using SNMP, but you can poll information and receive SNMP traps.

NOTE	Because SNMP traps use an unreliable transport protocol, called Unreliable Data Protocol (UDP), delivery of these messages cannot be guaranteed. However, SNMP Version 2 uses an Inform message to issue a notification that requires an acknowledgment from the SNMP manager. ONS 15454 does not support the Inform message, yet it can support SNMPv1 and SNMPv2c simultaneously.

SNMP Traps

SNMP traps allow a MSPP to report the occurrence of an event to one or more network-management systems; the ONS 15454 can send traps to up to 10 different SNMP-management systems.

The network-management system can use events that occur on an MSPP to carry out fault management, capacity management, performance management, or configuration management.

If a non-EMS architecture is being used, SNMP traps are needed to avoid excessive polling of MIB variables to monitor the health of the network, as discussed in the beginning of this chapter. Extensive SNMP polling results in the following:

- Additional configuration needed on the fault-management and performance-management software
- Increased traffic on the data communications network and SONET Data Communications Channel (DCC), especially in large networks
- Load on the fault-management and performance-management software
- Additional computing hardware required as a network grows

Designing and configuring the fault-management and performance-management software to use SNMP traps and limited polling eliminates the issues previously listed and enables faster notification of faults in the network.

ONS 15454 Traps

The ONS 15454 generates all alarms, conditions, and threshold crossing alerts (TCAs) as SNMP traps. These traps are defined in the specific Cisco MIB for the ONS 15454, called *CERENT-454-MIB*.

NOTE	ONS 15454 alarms are provisionable. Alarms that are configured to be suppressed using the CTC Alarm-Provisioning tool are reported as a "Nonreported," or "NR," severity condition. Conditions defined as NR are not reported as SNMP traps.

This MIB defines more than 400 different traps. The ONS 15454 supports additional traps that are defined in other MIBs. Examples of some important ONS 15454 traps follow:

- **entConfigChange (ENTITY-MIB-rfc2737)**—This trap reports changes in the ports and interfaces that occur through CTC or TL1.

- **risingAlarm (RMON-MIB-rfc2819)**—This trap is generated when an alarm entry crosses the rising threshold.

- **fallingAlarm (RMON-MIB-rfc2819)**—This trap is generated when an alarm entry crosses the falling threshold.

- **performanceMonitorThresholdCrossingAlert (CERENT-454 MIB)**—This trap is generated when a threshold is exceeded for a defined performance parameter.

The current alarms and conditions associated with these traps are accessible using SNMP get requests on the 15454 Alarm Table, called cerent454AlarmTable, defined in the CERENT-454-MIB, as shown in Figure 14-4.

Figure 14-4 *ONS 15454 SNMP Alarm Table*

```
Cerent454AlarmType ::= TEXTUAL-CONVENTION
                    --DISPLAY-HINT "d"
    STATUS current
    DESCRIPTION
            "This TC is used by the NMS to map the alarm type
            received by the trap and display the string
            shown in LHS."
    SYNTAX INTEGER {
        alarmUnknown                                              (1),
        alarmCutoffIsInManualMode                                 (10),
        failureDetectedExternalToTheNE                            (20),
        externalError                                             (30),
        excessiveSwitching                                        (40),
        terminationFailureSDCC                                    (50),
        incomingFailureCondition                                  (60),
        alarmIndicationSignal                                     (70),
        alarmIndicationSignalLine                                 (80),
        alarmIndicationSignalPath                                 (90),
        alarmIndicationSignalVT                                   (100),
        channelFailureAPS                                         (110),
        channelByteFailureAPS                                     (120),
        channelProtectionSwitchingChannelMatchFailureAPS          (130),
        channelAutomaticProtectionSwitchModeMismatchAPS           (140),
        channelFarEndProtectionLineFailureAPS                     (150),
        inconsistentAPSCode                                       (160),
        improperAPSCode                                           (170),
        nodeIdMismatch                                            (180),
        channelDefaultKAPS                                        (190),
        connectionLoss                                            (200),
        bipolarViolation                                          (210),
        carrierLossOnTheLAN                                       (220),
        ...                                                       ...
```

Figure 14-4 does not show the complete list of trap values because nearly 700 trap values are defined. You can use the 15454 Alarm Table to validate the traps that the ONS 15454 receives. Additionally, an SNMP manager can poll this alarm table for the current alarms and conditions instead of listening to SNMP traps from the ONS 15454.

Trap Example: Improper Removal

The ONS 15454 generates traps that contain a unique ID to identify the alarm. This ID is referred to as an object ID (OID). The trap contains information that identifies where the alarm originated. For example, an "Improper Removal" alarm is generated when a card from a 15454 shelf is unseated without deleting it from the 15454 database, as shown in Figure 14-5.

Figure 14-5 *Improper Removal of 15454 Card*

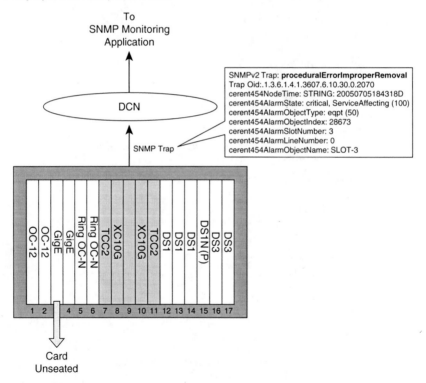

This SNMP trap is automatically issued to the IP addresses assigned as the SNMP manager in CTC. The trap message contains key information about the event that just occurred. This information is contained in the data field of the trap, which is formally called the *Variable Bindings* or *Varbind List*. In this example, the message identifies the following items:

- **Time of event**—This is the time stamp of the alarm, indicating the date and time when the alarm locally occurred. This is labeled as the CerentNodeTime in the trap message; Cerent is the original name of the ONS 15454. The time is displayed in the following format: YYYYMMDDhhmmss{S/D}. In this example, 20050705184318D is the time stamp, which converts to 6:43:18 PM EDT on July 5, 2005. If the last octet was S, this would indicate standard time.

- **Alarm state**—This is the severity of the alarm (critical, major, minor, and so on), indicating whether the alarm is service affecting or nonservice affecting. In this case, unseating a card produces a critical alarm, which is service affecting. This is labeled as cerent454AlarmState in the trap message.

- **Entity that raised the alarm**—The entity is the equipment that raised this alarm. This is labeled as cerent454AlarmObjectType in the trap message.

- **Slot associated with the alarm**—In this example, the card was unseated from Slot 3. This is labeled as cerent454AlarmSlotNumber in the trap message.

- **Port associated with the alarm**—If port information is not relevant to the alarm, the value is 0. This is labeled as cerent454AlarmPortNumber in the trap message.

- **Line associated with the alarm**—If line information is not relevant to the alarm, the value is 0. This is labeled as cerent454AlarmLineNumber in the trap message.

- **TL1 name**—This value provides more detailed information about what entity generated the trap. Entities can include Slot, Port, Synchronous Transport Signal (STS), Virtual Tributary (VT), and Bidirectional Line Switch Ring (BLSR). For example, this value provides STS or VT numbers if an STS or VT circuit triggers the trap. Or, if a BLSR ring event triggers a trap, the ring number is provided here. In this example, SLOT-3 is the value. If a VT circuit associated with Slot=6, STS=192, VT Group=7, and VT=4 raises a trap, the value is VT1-6-192-7-4.

NOTE   This TL1 name is identical to the access identifier (AID) value contained in every TL1 alarm or event generated from the ONS 15454. In this example, the AID is SLOT-3. Other AIDs that can be contained in the trap message are Facility (FAC), VT1, STS, and (VT Group) VTG. The following examples help you further understand the TL1 name. In these examples, *s* stands for "slot," *p* stands for "port," *sts* stands for "STS number," *vtg* stands for "VT group number," and *vt* stands for "VT number." Table 14-1 contains some other examples of other AIDs formats.

Table 14-1   *AIDs Format Examples*

Format	Example
FAC-s-p	FAC-5-1 for port 1 of OC12 in Slot 5.
SLOT-s	SLOT-16 for OC12 in Slot 16.
VT1-s-sts-vtg-vt	VT1-6-4-6-3 for VT in module located in Slot 6 (sts=4, VTGroup=6, vt=3).
STS-s-sts	STS-2-5 for STS object in module located in Slot 2 associated with STS 5 (that is, it corresponds to the second STS on Port 2).

The ONS 15454 also generates a trap for each alarm when the alarm clears, as shown in Figure 14-6.

Figure 14-6 *15454 Card Reseated*

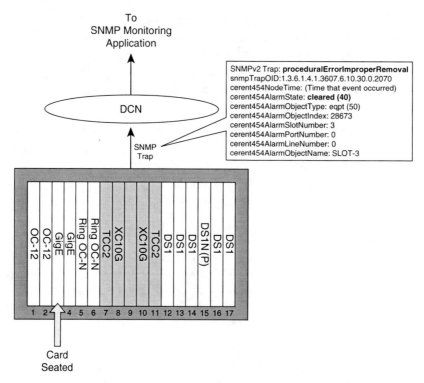

As you will notice, the trap generated is nearly identical to the previous trap, except that the alarm state has changed to "cleared" and the time of the event has a new time stamp.

Trap Example: Carrier Loss on G1K Port

Before talking about SNMP traps that are issued for a carrier loss on a G1K port, it is important to discuss a feature on the G1K card called link integrity. Link integrity means that the entire end-to-end Ethernet link fails if any point of the end-to-end path is not working. The Ethernet ports and the SONET circuit path interconnecting these ports must be up and working before Ethernet traffic can flow. In other words, the link integrity feature ensures that the transmit lasers on the G1K card at each end are not turned on until all SONET and Ethernet errors along the path are cleared. For example, if one of the G1K ports is manually disabled using CTC or is in a Carrier Loss state (CARLOSS) because the attached switch or router's Ethernet port is down, this port is considered to be in a failure state because the end-to-end Ethernet path is unavailable. As a result of this port failure, the

G1K card at each end disables its transmit laser, which prevents the port from going into service.

The G1K card can be set to Automatic In-Service state (AINS). AINS is supported in Release 5.0 and later. This means that the G1K port is initially in a state that suppresses alarm reporting, but traffic is carried and loopbacks are allowed. After a predefined period, referred to as the soak period, the port changes to in-service and alarm reporting is no longer suppressed.

Figure 14-7 shows a case in which a router's Ethernet port is shut down. As a result, Port 1 on the G1K card issues a CARLOSS alarm.

Figure 14-7   *Carrier Loss on G1K Port*

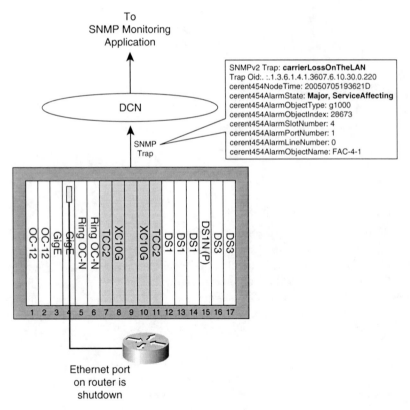

The trap name is carrierLossOnTheLAN, defined in the Cisco MIB. After the router's Ethernet port is activated, the CARLOSS is cleared on the G1K port and the same trap is sent, but this time with the Alarm state set to "cleared." Figure 14-8 shows a carrier loss cleared on a G1K port.

Figure 14-8 *Carrier Loss Cleared on G1K Port*

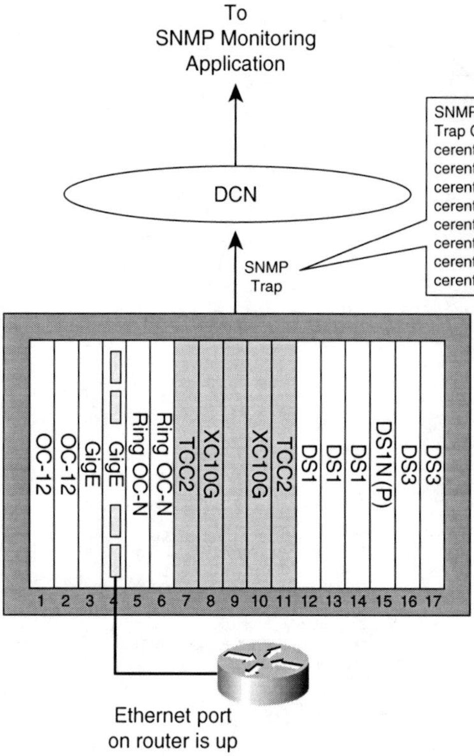

Trap Example: Protection Switch

A protection switch raises a condition called Working Switched To Protection (WKSWPR). This is the case for all protection types. This occurs when a loss of signal (LOS) or signal degrade (SD) occurs on an electrical or optical circuit. When a revertive protection group is established, a WKSWPR condition is raised when the working card switches to the protect card. As soon as the traffic is switched back to the working card, the WKSWPR is cleared. In the case of a nonrevertive protection group, a Switched Back to Working (WKSWBK) condition is issued when the traffic is manually switched back to the working card.

NOTE	1: *N* protection groups in the system are always revertive.

Figure 14-9 shows the SNMP trap that indicates a DS1 working card has switched to a DS1N protect card.

Figure 14-9 *Working Card Switch to Protection Card*

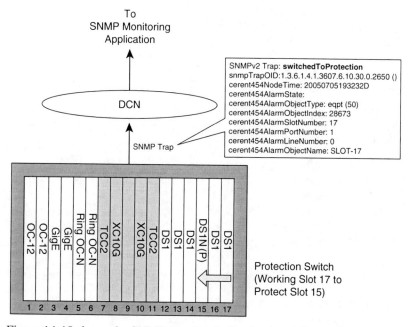

Figure 14-10 shows the SNMP trap that indicates the traffic has switched back to the DS1 working card.

Figure 14-10 *Traffic Switched Back to Working Card*

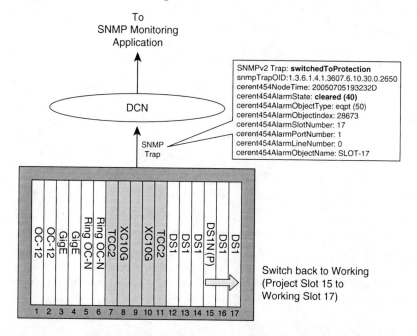

Trap Example: Audit Trail Full

To facilitate security monitoring, the ONS 15454 generates a trap when the audit log reaches 80 percent and also at 100 percent. The audit trail keeps track of user activities on the ONS 15454.

ONS 15454 MIBs

The ONS 15454 has its own enterprise-specific MIBs in addition to standard MIBs. Supported MIBs are listed at Cisco.com at http://www.cisco.com/public/sw-center/netmgmt/cmtk/mibs.shtml.

Refer to Cisco documentation to determine whether the MIB file is backward compatible. For example, if a customer has ONS 15454 Release 5.0 and ONS 15454 Release 4.6 running in the network, all MIB objects defined in Release 4.6 will still be supported in Release 5.0. If the MIB file is backward compatible, this allows the operations center to continue to monitor all the nodes in the network by polling the same MIB objects, even after certain 15454 nodes are upgraded to a newer software release.

NOTE   The version of the MIB file is indicated in the Description field at the beginning of the file. For example, the following lines were extracted from the Cerent-454-MIB:

DESCRIPTION

"This file can be used with R5.0 release."

MIB Object Example: Retrieving DS3 Port Name

The MIB object that identifies the DS3 port name is dsx3CircuitIdentifier. This MIB object is from the standard MIB called DS3-MIB-rfc2496. This corresponds to the port name "Customer ABC" on the DS3 card shown in Figure 14-11.

Up to 12 different port names can be assigned per DS3 card. Therefore, 12 different rows of the dsx3CircuitIdentifier exist in the MIB for every DS3 card. These port names can be retrieved using the **SNMP get** and **getNext** commands.

MIB Object Example: Retrieving Serial Numbers

Another standard MIB is the ENTITY-MIB (RFC 2737). This MIB has an objected called entPhysicalSerialNum that contains the serial numbers for each ONS 15454 card.

As an example, if a DS3 card is in Slot 2 in an ONS 15454 chassis, you need to translate Slot 2 to an index number that represents Slot 2 in the table. This is calculated as follows:

Table index = (Slot# × 4096) + 1

In this example, the table index would be $(2 \times 4096) + 1$, which equals 8193.

Figure 14-11 *DS3 Port Name Displayed in CTC*

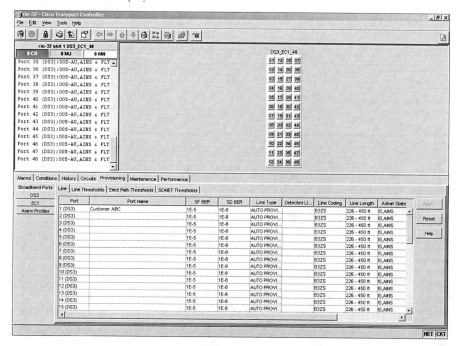

Setting Up SNMP on the ONS 15454

Setting up SNMP in the ONS 15454 is a simple task. First, assign the community names in the ONS 15454. The community name is like an assigned username, which is assigned and carried by each SNMP request message that the fault-management software issues. The community name in the SNMP request message must match the community name assigned in the ONS 15454. Community names are provisioned using CTC, as shown in Figure 14-12.

An SNMP request will be accepted if its community name matches any community name listed in the SNMP trap destinations table. In this example, the community name is public and the trap destinations are 10.1.1.40 and 10.1.1.50. All SNMP requests will be dropped if its community name does not match any of the community names identified in this table.

The second step is to assign the IP address of where the SNMP traps need to be delivered. Up to 10 trap destinations can be assigned.

Figure 14-12 *Provisioning Community Names Using CTC*

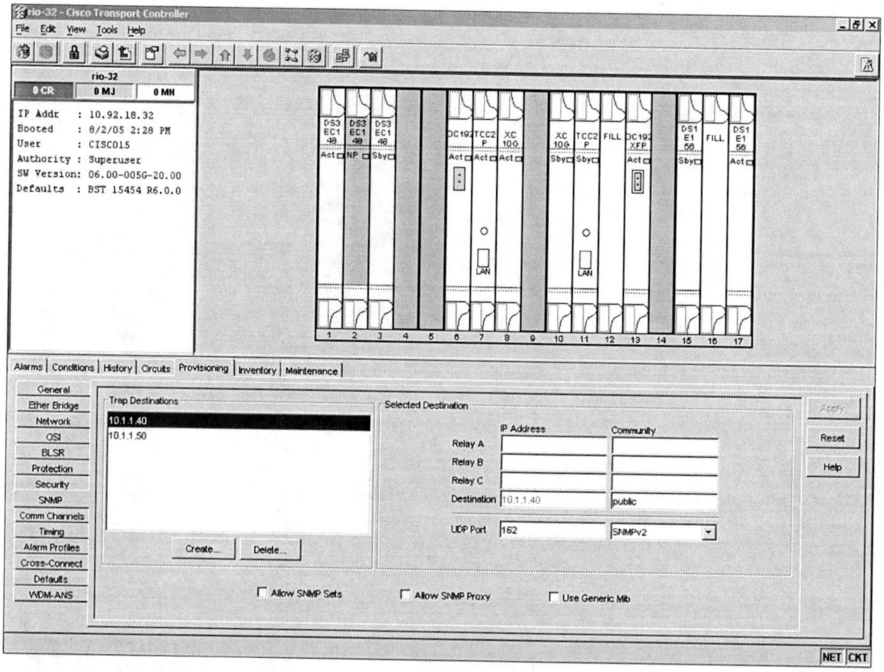

Loading MIBs

You must load the ONS 15454 enterprise-specific MIB files and standard MIB files as one of the first steps in setting up your fault-management and performance-management systems. You must follow two rules:

1 Ensure that the right versions of MIB files are loaded. If the release of the ONS 15454 is Release 5.0 or later be sure that the ONS 15454 MIBs are from the same version.

   Be aware that two versions of MIBs are supported on the ONS 15454: SNMPv1 and SNMPv2 MIBs.

2 Load the MIB files in the order specified in the MIB dependency table for the ONS 15454 release you are working with. Otherwise, the SNMP manager might not compile one or more of the MIB files.

Using TL1 for Fault Management

Transaction Language 1 (TL1) is a standard set of ASCII messages that an OSS application uses to manage a network element, such as MSPPs. Similar to SNMP, two key types of messages exist:

- TL1 commands

- TL1 autonomous messages

TL1 commands can configure the MSPP or retrieve information. Setting up an STS cross-connect is one example of using the TL1 command. You construct the TL1 command as follows:

```
ENT-CRS-<STS_PATH>
```

Retrieving the current status of all active alarms and conditions from an ONS 15454 is another example of using the TL1 command. You construct the TL1 command as follows:

```
RTRV-ALM-ALL
```

If these TL1 commands are successful, the ONS 15454 issues a response message that contains a number called the completion code. If a TL1 command fails, the ONS 15454 issues an error response that contains a number, called the DENY code.

TL1 autonomous messages are like SNMP trap messages: They are used to report alarms, conditions, and configuration changes. In other words, the OSS application does not have to request that these messages be sent; the ONS 15454 issues these messages without any intervention.

TL1 Versus SNMP

TL1 and SNMP have similar characteristics. Both TL1 and SNMP use autonomous messages to notify management systems of alarms and conditions from a network element, such as the ONS 15454. In addition, TL1 and SNMP both have mechanisms to retrieve the current alarms and conditions from a network element. TL1 uses the commands RTRV-ALM-ALL and RTRV-COND-ALL, and SNMP uses get/getNext request messages to retrieve alarms and conditions from 15454 MIB tables. In the case of the ONS 15454, the cerent454AlarmTable contains all the active 15454 alarms and conditions that an SNMP-management system can retrieve.

Thus, SNMP and TL1 can be used to monitor the health of the 15454 by using autonomous messages or retrieving alarms and conditions. One key difference, however, is that SNMP cannot provision the ONS 15454, whereas TL1 can provision electrical, SONET, Layer 1 Ethernet, and DWDM services on the ONS 15454.

TL1 has traditionally been used to manage telecommunications equipment, such as SONET add-drop multiplexers (ADMs), whereas SNMP has traditionally been used to manage data communications devices, such as Ethernet switches and routers. However, SNMP is now being used in telecommunications equipment, such as MSPPs. In certain large-scale service-provider networks, in which thousands of network elements (NEs) are deployed, TL1 is commonly used as the management protocol for transport SONET and DWDM networks; SNMP is commonly used as the management protocol for Ethernet and IP networks. In the latter case, some NEs are managed directly via SNMP and other NEs are managed using an EMS, which, in turn, interfaces with a newer generation of OSS. This architecture was discussed in the beginning of this chapter.

Everything in TL1 and SNMP can be accomplished in CTC. However, be aware that certain configuration tasks can be carried out only in CTC and cannot be set up by TL1. Automatic provisioning of an end-to-end circuit by selecting only the two endpoints of the circuits (for example, DS3 port in Node A and DS3 port in Node Z) can be accomplished only using CTC. The equivalent action in TL1 is setting up a series of cross-connects in each 15454 that the end-to-end circuit traverses.

Using TL1 for Ethernet Services

TL1 cannot provision certain parameters. These mainly consist of Layer 2 Ethernet configuration, such as setting up a resilient packet ring (RPR), virtual LAN (VLAN) IDs, and quality of service (QoS). TL1 does not support these parameters because no TL1 commands exist for configuring these parameters. Instead, a Layer 2 capable EMS like Cisco Transport Manager (CTM) can be used to set up these parameters.

The ML 100 and ML 1000 Ethernet cards are considered Layer 2 cards. These are the Ethernet cards that support RPR and QoS features used in setting up Multipoint Ethernet service. You can use TL1 on the ML Ethernet cards for card inventory, alarm management, Layer 1 provisioning, and the retrieval of Ethernet performance information. In addition, you can use TL1 to provision SONET circuits and transfer a Cisco IOS startup configuration file to the memory on the Timing Communications Control 2 (TCC2) card. Each ML card runs its own instance of Cisco IOS software and must have a Cisco IOS configuration file.

The G1K and the CE100 Ethernet cards are considered Layer 1 cards, on which QoS and VLANs parameters do not exist. These cards map the Ethernet frames coming into the Ethernet port directly onto a point-to-point SONET circuit. As a result, you can use TL1 to provision a point-to-point Ethernet service using G1K and CE100 cards.

Using CTC for Fault Management

Every alarm and event that the ONS 15454 generates is captured in the CTC alarm and conditions tables. A condition is a fault or status that the ONS 15454 detects. In CTC, the conditions window shows all conditions that occur (including those that are superseded) unless the root-cause filter is on, in the conditions table. For example, if loss of frame (LOF) and loss of signal (LOS) are present in the network, CTC shows both the LOF and LOS conditions in this window; LOS occurs when the port on the card is in service but no signal is being received.

The CTC alarm window would show only LOS because LOS supersedes and replaces LOF. Having all conditions visible can be helpful when troubleshooting the ONS 15454. Fault conditions include reported alarms and Not Reported or Not Alarmed conditions.

NOTE	Alarms that are suppressed are normally found on the conditions table in CTC. The conditions table is used to diagnose system problems, without any filtering of alarms or conditions. As a result, any alarm that normally would be suppressed is viewable in the conditions table as an NR severity condition.

CTC is a useful and practical tool that can be used to monitor a 15454 network. Keep in mind, however, that it is not intended as a tool to monitor a network of many 15454s. CTC is typically used to view and monitor one ring of 15454s at a time. It is a useful tool when troubleshooting is required on a 15454 ring.

SNMP traps can be correlated with the associated alarm generated in the CTC AlarmTable. The value of TL1 name (labeled as cerent454AlarmObjectName in the trap) which is also called the AID (as discussed in the trap section) is identical to the column called Object in the Alarm window in CTC. Therefore, you can use CTC to look into the cause of the alarm further when the SNMP-monitoring application receives a trap. Figure 14-13 shows where the object is located in the Alarm window. In this example, a CARLOSS alarm is present on a G1K. The Object field has the value FAC-4-1, which translates to Port 1 on the card in Slot 4.

Figure 14-13 *CTC Alarm Objects*

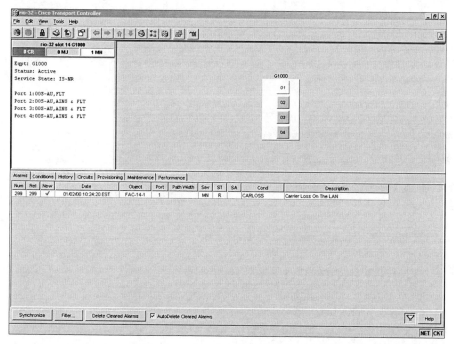

CTC supplies a graphical representation of the 15454 network topology, with current standing alarms and conditions, and depicts the current 15454 shelf configuration. CTC is an easy-to-use application because it automatically discovers all DCC-enabled 15454s that are interconnected to the 15454 node that CTC logs into. The DCC is created by using the bytes in the SONET signal's overhead to create a communications channel among multiple 15454s. CTC uses this communications channel, along with the built-in routing protocol of the 15454, to discover each 15454. After the network map of all the 15454s is created in CTC, the user can log into any of the 15454s identified in the network map and view all the alarms, as shown in Figure 14-14.

Figure 14-14 *CTC Alarm Table in Network Map View*

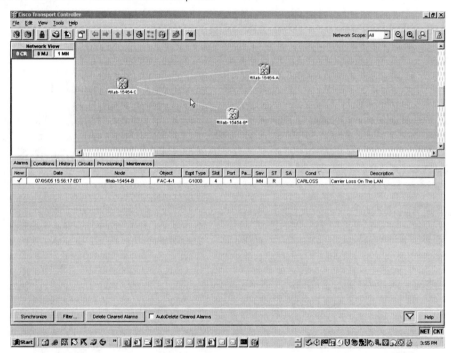

The user can then drill down into a single 15454 node and view the alarms associated with that particular 15454. This allows you to quickly filter out alarms from other 15454 nodes in the network. You can then filter again by double-clicking the 15454 card, which will show only the alarms associated with that card.

MSPP Performance Management

Why care about monitoring the performance of services? Services, especially voice and video, are impacted when performance degrades to a certain level. For example, the quality of voice calls might begin to degrade or the voice call might even drop. Digital video transmission might have pictures that are distorted or even frozen. This is unacceptable in a service-provider environment, especially when SLAs have been set up with customers. Not meeting the service level specified in this agreement can cause monetary damages.

A Threshold Crossing Alert (TCA) is a key component of performance management. You use TCAs to detect early performance degradation on the services that the ONS 15454 provides. These services include Ethernet, storage, optical, electrical, and DWDM services.

The ONS 15454 uses thresholds to set error levels for performance-monitoring (PM) parameters. The ONS 15454 accumulates the counters that keep track of each PM parameter. For example, one PM parameter is the DS3 line code violations, DS3 CV-L, which indicates the number of coding violations occurring on the DS3 line. In this case, the error level is defined in CTC to determine how many DS3 line code violations need to occur before a TCA is issued. Each PM parameter's TCA value needs to be set up in CTC. When the threshold is triggered, the ONS 15454 sends out the TCA as an SNMP trap and TL1 message. In addition, the TCA is displayed in CTC.

NOTE   ONS 15454 contains a default value for every PM parameter in the ONS 15454. You must change the threshold if the default value does not satisfy your requirements. For example, each service provider has its own set of service-level requirement to meet the SLA established with its customers. You might have to lower the default thresholds to have a high QoS on the line.

The SNMP trap name is performanceMonitorThresholdCrossingAlert. The SNMP trap contains all the details associated with a TCA displayed in CTC. In addition to the normal SNMP trap information (slot, port, and so on), the details associated with a TCA are added in the trap message:

- **Threshold monitor type**—Indicates the PM metric being monitored over a specified time interval. Examples include CV (that is, coding violations for DS3 card), SES (that is, severely errored seconds for DS3 card), laserBiasMax (that is, laser power level for DWDM card), and gfpStatsRxFrame (that is, number of received frames for storage card). This is a small sample of monitor types; the ONS 15454 supports many more.

- **Threshold location**—Indicates whether the TCA value is local or remote. In other words, indicates whether the threshold is being monitored on the near end or far end of the circuit.

- **Threshold period**—Indicates the sampling period. This can be 1 minute, 15 minutes, or 1 day.

- **Threshold set value**—Identifies the threshold count value for a current time period. The time period can be 15-minute intervals or a 1-day interval. In other words, this is the number of errors that must be exceeded before the threshold is crossed.

- **Threshold current value**—Identifies the counts for the current 15-minute interval for each monitored performance parameter (for example, *DS3 CV-L*). This is the value at the time of raising the TCA.

- **Threshold detect type**—Identifies the type of object being monitored—Monitor (MON) or Termination (TERM), as defined by TL1.

Ethernet Performance Monitoring

Monitoring Ethernet traffic is similar to monitoring optical, electrical, and DWDM traffic. You also establish thresholds for performance parameters being monitored for Ethernet. Ethernet traffic has its own set of performance parameters that thresholds can be established for. These performance parameters are defined in MIB-II and the Remote Network Monitoring (RMON) MIB.

What Is RMON?

RMON enables you to monitor Ethernet traffic on an ONS 15454 network. You can use RMON to view the Ethernet traffic flow in the network to assist in proactively monitoring the service.

RMON is an MIB that defines a group of Ethernet performance statistics. Monitoring software can gather these standard sets to allow the organization to keep track of the health of the network. These Ethernet statistics can be used not only to provide performance monitoring, but also to analyze the network utilization for capacity planning and troubleshooting reasons.

The RMON MIB contains different groups. A vendor can implement one or more of these groups on its communications equipment. These RMON groups are supported on the ONS 15454:

- Ethernet Statistics
- History Control
- Ethernet History
- Alarm
- Event

NOTE   The ONS 15454 supports the High-Capacity Remote Monitoring Information Base (HC-RMON MIB). The HC-RMON MIB enables 64-bit statistics for the etherStatsHigh CapacityTable and the EtherHistoryHighCapacityTable. The etherStatsHighCapacityTable is an extension of the etherStatsTable that adds 16 new columns for performance-monitoring data in a 64-bit format.

For example, the Cisco 15454 supports the alarm group. The alarm group is one section of the MIB that contains a table. This alarm table contains a list of alarm thresholds that can be configured to program the 15454 to send an alarm message when this threshold is reached in a 15-minute interval or in a 24-hour interval, for example. This interval is a configurable parameter and can be specified as small as 15 seconds. As an example, you can configure the alarm group to send out a trap if the number of incoming octets (ifInOctets) for an interface is more than 8000 over a period of 15 seconds.

The RMON MIB also controls the sampling of Ethernet metrics. The sampling period can be 1 minute, 15 minutes, 1 hour, or 1 day. The RMON MIB stores this information in a table, called the etherHistoryTable, which OSS applications can retrieve. This information is stored in the table on a per–Ethernet interface basis. For example, if the number of Ethernet frames is being counted every 15 minutes on the first port of a G1K card in slot 1, these samples are stored in the table and associated with Port 1 in Slot 1. In another scenario, you might want to set the sampling period to 1 and stop sampling at 60. In this case, the RMON MIB creates 60 entries in the table.

Creating a Threshold in RMON

Two threshold values must be specified when creating a RMON threshold: the rising threshold and the falling threshold. Other parameters must be configured as well:

- **Slot**—This identifies the Ethernet card.
- **Port**—This identifies the port on the Ethernet card.
- **Variable**—As an example, ifInOctets identifies the total number of octets received on the interface.
- **Alarm type**—This identifies whether the event will be triggered by the rising threshold, the falling threshold, or both the rising and falling thresholds.
- **Sample type**—This is either Relative or Absolute. Relative restricts the threshold to use the number of occurrences in the user-set sample period. Absolute sets the threshold to use the total number of occurrences, regardless of time period.
- **Sample period**—The sampling period can be 1 minute, 15 minutes, or 1 day.
- **Rising threshold**—For a rising type of alarm, the measured value must move from below the falling threshold to above the rising threshold. For example, if a network is running below a rising threshold of 1000 collisions every 15 minutes, and a problem causes 1001 collisions in 15 minutes, the excess occurrences trigger an alarm.
- **Falling threshold**—A falling threshold is the counterpart to a rising threshold. When the number of occurrences is above the rising threshold and then drops below a falling threshold, it resets the rising threshold. For example, when the network problem that caused 1001 collisions in 15 minutes subsides and creates only 799 collisions in 15 minutes, occurrences fall below a falling threshold of 800 collisions. This resets the rising threshold so that if network collisions again spike over 1000 per a 15-minute

period, an event again triggers when the rising threshold is crossed. An event is triggered only the first time a rising threshold is exceeded.

The ONS 15454 sends an RMON TCA every time a threshold is reached. This alarm is displayed as rmonThresholdCrossingAlarm in CTC. Figure 14-15 shows SNMP traps being sent when rising and falling thresholds are reached.

Figure 14-15 *Rising and Falling Thresholds in RMON*

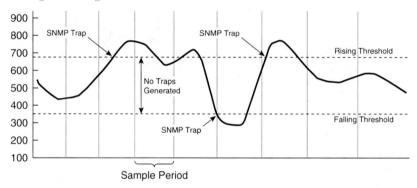

The ML Series cards have additional RMON capabilities, compared to the Layer 1 Ethernet cards. You can set up a customized threshold-crossing trap by configuring the RMON MIB to monitor an object in any other MIB used in the ONS 15454. This must be accomplished in Cisco IOS. When triggered, the ONS 15454 generates an SNMP trap to all defined trap destinations. For example, perhaps you want to monitor the interface errors on Port 1 on the ML 1000 card in Slot 1. The first step is to set up the alarm in RMON. The following command is issued in IOS:

```
# rmon alarm 10 ifEntry.14.1 200 delta rising-threshold 500 10 falling-threshold 0
owner config
```

This IOS command configures RMON to monitor ifEntry.14.1 once every 200 seconds. If the ifEntry.14.1 value shows a MIB counter increase of 500 or more, the alarm is triggered. As a result, this alarm triggers event number 10, which is configured with the **rmon event** command as follows:

```
# rmon event 10 log trap communityStr description "High If Errors" owner config
```

This event generates a trap called High If Errors and logs the event in a table.

Using the RMON to Monitor Ethernet Performance

You can use the RMON MIB to monitor the 15454 Ethernet cards that are discussed in Chapter 7, "ONS 15454 Ethernet Applications and Provisioning":

- ML 100 and ML 1000 cards
- CE 10/100 card
- G1K cards

In addition to the Ethernet cards, you can use the RMON MIB on other ONS 15454 cards, such as the Fibre Channel (FC) card and many of the DWDM Transponder-based cards.

NOTE RMON must be set up in the ML 100 and ML 1000 by using IOS commands. Note that the ML 100 and ML 1000 cards have SNMP agents. The ONS 15454 forwards SNMP requests messages to the ML cards through a proxy agent and uses the slot identification of the ML cards contained in the SNMP message to distinguish the request from the ONS 15454 SNMP request.

Figure 14-16 shows the table used in CTC to define RMON thresholds.

Figure 14-16 *RMON Configuration Table in CTC*

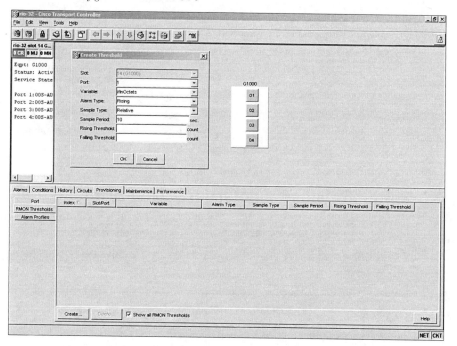

Multipoint Ethernet Monitoring

Layer 1 Ethernet refers to point-to-point Ethernet service using the G1K or CE 10/100 cards. As discussed in the previous section, RMON can be used to monitor these cards. The parameters that are monitored correspond to different Ethernet statistics, such as the number of Ethernet frames received on a port.

You can also monitor performance on other variables, such as QoS. QoS is considered Layer 2 information and can be used on Multipoint Ethernet service using the ONS 15454 ML 100 and ML 1000 cards. For example, you might want to monitor per-CoS packet statistics on the ML Series card interfaces when CoS accounting is activated. In other words, a network operations center (NOC) might want to monitor the number of Ethernet frames carrying voice packets on an interface. Monitoring the number of frames with a CoS value of 5 (this assumes that this is the CoS assigned to voice traffic) enables the ONS 15454 to keep track of the amount of Ethernet traffic on a per-service basis.

CoS-based traffic utilization is available on a per-port or per-VLAN basis. IOS commands (native language of ML Ethernet cards) or SNMP can be used to gather this information.

Using Local Craft Interface Application Versus EMS

When do you use an EMS solution over only the local craft interface application? In the case of the ONS 15454, CTC is a powerful application that can be used to monitor and manage multiple 15454 rings. In large network deployments, with hundreds or thousands of deployed 15454s, a more automated approach of provisioning and monitoring is needed. The CTM is capable of managing thousands of 15454s. Large service providers use application programming interfaces (APIs), referred to as NBIs, to enable their OSS to have an interface to the MSPP network. APIs use application protocols, such as Extensible Markup Language (XML) and CORBA, so that OSS can instruct the EMS to carry out provisioning and monitoring tasks that the OSS requires. This NBI must sustain a high rate of messages exchanged between the EMS and OSS.

Other capabilities found only in CTM and not in CTC include:

- Capability to provide end-to-end provisioning on multiple DCC domains
- An Oracle database to store faults, PM statistics, and audit trail information from a large number of 15454s for many days
- Capability to store many NE database instances and archive the NE database on a regular basis
- Capability to automatically store audit trail information in its database
- Provisioning of Layer 2 Ethernet parameters, such as RPR and QoS

Chapter 15 discusses other advantages and details of CTM.

It is important to note that CTC is used by companies such as service providers that have large-scale 15454 networks managed by CTM. CTC is an excellent tool for service-provider technicians who need to log into individual 15454 nodes or 15454 rings for maintenance, local provisioning, and troubleshooting tasks. CTC also can quickly diagnose problems from a remote location.

Summary

This chapter is an overview of different network-management architectures to monitor fault and performance on an MSPP network. Also discussed were the different management protocols used in the ONS 15454. TL1 and SNMP are both supported on the ONS 15454, highlighting that TL1 can provision and monitor the ONS 15454, whereas SNMP can only monitor, not provision, the ONS 15454.

Whether SNMP or TL1 is used, it is important for you to define what information needs to be monitored to meet your fault and performance objectives. If SNMP will be used, the MIB objects and traps must be determined.

Many large service providers are trying to migrate away from their heritage OSS by adding a newer set of OSS that is capable of monitoring not only SONET rings, but also Ethernet services running over an MSPP network. This means that TL1 is still being used to monitor Layer 1 services, such as SONET and DWDM, and the EMS manages Layer 2 Ethernet services.

This chapter covers the following topics:

- Overview of Management Layers
- Why Use an EMS?
- Using the ONS 15454 Element Management System
- Ethernet Management
- Integrating to an OSS Using the Northbound Interface

Large-Scale Network Management

Large-scale network management means that an operational support system (OSS) can manage thousands of network elements (NEs). For instance, some of these systems can manage more than 10,000 NEs. An OSS consists of various systems, including an element-management system (EMS), a network-management system (NMS), and a performance-monitoring system, to ensure that the correct level of service is being provided to end users.

An OSS supports daily network operations functions, such as adding electrical drops on a customer Multiservice Provisioning Platform (MSPP) private ring. An OSS uses fault, configuration, performance, and accounting information to ensure that the customer service is performing properly. This means, for example, that if a point-to-point Gigabit Ethernet (GigE) service begins to degrade, the OSS detects the start of this degradation and generates an alarm that a network problem exists.

Large service providers typically implement their own unique OSS. The different systems that make up an OSS should have standard interfaces. This is necessary for the different OSS components to communicate with each other. For example, many network manufacturers offer an EMS with their MSPP. Service providers expect a vendor's EMS to integrate with their existing OSS using standard interfaces, such as Tele-Management Forum (TMF) 814. TMF 814 uses a set of defined Common Object Request Broker Architecture (CORBA)-based messages to enable an OSS to provide end-to-end management and provisioning of multivendor networks. For example, the TMF standard enables an NMS to create and manage end-to-end connections and devices across TMF-compliant EMSs.

Large service providers, such as incumbent local exchange carriers (ILECs), desire certain functions as part of their OSS. Reasons to have these functions include being able to turn up new services for customers in a timely manner, to ensure that the performance of the service does not differ from the contractual agreement that is in place between the customer and the service provider, to properly update the network in case of an outage, and, of course, to stay competitive with other service providers.

Examples of these functions are listed here:

- **Flow-through provisioning**—This enables a new service order (for example, Ethernet point-to-point service) to flow through the OSS automatically, without manual intervention.

- **Policy management**—This associates network events with policy rules established by the service provider. As an example, bandwidth-on-demand service might be associated with certain policy rules. A private ring customer might want to add bandwidth during certain weeks within the year. The customer network management (CNM) application might enable the customer to perform this through software application. The rules for adding the extra bandwidth can be defined within the policy rules, which would be automatically checked when the user attempts to add this bandwidth. In this case, policy rules defines when the additional bandwidth can be added and how long the bandwidth can be used.

- **CNM**—CNM provides service provider customers the ability to monitor and activate their own services. The CNM application receives the information from other OSS components, such as the NMS, which, in turn, gets information from the EMS. This information, such as alarms from an MSPP OC-48 ring, is conveyed to the customer through the user interface on the customer's CNM application.

- **High availability (HA)**—Ensuring that the OSS components have high availability is important in a service provider environment. The daily operations that support customer services, such as Ethernet over Synchronous Optical Network (SONET), must never stop. In many cases, a service-level agreement (SLA) is established between the customer and the service provider when services are sold. This contract contains metrics that can be measured and that the service provider must meet. A service provider might have to pay substantial fees to the customer if these metrics are not met. As a result, it is important that the multiple systems that make up an OSS be capable of recovering from a failure. Local and geographical redundancy is often used to accomplish this.

- **Number of NEs**—The *number* of NEs, such as MSPPs, is important when deploying an OSS, such as an EMS. Most service providers require a single EMS server to support more than 1000 NEs.

- **Data discovery**—An EMS must automatically discover the configuration of the NE, as well as all NEs that make up the network. For example, Cisco's optical EMS, called Cisco Transport Manager (CTM) automatically discovers 15454s on a ring when the EMS begins communicating with the Gateway Network Element (GNE). This discovery includes detecting changes in configurations on the NE, as well as discovering new NEs and interface cards. For example, when you insert a GigE card into an MSPP node, the EMS detects this event and updates its inventory in near-real-time.

- **Alarm correlation and suppression**—Alarm correlation and suppression might exist in the EMS, or in the NMS, or even in the NE. In fact, alarm correlation and suppression might exist in all these systems. Alarm correlation detects problems quickly and helps identify the root problem faster. This allows quicker resolution with less effort and faster mean-time-to-repair (MTTR).

- **OSS and network being synchronized**—The OSS should always represent the current state of the network. In other words, all the databases residing in an OSS (such as EMS and NMS) must be synchronized with the network. If this is not the case, applications such as CNM will not be useful.

Overview of Management Layers

In this section, you learn about the layers of management functions that must exist in an OSS.

Network management is defined as a collective set of management functions that can be applied to a network, such as an MSPP interoffice (IOF) ring. These functions are commonly referred to as *FCAPS:*

- Fault management
- Configuration management
- Accounting management
- Performance management
- Security management

These functions exist as part of the telecommunications management network (TMN) framework. The TMN layers are the framework that defines how management is logically separated in an OSS. Developed within the International Telecommunications Union (ITU), the TMN framework breaks down the management functions into different layers that represent a hierarchical structure. Figure 15-1 shows the TMN framework.

Figure 15-1 *TMN Framework*

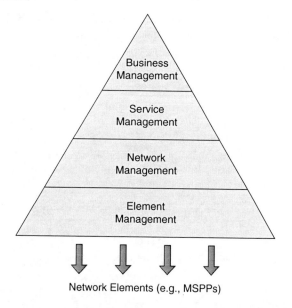

Network Elements (e.g., MSPPs)

The TMN framework has four layers:

- **Element Management Layer (EML)**—The bottom of the pyramid, referred to as the EML, is responsible for managing each individual element of the network. The EML communicates management data, such as alarms, collected from the network elements to the next-highest layer: Network Management.

- **Network Management Layer (NML)**—The NML has the responsibility of ensuring end-to-end connectivity, network reliability, and capacity planning. It carries out these responsibilities by having access to the management data that the EML provides.

  The NML has complete visibility of the entire network, which consists of different technologies—for example, Asynchronous Transfer Mode (ATM), digital subscriber line (DSL), SONET, and so on—or a subsection of the network, such as all the MSPPs. The NML is also responsible for collecting any failures or performance degradations in the network that violate an SLA or help to provide advance warning of an SLA violation. Failures can be reported as alarms, error rates, and performance levels. Thus, the NML has the responsibility to report service-affecting problems to the Service Management Layer (SML). For example, an OSS that supports NML must provide the appropriate performance-management and fault-management functions so it can provide information that applications at the SML require (for example, SLA-monitoring applications and customer network management applications).

- **Service Management Layer (SML)**—The SML is responsible for managing the services within MSPP networks, such as Ethernet over SONET. The SML normally includes the contractual aspects of services that are provided to the end customers. Customer network management and service-layer management are supported at this layer.

- **Business Management Layer (BML)**—The BML's responsibilities include billing and company policies. Policy-management applications reside at this layer.

It is important to realize that, in many implementations of the Telecommunications Management Network (TMN) framework, management data might not be passed from a layer to the layer directly above it in all cases. Management data might pass from one layer to any layer above it, or to another system within the same layer, if required. For example, an EMS that supports EML functions might pass MSPP alarms data to a customer network management application residing at the SML. Each service provider or large enterprise that deploys its own OSS defines where the TMN functions will reside, keeping in mind the importance of not duplicating functionality. Duplication increases the complexity and cost of integrating different OSS components.

Furthermore, in certain OSS deployments, an application might support functions that span multiple layers of the TMN model. For example, an EMS might support both EML and NML functions within the EMS system. This EMS might support the correlation and aggregation of statistics collected from the MSPP network, or it might not provide this correlation and might leave this function for a NMS. CTM can manage individual 15454s in an interoffice facility (IOF) ring, for example, but it can also provision and maintain end-to-end circuits traversing one or more 15454 rings.

Why Use an EMS?

In this section, you learn about the main functions of an EMS. First, let's answer the question "Why do you need an EMS?" A well-designed EMS can have a profound impact on decreasing operational expenses, thereby increasing profitability. A well-designed EMS increases the efficiency of network resources and reduces the cost of operating the network by automating tasks and eliminating the need for manual intervention.

A well-designed EMS includes capabilities for the following:

- **New services**—A legacy OSS that is being used by an ILEC is not capable of managing new services, such as Layer 2 Ethernet services, where a shared Ethernet local-area network (LAN) can be established across a SONET ring. As discussed in Chapter 7, "ONS 15454 Ethernet Applications and Provisioning," MSPPs can support these new services. However, the legacy OSS cannot support provisioning of these new services. This results in ILECs deploying EMSs and other OSS components to offer these new services.

- **OSS integration**—An EMS should be capable of integrating with an OSS using standard interfaces such as TMF 814. These interfaces allow the flow-through provisioning discussed earlier in this chapter and forward alarms to the upper-layer OSS. Because the EMS supports a standard northbound interface, the OSS does not have to connect to each NE; it establishes only one connection to the EMS. Many service providers, are deploying EMSs.

- **A-to-Z circuit provisioning**—This method of creating a circuit provides tremendous cost savings over the traditional node-by-node cross-connect approach. This capability enables new services to be quickly turned up.

- **Bulk software download and scheduling**—This increases the efficiency to perform software upgrades of the network, which, in turn, makes it easy to gain access to new features of the latest software version supported by the NE.

- **Centralized NE audit trail**—This allows for a centralized repository of all user operations on the NEs. An EMS should keep an audit trail log of the configuration changes and security violations on the NE.

- **Centralized management of NE user accounts**—Administering user accounts on the NE, such as an MSPP, can be done from a centralized location using an EMS. The EMS should be capable of adding, modifying, and deleting user accounts across multiple MSPPs. For example, if a service provider hires a new technician who requires access to 100 MSPPs, centralized management saves a tremendous amount of time over setting up the account on each MSPP one at a time.

- **Automatic daily memory backups**—This eliminates the need to manually initiate backups from the NE.

- **Alarm profile management**—This allows alarm severities to be customized based on the policies within the service provider policies.

Using the ONS 15454 Element Management System

CTM is an EMS that manages a full line of optical products, including the ONS 15454. CTM provides fault-, configuration-, performance-, and security-management (FCPS) functions for thousands of 15454s using a Java-based graphical user interface (GUI) and the TMF 814 interface to provide connectivity to a service provider's OSS. As an example, CTM enables a user to monitor alarms from all the 15454s in the network using an alarm table that displays the highest alarm severity, NE ID, probable cause, physical location, and date/time stamp.

CTM uses an alarm table to identify standing conditions on the ONS 15454 and automatically moves the alarm to an alarm history table when it clears. CTM also enables a user to configure 15454 shelf, card, interface, and port parameters. The EMS functions that CTM supports include these:

- A-to-Z circuit provisioning
- Bulk software download and scheduling
- Centralized NE audit trail
- Centralized management of NE user accounts
- Automatic daily memory backups
- Alarm profile management
- Ethernet Layer 2 management
- High availability (local and geographical redundancy)
- Standard interfaces for OSS integration [CORBA, TL1, and Simple Network Management Protocol (SNMP)]

CTM Architecture

CTM uses a client/server architecture. This means that there are two software components to CTM: client software and server software. Figure 15-2 shows the architecture for CTM.

Figure 15-2 *CTM Architecture*

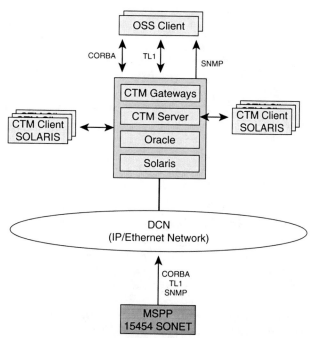

The client software runs on Windows-based computers or Solaris workstations; the server software uses a dedicated Sun Solaris server. The client uses CORBA as the application protocol to communicate to the CTM server. CTM also communicates to a higher-layer OSS by using the protocols CORBA, TL1, and SNMP. It is important to note that the CORBA interface manages all the services available on the MSPP, such as shared Ethernet services.

CTM communicates to the MSPP and other Cisco optical NEs by using Transmission Control Protocol/Internet Protocol (TCP/IP) communications. This requires CTM to have IP connectivity to each MSPP to perform its management functions. Various types of networks support connectivity between CTM and the MSPPs. Typically, service providers use a network of routers and Ethernet switches. Figure 15-3 shows an example of a network connecting CTM to an MSPP ring.

CTM needs to know only the IP address of the target MSPP, normally the Gateway Network Element (GNE); it is not concerned with how the IP packets get delivered to the MSPPs. Figure 15-3 shows how CTM connects to the GNE using a network of routers and switches. CTM must also communicate with the other MSPPs on the ring. These NEs are called External Network Elements (ENEs). CTM communicates directly with each ENE to carry out FCPS functions described earlier in this chapter. CTM uses the SONET data communications channel (DCC) link or the optical supervisory channel (OSC) for dense wavelength-division multiplexing (DWDM) to communicate with ENEs.

Figure 15-3 *CTM Network Connectivity Example*

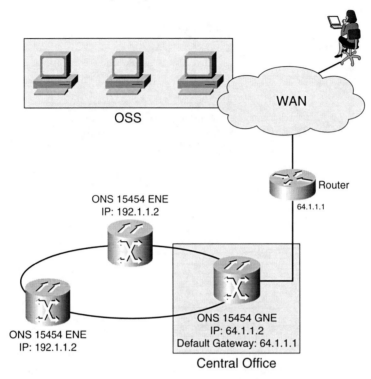

CTM carries out a regular health poll check that continuously verifies that the communications path to each individual 15454 is available. If a disruption occurs on this path, CTM resynchronizes with the NE's alarm, inventory, and configuration state after connectivity is re-established.

System Management Capabilities

CTM is capable of monitoring its own health by generating thresholds on certain possible bottlenecks. This information can be used when a service provider wants to know when the CTM server is beginning to reach the maximum number of NE support with its current hardware configuration. A service provider using high availability can force the redundant server to be used while the active server is upgraded or replaced. CTM generates alarms for the following performance bottlenecks:

- Disk use
- Central processing unit (CPU) use
- Memory use (random-access memory [RAM] and Swap)
- Circuit-creation time

CTM also monitors and reports internal alarms for loss of communication to 15454s, memory backup failures, and login security violations. These alarms help the user to quickly isolate fault conditions within CTM.

Fault-Management Capabilities

CTM detects, isolates, and reports faults on MSPP rings. CTM uses these capabilities, collectively referred to as fault management, to maximize the availability of network services. CTM discovers and helps to localize the faults in a 15454 network. Figure 15-4 shows the alarm table that displays the faults generated by the 15454.

Figure 15-4 *CTM Alarm Table*

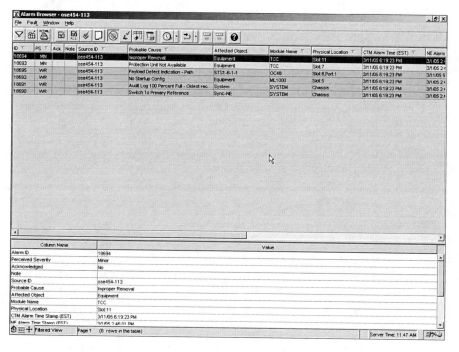

This alarm table is called an alarm browser, and each alarm is sorted by severity. If CTM is integrated into a higher-layer OSS through either TL1 or CORBA, these same alarms are forwarded to the OSS. CTM's fault-management capabilities include these:

- Fault resynchronization after connection loss
- Storage of all alarms/events in the database
- Forwarding of all alarms on the northbound interfaces
- Fault status display on most of CTM's user interfaces

- Six severities: critical, major, minor, warning, cleared, and indeterminate
- Recording of both CTM time stamp and 15454 time stamp
- Real-time alarm table and historical alarm log
- Sorting, filtering, and exporting of alarm window contents
- Alarm suppression

Fault management and performance management are closely tied together. Certain performance-monitoring (PM) data is forwarded to the alarm window in CTM and onto CTM's northbound interfaces. This is important to ensure that the degradation of service is identified as quickly as possible. Performance management is discussed in the section "Performance-Management Capabilities."

Configuration-Management Capabilities

Configuration management deals with network discovery and the provisioning of NE resources and services. CTM performs network discovery. This means that CTM can discover changes in the network autonomously. As an example, CTM discovers new 15454s and updates its database and user interfaces with the new network information; this can include both physical topology and logical circuits. As a result, the network map view in CTM dynamically updates with new 15454s if a new 15454 is added to an existing ring. In addition, CTM updates its database and user interfaces (for example, alarm table and inventory table) with changes such as inventory, alarms, service and common control cards, and protection mechanisms in any one of the 15454s in the network that might contain up to as many as 3000 NEs.

CTM can configure 15454 shelves, cards, interfaces, and port parameters. A useful feature that CTM supports and service providers commonly use is scheduling tasks. Software downloads to update thousands of 15454s can be scheduled at a specific time. It is important to note that there are no management or service interruptions during the software-download process. Keep in mind, however, that downloading software to a 15454 only stores the software in memory; it does not activate it. The software must be activated one by one.

CTM can also schedule other tasks, such as automatically backing up memory for the 15454s once a day. The Job Monitor table in CTM provides information about scheduled tasks. This table identifies the username of the person who entered a specific task, the time that the task began, and the time that the task ended.

Performance-Management Capabilities

A key aspect of performance management is monitoring SLAs. In many cases, a service provider has a contractual obligation to its customer to provide a consistent service delivery or a certain level of SLA. This service could be a point-to-point GigE service using an underlying MSPP ring to deliver this service. In some situations, customers require compensation in case of service delivery failure. Therefore, performance-management mechanisms must be accurate and traceable. Collecting, filtering, and correlating performance data collected from the NEs is part of performance management. The EMS aids in carrying out these tasks.

Performance monitoring permits users to view the performance of a service on an MSPP ring. In a service-provider environment, the user can be either the service-provider customer or a network administrator within the service-provider organization. Several service providers enable the end customer to view this data and the status of their network, which is referred to as customer network management (CNM). This is being offered for traditionally TDM services, such as DS1 and DS3 services. Service providers are beginning to offer these same services with MSPP rings.

An EMS should support CNM applications by collecting the appropriate metrics. As an example, an EMS can simply pass the raw PM data collected from the MSPP to higher-layer OSS components. Those components, in turn, pass this data to the CNM application. The EMS can massage this PM data by correlating, aggregating, and filtering it before passing it to a higher-layer OSS. Alternatively, the service provider might decide to simply pass the raw PM data directly to the CNM application and let the end customer analyze the data and create meaningful reports and graphs to help monitor the service provided by the service provider.

Before collecting this data, the first step is to identify the data that the MSPP network must collect. This particular task should be allocated a lot of time to ensure that the correct information is collected from the network so that it can be used to help in accurately monitoring the service.

NOTE

The service provider can use other sources of information resident in another OSS to help determine the accuracy of the information collected.

However, keep in mind that the NEs must have the data available to accomplish this. The ONS 15454 from Cisco provides detailed data that it makes available for CTM to collect. In addition, the 15454 can be set up so that autonomous messages can be sent to identify when certain performance thresholds are crossed. These are commonly referred to as threshold-crossing alerts (TCAs). For more information on TCAs, see Chapter 14, "Monitoring Multiple Services on a Multiservice Provisioning Platform Network."

Because PM data must be collected to help the service provider validate that the service is operating correctly, the collecting intervals for PM data should not be missed. CTM recovers from any gaps in PM data collection that can occur from communication failure between CTM and the ONS 15454. In this event, when CTM detects a gap in the PM data, it retrieves the missing data from the PM history registers on the 15454.

CTM can collect PM data from the 15454s. The user interface called the PM Query Wizard performs this task. The type of PM data collected is defined within this wizard and includes the following options:

- PM type (for example, SONET Section, SONET Line, SONET VT, SONET STS, DS1, DS3, or Ethernet)
- Interval of 15 minutes or 1 day
- Near-end or far-end
- Time period
- Module types, as applicable (for example, Ethernet cards only)
- Physical location

CTM forwards all alarms, including TCA events that are received from the 15454s, to the CORBA northbound interface. In addition, CTM displays all TCAs in the Alarm Log Table. Figure 15-5 shows the PM Query Wizard.

Figure 15-5 *PM Query Wizard*

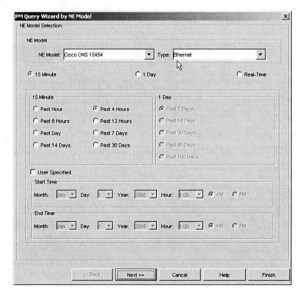

Security-Management Capabilities

More than ever, security management is a high priority for service providers. Security mechanisms must be built into the network along with the OSS components to provide effective security to protect revenue-generating services. Support for security management within the EMS is no exception.

An EMS must be capable of preventing and detecting inappropriate use of network resources and services. CTM accomplishes this task by controlling user access and NE access, and keeping an audit of actions performed on the CTM server and on the NE. CTM implements this task at two levels:

- CTM application level
- NE application level

CTM application-level security functions include the following:

- Uses a wizard to add and modify CTM user accounts
- Requires users to select a password that contains alphabetical, numerical, and special characters
- Enforces password aging
- Uses automatic screen lock and automatic logout after client inactivity period
- Initiates user lockout after a certain number of failed login attempts
- Enables an administrator to force a user to log out
- Creates custom user security profiles to restrict users to access only certain CTM capabilities and access certain 15454s
- Keeps an audit log table that contains user activities and violations during a certain time period

The Audit Log table in CTM contains the following information:

- **Source ID**—Displays the source of the event. For example, events performed by the CTM server show CTM as the source ID; otherwise, it will be the IP address of the CTM client.
- **Username**—Displays the name of the user performing the logged event.
- **Service name**—Displays the name of the service that generated the audit log entry.
- **CTM time stamp**—Displays the date and time at which the event occurred on the CTM server.
- **Category**—Displays the name of the category where the event occurred. Events belong to one of the following categories: GateWay/TL1, CTM Server Administration, CTM Server Connectivity, CTM Server Security, CTM Server Topology, CTM Server Circuit, or CTM Server Network.
- **Message**—Describes the event that occurred.

Creating custom user security profiles allows CTM to be customized to a particular service provider environment. By default, CTM supports five user levels: SuperUser, SysAdmin, NetworkAdmin, Provisioner, and Operator. In today's environment, an EMS needs to provide further granularity in terms of which CTM functions a user can use. CTM supports this additional granularity. CTM defines a list of operations that can be assigned to a user. After a user profile is created, this profile can be assigned to a new CTM user. As an example, a user profile can be created to restrict a CTM user to provision circuits of only a certain size and to provision only Layer 2 Ethernet services. In addition, you can restrict users to carry out certain capabilities relating to Layer 2 Ethernet management. Any one of the following capabilities can be assigned to a custom user profile:

- Enable a contiguous series of circuits interconnecting ML cards to become a resilient packet ring (RPR) topology
- Launch the IOS Users table and create, modify, and delete users
- Create, edit, or delete L2 services on an L2 topology
- Launch the Manage VLANs table, and create and delete VLANs
- Create, edit, and delete L2 quality of service (QoS) profiles on an L2 service
- Define custom QoS settings

CTM also supports NE application security management. Centralized management of 15454 user accounts is one important NE security-management function. When an administrator creates a new user in CTM user, a username and password can be entered to give the user access to the 15454, thus eliminating the need for the administrator to create these accounts individually on each of the 15454s. This is a significant savings in time, especially when there are thousands of 15454s in the network.

Another important NE application-level security function is uploading, storing, and viewing 15454 audit logs to enable tracking for all user actions that are performed locally at the NE. CTM logs all requests and any 15454 notifications, such as adding a new service card. This information can be used for auditing purposes later.

The Audit Trail table in CTM displays ONS 15454 activities and violations related to security. This table contains the following fields:

- **NE ID**—The identification of the selected 15454
- **Sequence Number**—A 15454-generated record ID
- **Time Stamp**—Date and time
- **Description of Operation**—Description of the audit trail operation
- **Status of Operation**—Status of the audit trail operation (Passed, Failed, or Aborted)
- **NE Username**—The 15454 user ID

High Availability

EMS high availability (HA) supports a protection scheme that can protect against a failure on the CTM server. This is an essential requirement with large service providers, especially incumbent local exchange carriers (ILECs). CTM supports HA at two levels: local redundancy and geographic redundancy.

The local redundancy solution is implemented with two identical Sun servers to provide automatic failover in case the primary Sun server fails. CTM has software that is always running, called an HA agent, to determine when the primary Sun server is no longer operational. When the primary server fails, the HA agent activates the secondary Sun server and acts as the primary server. If this scenario occurs, all CTM clients automatically reconnect to the secondary server. This 1:1 protection configuration is nonrevertive. This means that the secondary server does not automatically switch back to the primary server when the primary server becomes operational.

CTM also supports geographical redundancy, which consists of two 1:1 protected CTM servers that are located in two different sites. The strongest type of redundancy is a combination of local and geographical redundancy. Many service providers implement this type of redundancy to respond to local or regional catastrophes, such as hurricanes and tornadoes. Special software is used to keep the two CTM servers at the different sites synchronized. In addition, this software communicates with the HA agent at each site to determine when to fail over to the other site.

CTM clients and the OSS higher-layer system interface to a virtual IP address so that the switching from the primary to secondary server is transparent. Figure 15-6 shows CTM using an HA configuration.

Ethernet Management

With the advent of new services being offered by MSPPs, it is necessary to provision and monitor these new services. As a result, it is important for the EMS to carry out these functions because a legacy OSS that has traditionally managed DS1, DS3, and SONET services cannot manage data services, such as shared Ethernet LAN services. The ONS 15454 implements shared Ethernet services, as discussed in Chapter 7.

Shared Ethernet service can be provisioned by CTM either by the using CTM's user interfaces or by CTM's CORBA northbound interface. CTM uses a configuration wizard that guides you step-by-step in setting up a shared Ethernet LAN service. The wizard supports provisioning advanced QoS mechanisms that are supported by the 15454's Ethernet cards.

Figure 15-6 *CTM HA Architecture*

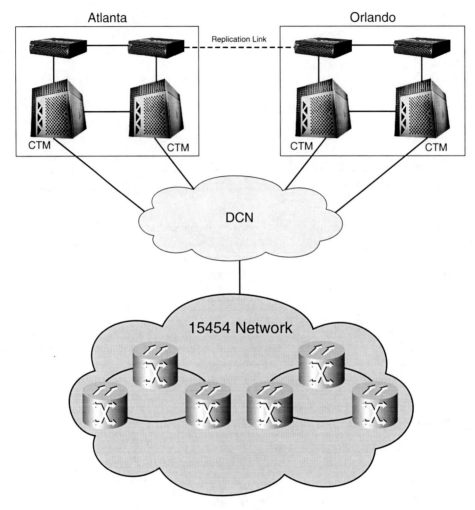

Layer 1 Provisioning

CTM is capable of separating Layer 1 circuit provisioning from Layer 2 Ethernet service provisioning. This means that TL1 messages from an external system, such as a legacy OSS, can set up the cross-connects in the 15454s to create a point-to-point Ethernet circuit or to create the SONET circuits for a shared Ethernet ring. A shared Ethernet ring is a contiguous series of ML Ethernet cards interconnected by SONET circuits that is used to create an Ethernet LAN service so that multiple users can share the bandwidth on the MSPP ring. Figure 15-7 shows five ML cards interconnected by SONET circuits forming an Ethernet ring topology.

Figure 15-7 *Shared Ethernet Ring Using ONS 15454 ML Cards*

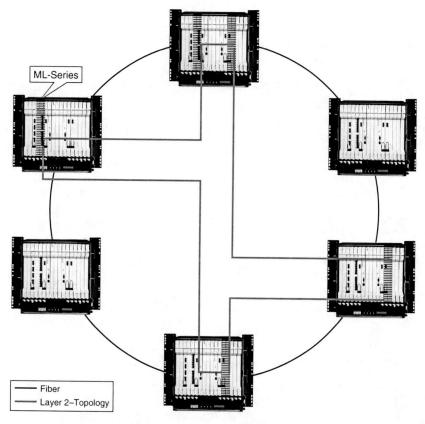

Layer 1 service provisioning is the first step required to set up a shared Ethernet service. ILECs use their existing legacy OSS to provision cross-connects on the MSPP to set up the SONET circuits. When this is accomplished, the EMS should be capable of discovering that the Ethernet cards have been stitched together with the series of cross-connects issued by the legacy OSS. CTM supports this discovery. Whenever CTM discovers a ring of Ethernet cards, CTM creates an entry in a table. This entry is called a Layer 2 topology.

The next step required to create a shared Ethernet LAN service is to configure RPR on the individual ML cards. This is accomplished in CTM by simply highlighting the entry in the Layer 2 Topology table and then selecting Upgrade L2 Topology. This action tells CTM to issue Cisco IOS Software commands to the ML card to set up a RPR. At this point, a shared Ethernet ring topology is created using RPR; now VLANs can be set up.

NOTE   15454 Ethernet cards can be added or removed in an Ethernet ring after a Layer 2 Ethernet service has been established, without affecting service. CTM uses a wizard to guide a user in adding or removing ML Ethernet cards.

Layer 2 Provisioning

CTM uses another setup wizard called an L2 Service Wizard to create a virtual LAN (VLAN) on the Ethernet ring or point-to-point circuit. When using this wizard, you do not need to issue any specific command-line interface commands on the ML card. All VLAN and QoS provisioning is accomplished using the L2 Service Wizard.

In summary, the following steps are required to set up a shared Ethernet LAN service on the 15454:

Step 1 Create the underlying Layer 1 circuit by using Cisco Transport Controller (CTC), CTM, or TL1.

Step 2 If a legacy OSS is used to set up cross-connect using TL1, CTM discovers those cross-connects and stitches them into an L1 circuit to form either a point-to-point or a ring topology.

Step 3 The discovered L2 topology must be enabled when circuits are set up using TL1.

Step 4 VLANs and QoS configurations can be applied to a wizard in CTM. This is referred to as creating a Layer 2 service.

As part of setting up Step 1, CTM can provision bandwidth settings on a RPR. Figure 15-8 shows the bandwidth settings.

Figure 15-8 *Bandwidth Settings for Layer 2 Topology*

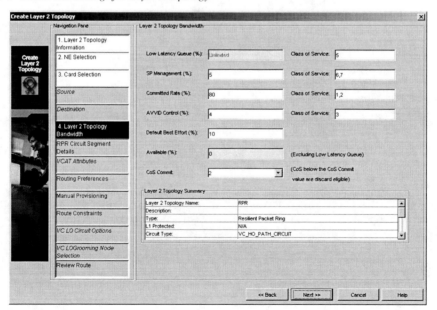

These settings are defined as follows:

- **Low-latency queue (LLQ) bandwidth**—Always set to unlimited and is not configurable. Class of service (CoS) associated with the LLQ must be selected. This is normally configured to a value of 5.

- **Service-provider management bandwidth**—Defines the bandwidth percentage used for service-provider management traffic. CoS associated with service-provider management must be selected. This is normally configured to a value of 6 or 7.

- **Committed Rate bandwidth**—Defines the bandwidth percentage used for the Committed Information Rate (CIR) traffic class. The sum of the percentage allocation for all traffic classes cannot exceed 99 percent. The bandwidth range for each traffic type has a minimum range of 1 percent and a maximum range of 99 percent. CoS associated with CIR traffic must be selected. This is normally configured to a value 1 or 2.

- **Avvid control bandwidth**—Defines the bandwidth percentage used for the voice control message, such as messages used for setting up a Voice over IP (VoIP) call. The sum of the percentage allocation for all traffic classes cannot exceed 99 percent. The bandwidth range for each traffic type has a minimum range of 1 percent and a maximum range of 99 percent. CoS associated with Avvid Control bandwidth traffic must be selected. This is normally configured to a value of 1 or 2.

- **Default best-effort bandwidth**—Defines the bandwidth percentage used for the best-effort traffic. The sum of the percentage allocation for all traffic classes cannot exceed 99 percent. The bandwidth range for each traffic type has a minimum range of 1 percent and a maximum range of 99 percent.

- **Available bandwidth**—Displays the bandwidth available for creating Layer 2 service on the topology, excluding the bandwidth assigned to the LLQ.

- **CoS commit**—Set when applying the base card configuration and ensuring that the value is the same on all the cards in the topology. All CoS below this value are eligible for discard.

The steps necessary in creating a Layer 2 service (Step 4 in the shared Ethernet LAN service setup process on the 15454) are as follows:

Step 1   Select an L2 topology from the L2 Topology table.

Step 2   Set a service provider VLAN ID. This VLAN ID, sometimes referred to as the outer tag, enables customer traffic to share the same RPR with other customers.

Step 3 Specify a customer ID and service ID. These are strings that a user can assign to an L2 service while provisioning. These are local to CTM and are not configurable on the NE.

Step 4 Specify the service drops. The user must specify at least one drop per ML card for a point-to-point L2 topology, and at least two drops on different cards for an RPR L2 topology.

Step 5 Select whether to specify one set of QoS parameters to all User-to-Network Interface (UNI) ports, or to specify QoS parameters for each UNI port individually.

Step 6 For each service drop, the following parameters must be set:

— Port status

— Port type: UNI or Network-to-Network Interface (NNI)

— Connection type: Dot1Q, QinQ, or Untagged

— Port VLAN ID

— Enable/Disable Interface

— Enable/Disable PM on the Interface

Step 7 Specify the QoS parameters for Committed Information Rate (CIR)/Peak Information Rate (PIR) or Best Effort. Predefined QoS profiles can be selected. Predefined profiles can be set up via another wizard in CTM. This allows a user who is not familiar with QoS to easily set up a Layer 2 service.

NOTE   When setting the connection type to QinQ, CTM automatically configures L2-tunneling stp, cdp, and vtp, and mode Dot1Q-tunnel. In addition, Spanning Tree Protocol (STP) is not configurable and is always set to Disabled. The reason is that STP is not needed for RPR because RPR provides a much faster ring protection (less than 50 ms restoration time) than STP.

Figures 15-9 through 15-12 show how to set up a Layer 2 service. Each user interface screen identifies which Layer 2 provisioning steps it accomplishes.

Figure 15-9 *Layer 2 Ethernet Wizard (Steps 2 and 3)*

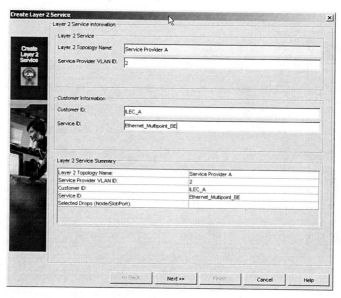

Figure 15-10 *Layer 2 Ethernet Wizard (Step 4)*

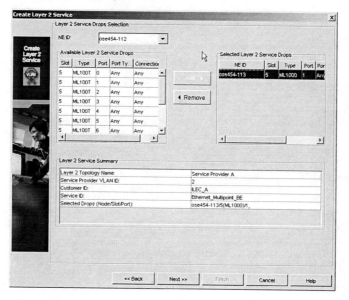

Figure 15-11 *Layer 2 Ethernet Wizard (Steps 5 and 6)*

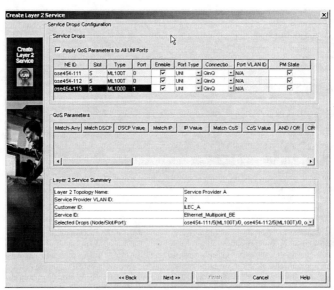

Figure 15-12 *Layer 2 Ethernet Wizard (Step 7)*

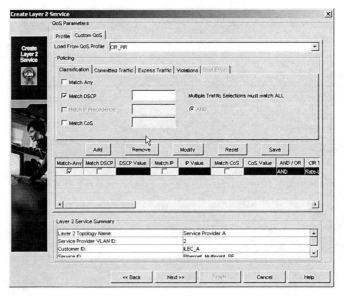

After a L2 service is created, it is identified in the Layer 2 Ethernet service table in CTM, as shown in Figure 15-13.

Figure 15-13 *Layer 2 Ethernet Service Table*

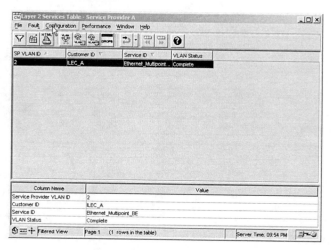

At this point, the following parameters associated with the Layer 2 service can be modified:

- Customer ID
- Service ID
- Ingress QoS parameters
- Port status

It is also important to understand that Ethernet service drops can be added or deleted from existing Layer 2 services without affecting the existing services on the Ethernet ring.

Integrating to an OSS Using the Northbound Interface

As explained earlier in this chapter, CTM provides a standard interface to integrate into a higher-layer OSS, such as an NMS. The interface addressed in this section is the CORBA Northbound Interface (NBI), also referred to as the CTM GateWay. The CTM NBI enables a service provider to integrate CTM with an existing OSS. This has recently become important because many large service providers, including many ILECs, have implemented or are implementing a new OSS to provision and monitor Layer 2 and Layer 3 services. It is worth noting that ILECs have traditionally used Telcordia legacy systems that are not capable of provisioning these new services. Figure 15-14 shows the northbound interfaces required on an EMS to support flow-through provisioning for Ethernet Layer 2 services.

Figure 15-14 *Separation of Layer 1 and Layer 2*

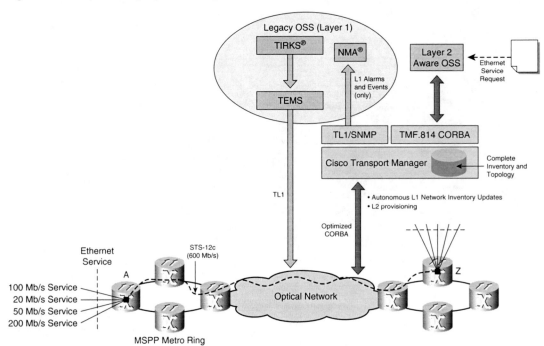

The CORBA NBI conforms to TMF 814 with additional extensions, allowing the interface to provide Layer 2 Ethernet provisioning, fault, and inventory management. CTM's CORBA NBI supports most of the Ethernet management functions discussed in the previous section:

- Creating an L2 topology—point-to-point or shared ring
- Creating an underlying SONET trunk subnetwork connections
- Creating VLANs and provisioning QoS

Summary

Many service providers require an EMS when deploying MSPP networks and offering Layer 2 Ethernet services. This is especially true for some ILECs that cannot use their traditional legacy OSS for managing shared Ethernet services supported by MSPP rings. CTM supports Layer 2 Ethernet services with a set of user interfaces and a CORBA NBI. This enables a service provider, such as an ILEC, to support Layer 2 Ethernet provisioning using CTM's GUI while using Telcordia's OSS to provision Layer 1 SONET circuits on an MSPP ring. By using CTM, you do not have to use the Cisco IOS Software commands to configure ML card parameters such as MTU size, or provision Layer 2 Ethernet services on a 15454 RPR; instead, you can use CTM's GUI to provision these parameters.

The user can use CTM to add ML cards to an existing RPR topology as well. This scenario is common in customer private MSPP rings offered by service providers when customers need to expand their Ethernet drops at their locations on the private ring. In this case, CTM automatically steps the user through a series of user interface windows to add or delete an ML card from the RPR topology.

When the service provider gains experience in offering Layer 2 Ethernet service, it can begin to integrate CTM into its OSS (for example, a next-generation OSS) by using the CORBA NBI. Service providers that do not have a legacy OSS or that already have a complete next-generation OSS can immediately use CTM's GUI or CORBA NBI to provision Layer 1 SONET and Layer 2 Ethernet services. When a next-generation OSS is operational, the service provider can support flow-through provisioning and support CNM by using the CORBA NBI to interface CTM with the next-generation OSS.

INDEX

Numerics

A

C

E

G

H

I

J–K

L

T

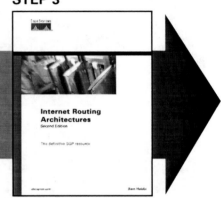

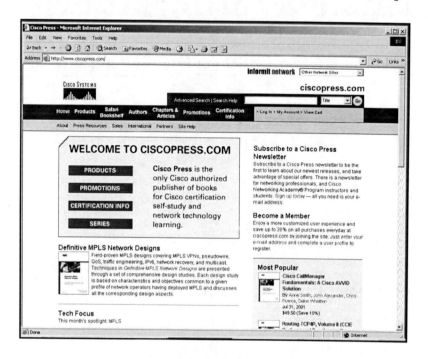

BOOKS ONLINE

ENABLED

THIS BOOK IS SAFARI ENABLED

INCLUDES FREE 45-DAY ACCESS TO THE ONLINE EDITION

The Safari® Enabled icon on the cover of your favorite technology book means the book is available through Safari Bookshelf. When you buy this book, you get free access to the online edition for 45 days.

Safari Bookshelf is an electronic reference library that lets you easily search thousands of technical books, find code samples, download chapters, and access technical information whenever and wherever you need it.

TO GAIN 45-DAY SAFARI ENABLED ACCESS TO THIS BOOK:

- Go to **http://www.ciscopress.com/safarienabled**
- Complete the brief registration form
- Enter the coupon code found in the front of this book before the "Contents at a Glance" page

If you have difficulty registering on Safari Bookshelf or accessing the online edition, please e-mail customer-service@safaribooksonline.com.